A Textbook of ALGAE

A Textbook of ALGAE

A.V.S.S. SAMBAMURTY
Department of Botany
Sri Venkateswara College
South Campus, Delhi University
New Delhi

I.K. International Pvt. Ltd.
New Delhi • Bangalore • Mumbai

Published by

I.K. International Pvt. Ltd.
S-25, Green Park Extension
Uphaar Cinema Market
New Delhi - 110 016 (India)
E-mail : ik_in@vsnl.net

Branch Offices:

A-6, Royal Industrial Estate, Naigaum Cross Road, Wadala
Mumbai - 400 031 (India)
E-mail : ik_mumbai@vsnl.net

G-4, "Embassy Centre", 11 Crescent Road, Kumara Park East
Bangalore - 560 001 (India)
E-mail : ik_bang@vsnl.net

ISBN 81-88237-44-2

© 2006 I.K. International Pvt. Ltd.

10 9 8 7 6 5 4 3 2

All rights reserved. No part of this publication may be reproduced, stored in a retrieval system, or transmitted in any form or any means: electronic, mechanical, photocopying, recording, or otherwise, without the prior written permission from the publisher.

Published by Krishan Makhijani for I.K. International Pvt. Ltd., S-25, Green Park Extension, Uphaar Cinema Market, New Delhi - 110 016 and Printed by Rekha Printers Pvt. Ltd., Okhla Industrial Area, Phase II, New Delhi - 110 020.

*Dedicated
to*
Sri Sai Baba of Shirdi

Preface

A Textbook of Algae has been designed for students of B.Sc. (Gen.) and B.Sc. (Hons.) of all Indian universities. Topics included are selected from syllabi of different universities. Many general topics are given in the first few chapters like: Life Cycles, Structure and Reproduction, Fossils of Algae, Chemical Constituents, Economic Importance of Algae, Ecology of Algae, Commercial utilization of Algae are all dealt with in detail with examples. After the general topics, type study of several algal forms are given class-wire, like: Chlorophyceae, Bacillariophyceae, Xanthophyceae, Phaeophyceae, Rhodophyceae and Myxophyceae. The book is amply illustrated with diagrams wherever necessary. Almost all the important algal forms have been discussed from their structure, reproduction, life cycle and evolutionary tendencies.

The author greatly acknowledges Blackwell Scientific Publications, Oxford, for certain diagrams and tables.

Several techniques in algae, glossary of algal terms and life cycles of different algae are included at the end of the book as Appendices.

I thank my daughter Jaishrie for typing and indexing the book.

Suggestions for improvement of the book are always welcome.

Author

Contents

Preface	v
1. Classification of Algae	**1**
Introduction	1
History of Algae	2
Classification	4
Classification by F.E. Fritsch (1945)	5
Classification by Michael J. Wynne and Gerald T. Kraft (1981)	11
Characteristic Features of Different Classes of Algae	15
2. Habitat and Habit of Algae (Ecology) of Algae	**22**
Types	22
Size of Algae	23
Form of Algae	23
Reproduction in Algae	23
Algal Ecology	24
Cyanophages or Phycophages	29
Extracellular Products of Algae	29
Fossil Algae	30
3. Commercial Cultivation of Algae	**32**
Introduction	32
The Substrate	43
4. Chemical Constituents of Seaweeds	**47**
Photosynthetic Pigments	47
Steroids and Other Triterpenoids	51
Other Terpenoids	53
Fatty Acids, Lipids, Hydrocarbons, Acetylenes, and other Low Molecular Weight Lipophilic Metabolites	59
Nonterpenoid Phenols	61
Vitamins	63
Nitrogenous Compounds	64
Polysulphides and Polyphosphates	65
5. Economic Importance of Algae	**67**
Economic Importance	67
Chemical Constituents of Algal Seaweeds	71

x Textbook of Algae

6. **Cytology and Ultrastructure** — 74
   - *Chlorophyceae* — 74
   - *Phaeophyceae* — 76
   - *Algal Cytology* — 78

7. **Thallus Organization** — 80
   1. *Unicellular Habit* — 80
   2. *Colonial Habit* — 80
   3. *Filamentous Habit* — 80
   4. *Siphoneous Habit* — 81
   5. *Pseudoparenchymatous and Parenchymatous Habit* — 81

8. **Life Cycles: Phylogeny—Sexuality** — 90
   - *Life Cycles* — 90
   - *Phylogenetic Relationships in Algae* — 92
   - *Algae and the Origin of Land Flora* — 92
   - *Sexuality in Algae* — 93
   - *Origin and Evolution of Sex in Algae* — 94

9. **Class: Chlorophyceae: General Characters and Type Study** — 98
   - *Chlorophyceae* — 98
   - *Evolution of Thallus Organization in Chlorophyceae* — 101
   - *Chlamydomonas* — 102
   - *Sphaerella (Haematococcus)* — 108
   - *Gonium* — 109
   - *Pandorina* — 111
   - *Eudorina* — 112
   - *Volvox* — 113
   - *Evolutionary Tendencies in the Order Volvocales* — 117
   - *Chlorella* — 119
   - *Hydrodictyon* — 120
   - *Ulothrix* — 121
   - *Ulva* — 124
   - *Cladophora* — 127
   - *Stigeoclonium* — 129
   - *Fritschiella* — 133
   - *Trentepohlia* — 137
   - *Draparnaldia* — 141
   - *Draparnaldiopsis* — 142
   - *Coleochaete* — 143
   - *Oedogonium* — 147
   - *Zygnema* — 150
   - *Spirogyra* — 153
   - *Desmids* — 155
   - *Chara* — 160
   - *Bryopsis plumosa* — 167

10. Class: Xanthophyceae — 169

Family 1. Botrydiaceae — *169*
Botrydium — *169*
Vaucheria — *175*

11. Bacillariophyceae — 203

Bacillariophyceae (Diatoms) — *203*

12. Class: Phaeophyceae (Brown Algae): General Characters and Type Study — 210

General Characters of Phaeophyceae — *210*
Ectocarpus — *212*
Dictyota — *216*
Sargassum — *219*
Laminaria — *224*
Fucus — *233*

13. Class: Rhodophyceae (Red Algae): General Characters and Type Study — 250

Salient Features of Rhodophyceae — *250*
Batrachospermum — *252*
Polysiphonia — *255*
Porphyra — *262*
Ceramium rubrum — *264*

14. Class: Myxophyceae (Blue-Green Algae): General Characters and Type Study — 273

General Characters of Myxophyceae — *273*
Heterocysts in Cyanophyceae — *276*
Gloeocapsa — *277*
Rivularia — *278*
Oscillatoria — *279*
Nostoc — *280*
Gloeotrichia — *282*
Scytonema — *283*

Appendix I : Life Cycles of Algae — 288

Life Cycle of Pandorina — *288*
Life Cycle of Volvox — *289*
Life Cycle of Hydrodictyon — *289*
Life Cycle of Ulothrix — *290*
Isomorphic Alternation of Generations in Ulva — *290*
Life Cycle of Cladophora glomerata — *291*
Oedogonium Life History of Nannandrous Species — *291*
Isomorphic Alternation of Generation in Cladophora — *292*
Life Cycle of Macrandrous Species Oedogonium — *292*
Life Cycle of Coleochaete — *293*
Life Cycle of Zygnema — *293*
Life Cycle of Desmids — *294*
Life Cycle of Vaucheria — *294*

xii Textbook of Algae

    Life Cycle of Ectocarpus   295
    Life Cycle of Sargassum   295
    Life Cycle of Dictyota   296
    Life Cycle of Batrachospermum (Haplobiontic)   296
    Life Cycle of Polysiphonia (Diplobiontic type)   297

Appendix II : Techniques in Algae   298
    Phytoplankton   298
    Preservation   298
    Preparation of Herbarium Sheets   299
    Staining Techniques in Algae   300
    Culture   301

Appendix III : Glossary of Algae   304

References   315

Index   317

1 Classification of Algae

INTRODUCTION

ALGAE are thallophytes (plants lacking roots, stems and leaves), which have chlorophyll *a* as their primary photosynthetic pigment and which lack a sterile covering of cells around the reproductive cells. Algae according to L.H. Tiffany (1958), "are among the most interesting plants of the world. The algae represent as wide a range in size, as found in any other group in the plant kingdom. Some of the giant kelps, which grow in marine waters, attain lenghts well over a hundred feet. The leaf-like blade may be a few feet in diameter. The surface area of a giant kelp is one the order of 6,000,000,000,000 times that of one of the minutest of the algae"!

Algae inhabit both fresh and salt waters. They flourish in ponds, in streams, in lakes, in warm springs; they hang from cliffs and cataracts. They, however, can also be found in almost any other environment on earth, from the algae growing in the snow of some American mountains to algae living in lichen associations on bare rocks, to unicellular algae in desert soils, to algae living in hot springs.

In matter of pigmentation perhaps no other group of plants exhibits so many different colours. Depending upon the type of predominant pigmentation, algae have been called as green algae, blue-green algae, yellow-green algae, brown algae, golden-brown algae, red algae etc. The chlorophylls are green, the carotenes are orange, the xanthophylls are yellow, and the phycobilins, are proteinaceous in composition, which consist of the blue phycocyanin and the red phycoerythrin. The blue greens perhaps, are the widest in the range of colour exhibition; red, rose, orange, yellow, green, blue, blue-green, purple, brown and violet.

In most habitats they function as the primary producers in the food chain, producing organic material from sunlight, carbon dioxide and water. Besides, forming the basic food, source for these food chains, they also form the oxygen necessary for the metabolism of the consumed organisms. Some algae particularly the reds and browns, are harvested and eaten as a vegetable, or the mucilages are extracted from the thallus for use as gelling and thickening agents.

Algae exhibit a striking phenomenon occurring in lakes during the late summer of the so-called 'water blooms' which are due mostly to the presence of the blue-greens. Such free floating algae are often termed as phytoplankton and is often classified as follows: the plankton of lakes is known as limnoplankton, that of ponds as *heleoplankton*, and that of rivers as *photoplankton*. G.M. Smith suggested *facultative plankton* for both benthic and pelagic forms.

Since late fifties of this century, major conceptual changes took place in algal studies, owing to the advent of successful preparation techniques for the electron microscope and, to a lesser extent, to new techniques of biochemical analysis of cell components and pathways. The increased resolution of electron microscope revealed the presence and structure of flagella, flagella hairs, flagellar swellings, eye spots, chloroplast, endoplasmic reticulum, thylakoid groupings, phycobilisomes, external scales, pit connections, scalariform vesicles, theca, projectiles, and nuclear structure and division, all characteristics important in the basic systematics of the algae.

Much work has been done in recent times on the ecology of algae, their role in the environment, the synthetic foods developed from edible algae, and certain unicellular algae, notably, *Chlorella* and *Chlamydomonas* have been put into the physiological studies apart from the genetic studies made on the unicellular alga, *Chlamydomonas*.

HISTORY OF ALGAE

"Many workers have been drawn to the study of algae for aesthetic and emotional reasons, and they have been attracted by their geometrical beauty and their sense of mystery." (Randhawa, 1960)

Early Work in Europe

The study of algae, their structure and reproduction is closely connected with the evolution and development of the microscope. September 7, 1674, the date on which Antoni von Leeuwenhoek reported a number of unicellular algae and flagellates, is considered as the date of birth of microbiology. However, almost a century elapsed before the study of algae was put on a scientific basis, and even in the last quarter of the 18th century, they were mostly designated as *Fucus, Corallina, Ulva,* and *Conferva*. Linnaeus (1754) gave the name of "*Algae*" to a group of plants amongst which he also included the Hepaticae. The first ones to draw attention were the marine algae on account of their size and structure. A.L. De Jussieu (1789) was the first to delimit the algae as known to us, but his characterisation of these plants was based on macroscopical features.

During that period most of the genera of the algae were described. Roth (1797-1805) named and described *Hydrodictyon, Batrachospermum* and *Rivularia*. Turner (1802) studied reproduction in *Fucus*.

H.F. Link (1820-33) studied the algal flora of Germany and described the genera *Tetraspora, Oedogonium* and *Spirogyra*. C.A. Agardh (1817-24), Professor in Lund University, Sweden, studied the algae of Scandinavia and established in 1824 the genera *Mougeotia* and *Zygnema*. Hassall (1842-45) described many species from Great Britain.

Thuret (1843-55) made valuable observations on motile spores and locomotory organs and described zoospore formation and reproduction in several genera. Braun (1835-55) made valuable observations on the development and taxonomy of *Chara*. Cienkowski (1855) described the reproduction in *Sphaeroplea*. Pringsheim (1855) discovered the mode of sexual reproduction in *Vaucheria, Oedogonium* and *Coleochaete*. Pringsheim watched the spermatozoids of *Oedogonium,* push into the receptive tips of the living eggs and saw the formation of the characteristic oospore wall. Pringsheim (1869) repeatedly observed the gradual fusion of the motile gametes of *Pandorina*. He gave detailed descriptions of growth, the formation of sexual organs, the mode of sexual union, and the development of oospores.

Areschoug (1866-84) described a number of new genera and species and made observations on reproduction in *Cladophora, Laminaria* and *Macrocystis*. Hertwig (1876) working on a species of sea urchin, showed for the first time that a significant feature of sexual reproduction was the fusion of the gamete nuclei. Schmitz (1879) observed a similar fusion of nuclei for the first time in *Spirogyra*.

The genus *Debarya* was established by Wittrock in 1872, *Mougeotipsis* by Palla in 1894, *Pleurodiscus* by Lagerheim in 1895, and *Temnogametum* by Wests in 1897. Mobius described a number of algae from Port Elizabeth, Brazil, Australia, and Java from 1888 to 1897. Pettkoff (1904-35) described a number of forms from Bulgaria. O. Borge (1894-1936) carried out extensive researches on the freshwater algae of Sweden.

G. Legerheim (1883-1902) studied the algal flora of Sweden and also examined collections from Ecuador, Abyssinia and India. E. Lemmermann (1891-1910) worked on the algae of Germany as well as on the collections received from China and Paraguay. O. Nordstedt (1873-97) described some new species from Argentina, Patagonia, Cameroons, New Zealand and Australia. Borzi (1878-94), De Toni (1888-1905) and Gomont (1893) carried on extensive studies on Cyanophyta. Schmitz (1883-89), Kjellman (1883), Kuckuck (1897-1912), and Phillips (1895-98) made important contributions to the study of marine algae. Strasburger (1897) worked out the life history of *Fucus* and Williams (1897-98) of *Dictyota*.

From among the British workers, the studies of G.S. West (1899-1916) and W. West (1889-1909) are classical. The Wests, father and son, not only made an exhaustive study of fresh-water algae discovering a number of new species but they also examined specimens sent to them by private collectors and leaders of expeditions to countries like Tanganyika, Victoria, Egypt, South-West Africa, Central Africa, Madagascar, West-Indies, Ceylon, Burma, Bengal and Madras. G.S. West's "Algae" published under the Cambridge Botanical Handbooks contains a very lucid account of the structure and reproduction of algae. F.E. Fritsch and F. Rich (1907-37) studied fresh-water of South Africa particularly from Natal, Transvaal, Cape Colony and Rhodesia. Yamanouchi (1909) described the life history of *Polysiphonia violacea*.

A good deal of interest has been taken in the U.S.A. in the study of algae. W.J. Hodgetts (1918-25) described conjugation in *Zygogonium erectorum* and also a number of new species of algae. E.N. Transeau and L.H. Tiffany made a substantial contribution. Transeau's work on the Zygnemaceae is of particular importance and so is that of Tiffany's on Oedogoniales.

Researches on algae in India and the neighbouring countries can be broadly divided into two periods: firstly, the period of pioneers (1806-1907) and professional botanists, mostly Englishmen working in the Botanical Survey of India or in the Universities in England, and secondly, from 1919 to date, the period of Indian Research Workers, mostly professors of Botany in Indian Universities, employees of the Botanical Survey of India, or amateurs. The macroscopic forms like Characeae and sea-weeds were the earliest to attract attention, and description of a new variety of *Chara polyphylla* (Braun) from Ceylon by Lebeck in 1798 marks the first milestone in the study of algae in India and the neighbouring countries. In 1806 Wildenow described *Chara zeylanica* from Tranquebar and Ceylon. Koening collected a number of algae which were later described by other workers. Many pioneers followed him to India in search of plants and among these can be mentioned Roxburgh (1832), Reynaud (1834), Wight (1834) and Belanger (1836).

In 1897, W. West and G.S. West described 45 species of desmids from Singapore. In 1902, Wests described 7 species of Rhodophyceae, 49 species of diatoms, 33 species of Myxophyceae, 246 species of desmids, and 84 species of Chlorophyceae from Ceylon. In 1907, Wests further recorded 58 species of diatoms, and 148 species of desmids. They also recorded 53 species of other green algae from Madras. F.E. Fritsch (1903-07) studied fresh-water and terrestrial algae of Ceylon in 1903.

Work of Indian Researchers

From 1919 onwards work on algae in India has largely been done by Indian researchers. Ghose (1919-32) carried on observations on the blue-green algae of Burma and Punjab. M.O.P. Iyengar from 1920 onwards published a number of papers on algae of South India, both fresh-water and marine. In collaboration with his students Balakrishnan, Desikachary, Kanthamma, Ramanathan and Subramaniam, apart from describing a number of new species and genera, he also worked on the life history of *Cylindrocapsa geminella* (1948) and described formation of gametes in a species of *Caulerpa*. He also described morphology and cytology of *Polysiphonia platycarpa*, sexual reproduction in *Dictyosphaerium* and life history and cytology of *Microdictyon tenius*. Along with Subrahmanyam he further described reduction division and auxospore formation of *Cyclotella meneghiniana*. His most important contribution is in the discovery of *Fritschiella tuberosa*, a very interesting terrestrial alga. Considering the volume and quality of his work, and the work he has been able to stimulate, he can rightly be called the 'father of modern algology of India'.

Bhardwaja (1928-36) made an effective contribution to our knowledge of the blue-green algae of Uttar Pradesh. He built up a school of algology at Benaras Hindu University. Out of his students, R.N. Singh (1938-59) published a series of papers on blue-green algae as well as Zygnemaceae, Oedogoniaceae and Chaetophorales of Uttar Pradesh. Singh also described the life history of *Fritschiella tuberosa* and *Draparnaldiopsis indica*. He also indicated the possibility of reclamation of saline usar lands in India by the cultivation of blue-green algae. V.P. Singh (1941) described some species of green algae from Chamba and Uttar Pradesh. C.B. Rao (1935-38) published a number of papers on the Myxophyceae and Zygnemaceae of UP and Madras.

Randhawa (1932-59) published a series of papers mostly on Zygnemaceae, Oedogoniales and Vaucheriaceae from the Punjab and Uttar Pradesh. He recorded 70 species of Zygnemaceae, including one new species of *Debarya*, 2 of *Mougeotia*, 2 of *Zygogonium*, and 11 of *Spirogyra*. In 1939 he described *Sirocladium kumaoense*, and a new genus and species from Kumaon, Himalayas, of which another new species was discovered from the Western Ghats in Bombay State in 1956. The number of terrestrial green algae discovered was truly remarkable. These included a terrestrial species of *Zygnema*, *Zygnema terrestre*, 2 terrestrial species of *Oedogonium*, a new species of *Oedogonium*, *Oedocladium himalayense*, and a new species of *Zygogonium kumaoense*. Apart from these, he described cyst formation in *Fritschiella tuberosa*, and a new type of *akinete* formation in *Vaucheria*.

S.R. Narayana Rao (1941-49) made useful observations on the fossil algae of India. R. Subrahmanyam (1945) made interesting observations on reproduction in some species of diatoms from South India. Allen (1925-42), Pal, Dixit, Kundu and Sundaralingam worked on the Charophytes of India and Burma. J.N. Misra (1937) described 5 species of *Zygnema* including two new species, and 5 of *Spirogyra*. S.C. Dixit (1937) recorded 2 species of *Spirogyra* and one of *Sirogonium*. Gonzalves and her students carried on useful observations on the algal flora of the Bombay State. A.K. Mitra and Saxena have been carrying on some interesting work on the culturing of the algae. Ramanathan (1939-46) described the life history of *Enteromorpha compressa* as well as sexual reproduction in *Carteria* sp. and *Dictyosphaerium indicum*.

Work on the cytology of algae has been published by M.B.E. Godward from London in modern times and from India by Prof. Y.S.R.K. Sarma and his students from Benares Hindu University.

CLASSIFICATION

Historically, the foundation of a real classification of algae was laid down by Linnaeus (1753) and by A.L. de Jussieu (1789). Later Vaucher (1803), Hodwig (1798), Roth (1797-1805), Lamouroux (1805-1816), Lyngbye (1820) and Harvey (1836) attempted to classify algae in a scientific way. Later on Kutzing (1861), Bohlin (1901), Luther (1899), Robenhorst (1869), Hassel (1902), Strasburger (1897), Kjellman (1883), Kuckuck (1912), Kylin (1906), Engler and Prantl (1912), and Blackman and Tansley (1902), brought forward several classifications of algae.

Pascher in 1931 published a very elaborate classification of algae who divided into eight classes viz., Chrysophyta, Phaeophyta, Pyrrophyta, Euglenophyta, Chlorophyta, Charophyta, Rhodophyta and Cyanophyta.

J.E. Tilden (1933) based on reserve food materials, pigmentation and flagellation classified algae into five classes viz., Chlorophyceae, Myxophyceae, Rhodophyceae, Phaeophyceae and Chrysophyceae.

F.E. Fritsch (1935) in his masterpiece on the *Structure and Reproduction of Algae*, Vol. I and Vol. II (1945) gave a sound classification of algae into eleven classes based on pigmentation, reserve food, flagellation and also mode of reproduction. They are Chlorophyceae, Xanthophyceae, Chrysophyceae, Bacillariophyceae, Cryptophyceae, Dinophyceae, Chloromonadineae, Euglenineae, Phaeophyceae, Rhodophyceae and Myxophyceae.

G.M. Smith (1938) classified algae in his *Cryptogamic Botany*, Vol. I into eight divisions and each division into several classes viz., Chlorophyta, Euglenophyta, Pyrrophyta, Chrysophyta, Phaeophyta, Cyanophyta and Rhodophyta.

Papenfuss (1955) produced eight Phyla and classified algae into Chlorophycophyta, Charophycophyta, Euglenophycophyta, Chrysophycophyta, Pyrrophycophyta, Phaeophycophyta, Schizophycophyta and Rhodophycophyta.

Chapman (1962) produced a new type of classification who divided algae into four major classes viz., Euphycophyta, Myxophycophyta, Chrysophycophyta, and Pyrrophycophyta. In Euphycophyta he included Charophyceae, Chlorophyceae, Phaeophyceae and Rhodophyceae which is an altogether different grouping.

G.W. Prescott (1969) divided algae into nine Phyla viz., Chlorophyta, Eulgenophyta, Chrysophyta, Pyrrophyta, Phaeophyta, Rhodophyta, Cyanophyta, Cryptophyta and Chloromonadophyta.

F.E. Round (1965) divided algae into eight Phyla viz., Cyanophyta, Chrysophyta, Chlorophyta, Euglenophyta, Pyrrophyta, Cryptophyta, Phaeophyta and Rhodophyta.

The classification proposed by F.E. Fritsch is followed in this book based on its practical value. Professor Felix Eugen Fritsch (1879-1954) was born on 26th April, 1879 at Hampstead, London. He graduated from London University and obtained his Ph. D. from Munich University, West Germany in 1902. In 1902 he was employed as Assistant in Botany Department in London University College (1902-1906) and later became Professor and Head of the Department of Botany in Queen Mary College, London in 1912. He retired as Professor Emeritus in 1948 from the same institution. He died on 23rd May 1954.

CLASSIFICATION BY F. E. FRITSCH (1945)

Class I: Chlorophyceae

Order I: VOLVOCALES
    Sub-order: (a) *Chlamydomonadineae*
        Family: Chlamydomonadaceae   Family: Sphaerellaceae
        Family: Polyblepharidaceae   Family: Phacotaceae
    Sub-order: (b) *Tetrasporineae*
        Family: Tetrasporaceae   Family: Palmellaceae
    Sub-order: (c) *Chlorodendrineae*
        Family: Chlorodendraceae

Order II: CHLOROCOCCALES
    Family: Chlorococcaceae
    Family: Chlorellaceae
    Family: Selenastraceae
    Family: Hydrodictyaceae
    Family: Eremosphaeraceae
    Family: Oocystaceae
    Family: Dictyosphaeriaceae
    Family: Coelastraceae

Order III: ULOTRICHALES
  Sub-order: (a) *Ulotrichineae*
    Family: Ulotrichaceae
    Family: Cylindrocapsaceae
    Family: Microsporaceae
    Family: Ulvaceae
  Sub-order: (b) *Prasiolineae*
    Family: Prasiolaceae
  Sub-order: (c) *Sphaeropleineae*
    Family: Sphaeropleaceae

Order IV: CLADOPHORALES
    Family: Cladophoraceae

Order V: CHAETOPHORALES
    Family: Chaetophoraceae
    Family: Coleochaetaceae
    Family: Pleurococcaceae
    Family: Trentepohliaceae
    Family: Chaetospaeridiaceae

Order VI: OEDOGONIALES
    Family: Oedogoniaceae

Order VII: CONJUGALES
  Sub-order: (i) *Euconjugatae*
    (a) *Mesotaenioideae*
      Family: Mesotaeniaceae
    (b) *Zygnemoideae*
      Family: Mougeotiaceae
      Family: Gonatozygaceae
      Family: Zygnemaceae
  Sub-order: (ii) *Desmidiaceae*
    Family: Desmidiaceae

Order VIII: SIPHONALES
    Family: Prtosiphonaceae
    Family: Derbesiaceae
    Family: Codiaceae
    Family: Chaetosiphonaceae
    Family: Vaucheriaceae
    Family: Caulerpaceae
    Family: Dasycladaceae
    Family: Valoniaceae
    Family: Phyllosiphonaceae

Order IX: CHARALES
    Family: Characeae

Class II: Xanthophyceae

Order I: HETEROCHLORIDALES

(a) Heterochloridineae
        Family: Heterochloridaceae
(b) Heterocapsineae
        Family: Heterocapsaceae
(c) Heterodendrineae
        Family: Mischococcaceae
(d) Heterorhizidineae
        Family: Heterorhizidaceae

Order II: HETEROCOCCALES
        Family: Halosphaeraceae   Family: Myxochloridaceae
        Family: Chlorobotrydaceae   Family: Chlorotheciaceae
        Family: Ophiocytiaceae

Order III: HETEROTRICHALES
        Family: Tribonemaceae   Family: Heterocloniaceae

Order IV: HETEROSIPHONALES
        Family: Botrydiaceae

Class III: Chrysophyceae

Order I: CHRYSOMONADALES
  Sub-order: *(a) Chrysomonadineae*
    (i) Chromulineae
        Family: Chromulinaceae   Family: Oicomonadaceae
        Family: Mallomonadaceae   Family: Cyrtophoraceae
    (ii) Isochrysideae
        Family: Isochrysidaceae   Family: Coccolithophoridaceae
        Family: Synuraceae
    (iii) Ochromonadeae
        Family: Ochromonadaceae   Family: Monadaceae
        Family: Lepochremonadaceae
    (iv) Prymnesieae
        Family: Prymnesiaceae
  Sub-order: *(b) Rhizochrysidineae*
        Family: Rhizochrysidaceae   Family: Lagyniaceae
    (v) Chrysocapsoneae
        Family: Chrysocapsaceae   Family: Nageliellaceae
        Family: Hydruraceae

Order II: CHRYSOSPHAERALES
        Family: Chrysophaeraceae   Family: Chrysostomataceae
        Family: Pterospermaceae

Order III: CHRYSOTRICHALES
        Family: Nematochrysidaceae   Family: Phaeothamnionaceae
        Family: Thallochiysidaceae

Class IV: Bacillariophyceae

Order I: CENTRALS
    (*a*) *Discoideae*
    (*b*) *Solenoideae*
    (*c*) *Biddulphioideae*
    (*d*) *Rutilarioideae*

Order II: PENNALES
      (*i*) *Araphideae*
    (*e*) *Fragilarioideae*
      (*ii*) *Raphidioideae*
    (*f*) *Eunotioideae*
      (*iii*) *Monoraphideae*
    (*g*) *Acanthoideae*
      (*iv*) *Biraphideae*
    (*h*) *Naviculoideae*
    (*i*) *Epithemoideae*
    (*j*) *Nitzshioideae*
    (*k*) *Surirelloideae*

Class V: Cryptophyceae

Order I: CRYPTOMONADALES
    (*a*) *Cryptomonadineae*
        Family: Cryptomonadaceae   Family: Nephroselmidaceae
    (*b*) *Phaeocapsineae*
        Family: Phaeocapsaceae

Order II: CRYPTOCOCCALES
        Family: Cryptococcaceae

Class VI: A. Dinophyceae (Peridinieae)

    (*a*) *Desmonadales*
        Family: Desmomonadaceae
    (*b*) *Thecatales*
        Family: Phorocentraceae
    (*c*) *Dinophysiales*
        Family: Dinophysiaceae   Family: Amphsioleniaceae

B. Dinokontae

I. *Dinoflagellatae*
    (*i*) *Peridineae*
      (*a*) *Gymnodinioideae*
        Family: Pronoctilucaceae   Family: Gymnodiniaceae
        Family: Polykrikaceae   Family: Noctilucaceae
        Family: Blastodiniaceae

      (b) *Amphilothaloideae*
      (c) *Kolkwitzielloideae*
      (d) *Peridinioideae*
          Family: Glenodiniaceae
          Family: Gonyauloceae
          Family: Ceratiaceae
          Family: Ceratocoryaceae
          Family: Protoceratiaceae
          Family: Peridiniaceae
          Family: Goniodomaceae
          Family: Podolampaceae
   (ii) *Dinocapsineae*
          Family: Dinocapsaceae
   (iii) *Rhizodinineae*
II. *Dinococcales*
          Family: Dinococcaceae
III. *Dinotrichales*
          Family: Dinotrichaceae   Family: Dinocloniaceae

Class VII: Chloromonadineae
Class VIII: Euglenineae
          Family: Euglenaceae   Family: Astaeiaceae
          Family: Peranemaceae

Class IX: Phaeophyceae
Order I: ECTOCARPALES
  (a) Forms Simple
          Family: Ectocarpaceae
  (b) Forms Haplostichous
          Family: Myrinemataceae   Family: Elachistaceae
          Family: Leathesiaceae   Family: Mesogloeaceae
          Family: Acrotrichaceae   Family: Spermatochnaceae
          Family: Splachnidiaceae
  (c) Forms Polystichous
          Family: Punctariaceae   Family: Asperococcaceae
          Family: Encoeliaceae   Family: Dictyosiphonaceae
Order II: TILOPTERIDALES
Order III: CUTLERIALES
Order IV: SPOROCHNALES
Order V: DESMARESTIALES
Order VI: LAMINARIALES
          Family: Chordaceae   Family: Laminariaceae
          Family: Lessoniaceae   Family: Alariaceae
Order VII: SPHACELARIALES
          Family: Sphacelariaceae   Family: Stypocaulaceae
          Family: Cladostephaceae   Family: Choristocarpaceae

Order VIII: DICTYOTALES
    Family: Dictyotaceae
Order IX: FUCALES
    Family: Fucaceae
    Family: Cystoseriaceae
    Family: Hormosiraceae
    Family: Ascoseriaceae
    Family: Himanthaliaceae
    Family: Sargassaceae
    Family: Durvilleaceae

Class X: Rhodophyceae
Sub-class I: Bangioideae
Order I: BANGIALES
    Family: Bangiaceae
    Family: Prophyridiaceae

Sub-class II: Florideae
Order II: NEMALIONALES
  (a) Forms Uniaxial
    Family: Acrochaetiaceae
    Family: Lemaneaceae
    Family: Bonnemaisoniaceae
    Family: Batrachosperroace
    Family: Naccariaceae
  (b) Forms Multiaxial
    Family: Thoreaceae
    Family: Chaetangiaceae
    Family: Helminthocladiaceae

Order III: GELIDIALES
    Family: Gelidiaceae

Order IV: CRYPTONEMIALES
    Family: Gloesiphoniaceae
    Family: Callymeniaceae
    Family: Dumontiaceae
    Family: Rhizophyllidaceae
    Family: Corallinaceae
    Family: Endocladiaceae
    Family: Grateloupiaceae
    Family: Cruoriaceae
    Family: Squamariaceae
    Family: Choreocolaceae

Order V: GIGARTINALES
  (a) No procarps
    Family: Calosiphoniaceae
    Family: Sebdeniaceae
    Family: Solieriaceae
    Family: Rhabdoniaceae
    Family: Nemastomaceae
    Family: Furcellariaceae
    Family: Rissoellaceae
  (b) Procarps present
    Family: Rhodophyllidaceae
    Family: Plocamiaceae
    Family: Gracilariaceae
    Family: Acrotylaceae
    Family: Gigartinaceae
    Family: Hypneaceae
    Family: Sphaerococcaceae
    Family: Mychodeaceae
    Family: Phyllophoraceae

Order VI: RHODYMENIALES
    Family: Champiaceae
    Family: Rhodymeniaceae

Order VII: CERAMIALES
    Family: Ceramiaceae
    Family: Delesseriaceae
    Family: Rhodomelaceae
- (a) Polysiphonieae
- (b) Lophothalieae
- (c) Bostrychieae
- (d) Rhodomeleae
- (e) Chondrieae
- (f) Laurencieae
- (g) Pterosiphonieae
- (h) Herposiphonieae
- (i) Lophosiphonieae
- (j) Polyzonieae
- (k) Amansieae

    Family: Dasyaceae

Class XI: Myxophyceae (Cyanophyceae)

Order I: CHROOCOCCALES
    Family: Chroococcaceae
    Family: Entophysalidaceae

Order II: CHAMAESIPHONALES
    Family: Dermocarpaceae
    Family: Chamaesiphonaceae
    Family: Endonemataceae
    Family: Siphononemataceae

Order III: PLEUROCAPSALES
    Family: Pleurocapsaceae
    Family: Hyellaceae

Order IV: NOSTOCALES
    Family: Oscillatoriaceae
    Family: Nostocaceae
    Family: Microchaetaceae
    Family: Rivulariaceae
    Family: Scytonemataceae

Order V: STIGONEMATALES
    Family: Pulvinulariaceae
    Family: Capsosiraceae
    Family: Nostochopsidaceae
    Family: Loefgreniaceae
    Family: Stigonemataceae

CLASSIFICATION BY MICHAEL J. WYNNE AND GERALD T. KRAFT (1981)

Division: *Rhodophyta*   **Class:** *Rhodophyceae*   **Sub-class:** *Florideophycidae*

Order: NEMALIONALES (including Bonnemaisoniales & Gelidiales)
    Family: Nemalionaceae
    Family: Helminthocladiaceae
    Family: Dermonemataceae
    Family: Bonnemaisoniaceae
    Family: Naccariaceae
    Family: Chaetangiaceae
    Family: Acrochaetiaceae
    Family: Gelidiaceae

Family: Gelidiellaceae
Family: Lemaneaceae

Family: Batrachospermaceae
Family: Thoreaceae

Order: CRYPTONEMIALES
Family: Corallinaceae
Family: Peyssonneliaceae
Family: Dumontiaceae
Family: Corynomorphaceae
Family: Hildenbrandiaceae
Family: Choreocolaceae
Family: Endocladiaceae

Family: Halymeniaceae
Family: Kallymeniaceae
Family: Weeksiaceae
Family: Gloiosiphoniaceae
Family: Pterocladiophyllaceae
Family: Pseudoanemoniaceae
Family: Tichocarpaceae

Order: GIGARTINALES
Family: Gigartinaceae
Family: Phyllophoraceae
Family: Gymnophloeaceae
Family: Binksiaceae
Family: Phacelocarpaceae
Family: Cystocloniaceae
Family: Dicranemaceae
Family: Mychodeophyllaceae
Family: Wurdemanniaceae
Family: Furcellariaceae
Family: Rissoellaceae
Family: Chondriellaceae
Family: Hypneaceae
Family: Rhizophyllidaceae

Family: Gracilariaceae
Family: Solieriaceae
Family: Calosiphoniaceae
Family: Caulacanthaceae
Family: Nizymeniaceae
Family: Acrotylaceae
Family: Mychodeaceae
Family: Plocamiaceae
Family: Sebdeniaceae
Family: Sarcodiaceae
Family: Sphaerococcaceae
Family: Cruoriaceae
Family: Polyideaceae
Family: Cubiculosporaceae

Order: RHODYMENIALES
Family: Rhodymeniaceae
Family: Lomentariaceae

Family: Champiaceae

Order: PALMARIALES
Family: Palmariaceae

Order: CERAMIALES
Family: Ceramiaceae
Family: Dasyaceae

Family: Delesseriaceae
Family: Rhodomelaceae

Sub-class: Bangiophycidae

Order: RHODOCHAETALES
Family: Rhodochaetaceae

Order: BANGIALES
Family: Bangiaceae
Family: Boldiaceae

Family: Erythropeltidaceae

Order: PORPHYRIDIALES
Family: Goniotrichaceae

Family: Phragmonemataceae

Family: Porphyridiaceae
Order: COMPSOPOGONALES
　　Family: Compsopogonaceae

Class: Phaeophyceae

Order: ECTOCARPALES (including Ralfsiales)
　　Family: Ectocarpaceae
　　Family: Ralfsiaceae
　　Family: Sorocarpaceae
Order: CHORDARIALES
　　Family: Myrionemataceae
　　Family: Leathesiaceae
　　Family: Acrotrichacae
　　Family: Chordariaceae
　　Family: Splachnidiaceae
　　Family: Elachistaceae
　　Family: Ishigeaceae
　　Family: Spermatochnaceae
　　Family: Notheiaceae
　　Family: Chordariopsidaceae
Order: SPOROCHNALES
　　Family: Sporochnaceae
Order: DESMARESTIALES
　　Family: Desmarestiaceae
　　Family: Arthrocladiaceae
Order: DICTYSIPHONALES
　　Family: Pogotrichaceae
　　Family: Buffhamiaceae
　　Family: Striariaceae
　　Family: Delamareaceae
　　Family: Coilodesmaceae
　　Family: Myriotrichiaceae
　　Family: Giraudyaceae
　　Family: Coelocladiaceae
　　Family: Punctariaceae
　　Family: Dictyosiphonaceae
Order: SCYTOSIPHONALES
　　Family: Scytosiphonaceae
　　Family: Chnoosporaceae
Order: TILOPTERIDALES
　　Family: Tilopteridaceae
　　Family: Masonophycaceae
Order: CUTLERIALES
　　Family: Cutleriaceae
Order: SPHACELARIALES
　　Family: Sphacelariaceae
　　Family: Cladostephaceae
　　Family: Choristocarpaceae
　　Family: Stypocaulaceae
Order: DICTYOTALES
　　Family: Dictyotaceae
　　Family: Dictyotopsidaceae
Order: LAMINARIALES
　　Family: Chordaceae
　　Family: Phyllariaceae
　　Family: Lessoniaceae
　　Family: Laminariaceae
　　Family: Arthrothamnaceae
　　Family: Alariaceae
Order: FUCALES
　　Family: Fucaceae
　　Family: Seirococcaceae

Family: Hormosiraceae
Family: Sargassaceae
Family: Himanthaliaceae
Family: Cystoseiraceae

Order: DURVILLAEALES
Family: Durvillaeaceae

Order: ASCOSEIRALES
Family: Ascoseiraceae

Division: *Chlorophyta*   Class: Chlorophyceae

Order: VOLVOCALES (including Tetrasporiales)
Family: Dunaliellaceae
Family: Volvocaceae
Family: Chlamydomonadaceae
Family: Palmellaceae

Order: CHLOROCOCCALES
Family: Chlorococcaceae
Family: Scenedesmaceae
Family: Characiosiphonaceae
Family: Oocystaceae
Family: Hydrodictyaceae

Order: ULOTRICHALES
Family: Ulotrichaceae
Family: Microsporaceae
Family: Cylindrocapsaceae
Family: Chaetophoraceae
Family: Schizomeridaceae

Order: TRENTEPOHLIALES
Family: Chroolepidaceae

Order: KLEBSORMIDIALES
Family: Klebsormidiaceae

Order: COLEOCHAETALES
Family: Coleochaetaceae

Order: ACROSIPHONIALES
Family: Acrosiphoniaceae

Order: ULVALES
Family: Percursariaceae
Family: Ulvaceae
Family: Monostromataceae

Order: PRASIOLALES
Family: Prasiolaceae

Order: SPHAEROPLEALES

Order: CLADOPHORALES (including Siphonocladales)
Family: Anadyomenaceae
Family: Valoniaceae
Family: Chaetosiphonaceae
Family: Boodleaceae
Family: Siphonocladaceae
Family: Cladophoraceae

Order: CAULERPALES (including Derbesiales, Dichotomosiphonales)
Family: Bryopsidaceae
Family: Derbesidaceae
Family: Udoteaceae
Family: Codiaceae
Family: Caulerpaceae
Family: Phyllosiphonaceae

Family: Dichotomosiphonaceae
Order: DASYCLADALES
    Family: Dasycladaceae   Family: Acetabulariaceae
Order: OEDOGONIALES
    Family: Oedogoniaceae
Order: ZYGNEMATALES
    Family: Zygnemataceae   Family: Desmidiaceae
Order: CHARALES
    Family: Characeae

CHARACTERISTIC FEATURES OF DIFFERENT CLASSES OF ALGAE

The primary classification of algae is based on five main criteria, namely: (see Table 1).

(1) Photosynthetic pigments, (2) the nature of reserve food, (3) the nature of cell wall, (4) the types of flagella, and (5) certain details of cell structure. The classification of algae depends on a combination of these characters and not on any single feature. The details of thallus structure and the process of reproduction are useful at the level of the classification of species and genus. But they are not specially significant in the primary classification. Fritsch recognised eleven classes in algae and their characters are listed below:

Class 1: Chlorophyceae

In these algae, the chloroplasts contain chlorophyll a, chlorophyll b, β-carotene and xanthophyll approximately in the same proportion as in higher plants. Starch is the reserve food. Pyrenodis are commonly surrounded by starch sheath. Cellulose is the main component of the cell wall. The motile cells have 2 or 4 equal flagella at the anterior end. Sometimes, flagella are several. Sexual reproduction is by isogamy, anisogamy or oogamy. The heterotrichous plant body is most highly evolved one. Plant body is not bulky and parenchymatous as in Brown and Red algae. Most of the members a haploid but some are diploid (Siphonales) while a few others show isomorphic alternation of generations between haploid and diploid plants. The class is better represented in fresh water than in salt water.

Class 2: Xanthophyceae

The class is smaller than Chlorophyceae. The chromatophores are yellow-green, as the carotenoids are in excess over chlorophylls. Chlorophyll e is restricted in this class. Chlorophyll b is absent. β-carotene and xanthophylls are abundantly present. Pyrenoids are lacking. Oil and fat are the storage products. The cell wall has a higher content of pectic material than in Chlorophyceae. The two flagella are of unequal length (heterokontae), the longer being pantonematic and the shorter acronematic. Sexual reproduction is rare and isogamous. It is more widely distributed in freshwater than in salt water.

Class 3: Chrysophyceae

The chromatophores are brown or orange coloured containing chlorophyll a, and carotenoids like fucoxanthin and diadinoxanthin. Fats and leucosin are the reserve foods. Motile cells have one or two flagella. When they are two, they may be equal or unequal. Cellulose cell wall is absent. Sexual reproduction is rare but it is isogamous.

TABLE 1 CLASSIFICATION OF ALGAE AND THEIR IMPORTANT CHARACTERISTICS

Division	Common Name	Pigments and and Plastid Organisation	Stored food	Cell wall[a]	Flagellar Number and Insertion[b]	Habitat[c]
Chlorophycophyta	Green algae	Chlorophyll a, b; a—and — carotenes + several xanthophylls; 2—6 thylakoids/band.[d]	Starch, resembling that of land plants, i.e., a^{-1} : 4-glucopyranosides= amylose and glucopyranosides with a^{-1} 4 and some 1 : 6 linkages.	Cellulose in = – 1 : 4—glucopyranoside, hydroxy-proline glycosides; or wall absent; siliceous and/or calcified in some[e].	1, 2—8, many, equal, apical	f.w., b.w., s,w., t., a.
Charophyta	Stoneworts	Chlorophyll a, b; a, and – carotenes + several xanthophylls; thylakoids variably associated	Starch, resembling that of land plants	Cellulose (= –1 : 4— glycopyranoside)	2, equal, subapical	f.w., b.w.
Euglenophycophyta	Euglenids	Chorophylla a, b; —carotene + several xanthophylls; 2—6 thylakoids/band, sometimes many	Paramylon (= –1 : 3 gluco-pryanoside) and facts	Absent	1—3 apical, subapical	f.w., b.w., s.w., a.
Phaeophycophyta	Brown algae	Chlorophyll a, c;—carotene+several xanthophylls including fucoxanthin; 3 thlakoids/band	Mannitol laminarin (=, 1 : 3— glucopyranoside, predominantly)	Cellulose, alginic acid, and sulfated mucopolysaccharides	2, unequal lateral	f.w. (rare) b.w., s.w.
Chrysophycophyta	Golden algae (including diatoms)	Chorophylla a, c (e in some); —carotenes + several xanthophylls; 3 thylakoids/band	Oil Chrysolaminarin (—1 : 3-glucopyranoside, predominantly)	Cellulose silicon, mucilaginous substances, and some chitin	1—2, unequal or equal, apical	f. w., b.w., s.w. t.a.
Pyrrhophycophyta	Dinoflagellates, in part	Chlorophyll a, c;—carotene+several xanthophylls; 2—3 thylakoids/band	Starch (probably like land plants), fats and oils	Cellulose or absent; mucilaginous substances	2, one trailing one girdling	f.w., b.w., s.w. a.
Rhodophycophyta	Red algae	Chlorophyll a, d, (in some); c—and r—phycocyanin, allo-	Floridean Starch (glycogen-like)	Cellulose, xylans, pectin, calcified in some	Absent	f.w., (some) b.w., s.w. (most)

contd.

TABLE 1 CONTD.

Division	Common Name	Pigments and and Plastid Organisation	Stored food	Cell wall[a]	Flagellar Number and Insertion[b]	Habitat[c]
		phycocyanin; c—and r—phycoerythrin; a+ —carotene+ several xanthophylls; 1 thylakoid/band	(glucopyranosides with a—1 : 4 and some 1 : 6 linkages)			

Class 4: Bacillariophyceae

Chromatophores are yellow or golden brown containing chlorophyll *a* and *c*, β-carotene and xanthophylls like fucoxanthin, diatoxanthin etc. Fat and volutin are the reserve foods. Pyrenoid-like bodies are present. The members are unicellular or colonial inhabiting fresh or marine waters. They are diploid and they are radially symmetrical (Centrales) or bilaterally symmetrical (Pennales). The cell wall is silicified and ornamented, consisting of two halves, the *epitheca* and *hypotheca*. Sexual reproduction involves formation of auxospores. The male gamete of centrales has a single pantonematic flagellum.

Class 5: Cryptophyceae

They have two large parietal chromatophores containing chlorophyll *a* and *b*, carotene, diatoxanthin, phycocyanin and phycoerythrin. Carbohydrates are the reserve foods. Motile cells are bifla gellate.

Class 6: Dinophyceae

Chromatophores are numerous, discoid, brown or dark yellow containing chlorophyll *a* and *c*, carotene dinoxanthin and peridinin Oil and carbohydrates are the reserve foods. Majority of the members are unicellular with a transverse and longitudinal flagella. They are plankton organisms, more widely represented in sea than in freshwaters.

Class 7: Chloromonadineae

The discoid chromatophores contain excess of xanthophyll. Pyrnoids are lacking and oil is the reserve food. The motile flagella have two equal flagella.

Class 8: Euglenineae

The cells have several green chromatophores. Photosynthetic pigments are chlorophyll *a* and *b*, β-carotene and xanthophylls (Neoxanthin, astraxanthin). The reserve food is paramylum. Euglenoid cells are naked. Flagella are 1—3, arising from an invagination at the anterior end.

Class 9: Phaeophyceae

Brown algae are restricted to sea except a few species. The degree of morphological complexity reached by these algae is greater than in any one of the previous classes. Usually, they are bulky and parenchymatous with complex external and internal differentiation. The simplest types are filamentous, some of which exhibit heterotrichy. Photosynthetic pigments are chlorophyll *a* and *c*, carotenes, and xanthophylls like fucoxanthin and diatoxanthin. Reserve foods are laminarin and mannitol. The characteristic constituents of cell wall include alginic acid and fucinic acid. The motile stages are pear-shaped with two unequal lateral flagella, one of which is unilocular and other plurilocular sporangia. Sexual reproduction ranges from isogamy to oogamy. The life cycles are of different types.

Class 10: Rhodophyceae

These are mostly marine. The simplest forms are filamentous but a majority attain a complex structure. The chromatophores are red containing the photosynthetic pigments r-phycoerythrin, r-phycocyanin together with chlorophyll *a* and *b* and carotenoid tetraxanthin. Reserve foods are floridein starch. Polysulphate esters are found in the cell walls. There are no motile stages with flagella. Sexual reproduction is oogamous. The carpogonium and spermatangia or antheridia are the sex organs. After fertilization, gonimoblast filaments and carposporangia are formed which are haploid or diploid.

Class 11: Myxophyceae or Cyanophyceae

Members of this class are abundant in freshwaters and in terrestrial habitats. Some are marine. They are unicellular, colonial or filamentous. Filaments are simple or may show false or true branching. The cells are *prokaryotic* without a nucleus, nucleolus, mitochondria and golgi bodies. Endoplasmic reticulum and chloroplasts are absent. The outer peripheral chromatoplasm contains photosynthetic pigments chlorophyll a, c-phycocyanin, c-phycoerythrin, myxoxanthin and myxoxanthophyll. The electron microscope reveals complex lamellar system in the chromatoplasm. Reserve foods are proteinaceous cyanophycin and cyanophycean starch. Flagellated structures are not formed. Sexual reproduction is lacking. The thallus has mucilage sheaths.

The primary classification of algae into different classes is based on (i) photosynthetic pigments, (ii) the type of reserve foods, (iii) nature of cell wall, (iv) the type of flagella and (v) some details of the cell structure. The details of somatic structure and the details of the process of reproduction are not particularly useful, although they are of importance in the classification of algae at the level of the lower taxa.

1. Photosynthetic Pigments in Algae

Algae vary in colour and it is often dependent on the environmental conditions. The exact nature of the pigment can be known by chemical analysis. The main pigments are chlorophylls, carotenoids and biliproteins.

Chlorophylls. On the basis of spectral properties, chlorophylls extracted from different algae have been recognized into different types *viz.*, Chlorophylls *a, b, c, d* and *e*. Chlorophyll *a* is found in all green cells except photosynthetic bacteria. Chlorophyll *b* found in higher plants is present in Chlorophyceae, Euglenineae and in no other class of algae. Chlorophyll *c* is widespread being present in Bacillariophyceae, Chrysophyceae, Cryptophyceae and Phaeophyceae. Chlorophyll *d* is present in Rhodophyceae and Chlorophyll *e* in zoospores of Vaucheria of Xanthophyceae.

Carotenoids. Carotenoids are of two kinds, carotenes and xanthophylls. β-carotene is present in most of the classes while α-carotene is found in the Caulerpales of Chlorophyceae, and to some extent in Rhodophyceae. In Charales (Chlorophyceae) lycopene and V-carotene are present. E-carotene is found in Bacillariophyceae.

Classification of Algae

There are several kinds of xanthophylls in algae, and some of them are exclusively found in particular algal groups:

Lutein	Chlorophyceae, Phaeophyceae, Rhodophyceae.
Fucoxanthin ⎫ Diatoxanthin ⎭	Chrysophyceae, Bacillariophyceae, Phaeophyceae.
Diadinoxanthin	Chrysophyceae, Bacillariophyceae, Pyrophyta.
Taraxanthin	Rhodophyceae.
Echinenone ⎫ Antheraxanthin ⎭	Euglenineae.
Myxoxanthin, ⎫ Myxoxanthophyll ⎭	Myxophyceae.

Biliproteins. These are water soluble pigments unlike chlorophylls and carotenoids which are soluble in lipid solvents. The free pigment cannot be extracted as it forms a complex with protein. So, these phycobilins are now termed as biliproteins to indicate the existence of pigment-protein complex. The bili-proteins are found in Rhodophyceae, Myxophyceae and Cryptophyceae. The bili-proteins are of two kinds, phycocyanin and phycoerythrin. They differ in Cyanophyceae and Rhodophyceae and therefore referred to as C-type and R-type.

The proportion of the various pigments changes in different groups and sometimes this may depend on the environmental conditions. Members of Chlorophyceae are green due to predominance of chlorophylls, those of Bacillariophyceae and Phaeophyceae are yellow-brown to brown, while members of Xanthophyceae are yellow-green due to excess of carotenoids. The Cyanophyceae are blue-green as C-phycocyanin and Chlorophylls occur in greater proportion. Rhodophycean members are red as r-phycoerythrin is found in greater proportion.

2. Reserve Foods

The nature of primary product of photosynthesis is same in all classes of algae but the nature of reserve food varies, affording an important criterion for classification. The reserve foods are mostly polysaccharides. In Chlorophyceae, the starch is like that of higher plants. Floridean starch and Cyanophycean starch are respectively stored in Rhodophyceae and Myxophyceae. Polysaccharides like laminarin in Phaeophyceae and leucosin in Bacillariophyceae are the reserve foods.

Other products like proteinaceous cyanophycin granules in Myxophyceae, floridoside and mannoglycerate in Rhodophyceae, mannitol in Phaeophyceae are also of taxonomic importance. A number of algae also accumulate fat. Bacillariophyceae accumulate higher proportion of fat than Chlorophyceae. Members of Myxophyceae resemble bacteria and differ from all other organisms in lacking sterols. In general, sterols vary with changes in the environment and therefore may not be useful in classification.

3. Cell Wall

The cellulose is the most common component of the cell wall. Alginic acid is found in the cell walls of Phaeophyceae and mucopeptide in the walls of Cyanophyceae. Bacillariophyceae have silicified walls.

4. Flagella

In Rhodophyceae and Cyanophyceae no flagellate structure are formed. In other classes, the nature of flagella, position and number are of importance. The fibrils of a flagellum typically show 9 + 2 structure. The flagella may be 2 or 4 and equal (Chlorophyceae) or may be only one as in Bacillariophyceae and Euglenineae. The flagella are of two kinds. The *acronematic* or whiplash type are smooth and the *pantonematic* or filmer type have longitudinal rows of hairs along the axis of the flagellum. In Chlorophyceae both are acronematic type, but in Cryptophyceae both are pantonematic. In Xanthophyceae and Phaeophyceae one is acronematic and the other is pantonematic type. The two flagella may be terminal as in Chlorophyceae or lateral as in Phaeophyceae.

5. Structure of the Algal Cell

There are two basic types of cells in the algae, *procaryotic* and *eucaryotic*. Procaryotic cells lack membrane bounded organelles (plastids, mitochondria, nuclei, Golgi bodies, and flagella) and occur in the Cyanophyceae. The procaryotic structure of the Cyanophyceae is described in detail under the class Cyanophyceae. The remainder of the algae are eucaryotic and have organelles.

A eucaryotic cell is often surrounded by a cell wall composed of polysaccharides that are partially produced and secreted by the Golgi body. The plasma membrane (plasmalemma) surrounds the remaining part of the cell; this membrane is a living structure responsible for controlling the influx and outflow of substances in the protoplasm. Locomotory organs, the flagella, propel the cell through the medium by their beating. The flagella are enclosed in the plasma membrane and have a specific number and orientation of microtubules. The nucleus, which contains the genetic material of the cell, is surrounded by a double membrane with pores in it. The contents of the nucleus are a nucleolus, chromosomes, and the background material or karyolymph. The chloroplasts have membrane sacs called thylakoids that carry out the light reactions of photosynthesis. The thylakoids are embedded in the stroma where the dark reactions of carbon fixation take place. The stroma has small 70 S ribosomes, DNA, and in some cases the storage product. Chloroplasts are surrounded by the two membranes of the chloroplast envelope. Sometimes chloroplasts have a dense proteinaceous area, the pyrenoid, which is associated with storage-product formation. Double membrane bounded mitochondria have 70 S ribosomes and DNA, and contain the respiratory apparatus. The Golgi body consists of a number of membrane sacs called cisternae stacked on top of one another. The Golgi body functions in the production and secretion of polysaccharides. The cytoplasm also contains large 80 S ribosomes and lipid bodies.

Cell Walls and Mucilages

In general, algal cell walls are made up of two components: *(i)* the fibrillar component, which forms the skeleton of the wall, and *(ii)* the amorphous component, which forms a matrix in which the fibrillar component is embedded. The most common type of fibrillar component is *cellulose*, a polymer of 1, 4 linked β-D-glucose. Cellulose is replaced by *mannan*, a polymer of 1, 4 linked β-D-mannose, in some Siphonaceous greens and in *Porphyra* and *Bangia* in the Rhodophyceae. Xylans of different polymers occur in some Rhodophyceae (*Porphyra, Rhodymenia*).

The amorphous mucilaginous components occur in the greatest amounts in the Phaeophyceae and Rhodophyceae, the polysaccharides of which are commercially exploited. *Alginic acid* is a polymer composed mostly of β—1, 4 linked D-mannuronic acid residues with variable amounts of L—guluronic acid. *Alginic acid* is present in the intercellular spaces and cell walls of the Phaeophyceae. *Fucoidin* also occurs in the Phaeophyceae and is a polymer of *a*-1, 2, *a*-1, 3, *a*-1, 4 linked residues of L—fucose sulfated at C-4. In the Rhodophyceae the amorphous component of the wall is composed of galactans or polymers of galactose, which are alternately β-1, 3 and β-1, 4 linked. These galactans include *agar* (made up of agaropectin and agarose), *carrageenan, porphyran, furcelleran,* and funoran. The amorphous polysaccharides of the Chlorophyceae are more complex, containing residues of D-galactose, L-arabinose, D-xylose, D-glucuronic acid, and L-rhamnose.

2 Habitat and Habit (Ecology) of Algae

TYPES

Based on their habitat three types of algae can be recognised, *viz.*, 1. Aerial and terrestrial algae; 2. Aquatic algae—(*a*) freshwater algae; (*b*) marine algae; 3. Algae of unusual habitat.

1. Aerial and Terrestrial Algae

These forms are often found epiphytic on trees like *Trentepohlia* and *Protococcus*. Some forms are found subterranean in soil which can withstand unfavourable conditions.

2. Aquatic Algae

Majority of the algae about 90 per cent are aquatic. They may be fresh water algae or marine algae. In the freshwater algae some are still water forms like *Oedogonium, Chara, Zygnema, Rivularia* etc. The running water forms among the freshwater algae include forms like *Cladophora* and *Vaucheria*.

The marine algae are those which live in sea water like *Sargassum, Dictyota, Ceramium, Gracilaria, Fucus, Laminaria* etc.

3. Algae of Unusual Habitat

They are found in different habitats like:
 (*a*) *Cryophytes* or *snow algae*, like *Haematococcus nivalis, Rapidonema, Chlamydomonas yellow-stonensis, Ancyclonema nordenskioldii, Protoderma*, etc. Some of these forms impart their own colour to the snow-fed mountains wherever they occur like red, pink, purple, yellow etc.
 (*b*) *Thermal algae,* which are found at very high temperatures as high as 85°C especially in hot springs.
 (*c*) *Halophytic algae* are found in water containing high concentrations like *Dunaliella, Stephnoptera, Chlamydomonas ehrenbergii* etc.
 (*d*) *Lithophytes* are found attached to stones and rocky areas, like *Rivularia, Gloeocapsa, Prasiola, Vaucheria,* Diatoms etc.
 (*e*) *Epiphytes* are algae found attached to other algae or higher plants like *Bulbochaete, Oedogonium, Coleochaete, Cephaleuros, Rivularia* etc.
 (*f*) *Aerophytes* are algae growing on leaves, bark or land animals termed respectively as epiphyllophytes, epihoeophytes, epizoophytes, like *Phormidium* and *Scytonema*.

Habitat and Habit (Ecology) of Algae   23

(g) *Symbiosis.* Many members of algae are found living in symbiotic association with plants or animals like *Nostoc* (on *Anthoceros*), *Anabaena cycadeae* (on *Cycas*, coralloid roots), *Anabaena* (on *Azolla*), on lichens (algae + fungi), *Chlorella* and *Azotobacter chroococcum*, on the nitrogen fixing nodules of Leguminosae plants.

Cladophora on snails and *Zoochlorella* within *Hydra* are examples of symbiosis on animals.

SIZE OF ALGAE

Size variations are enormous in algae from unicellular microscopic *Chlamydomonas* to huge tree like forms like *Laminaria, Postelsia palmaeformis,* and *Macrocystis pyrifera* which reach a height of 60 m.

FORM OF ALGAE

Like size, the structure of the algal organisation also varies enormously from simple, unicellular *Chlamydomonas* to highly differentiated plant like structure into leaf, stem and rhizoids as in *Sargassum, Laminaria* etc. Intermediate forms are simple filamentous colonial and branched algae like Oedogonium, Volvox, Polysiphonia, Batrachospermum etc.

REPRODUCTION IN ALGAE

Reproduction in algae takes place in three ways—1. Vegetative, 2. Asexual, and 3. Sexual.

1. Vegetative Reproduction

It may be simple fragmentation of the filament as in Chlorophyceae, Phaeophyceae and Rhodophyceae. Sometimes as in Myxophyceae it may be by hormogones which divide the filament into two segments. Cells of *Pleurococcus* divide into several parts.

2. Asexual Reproduction

It takes place by a variety of spores formed in different algae. They include: (a) Zoospores, (b) Palmella stage, (c) Aplanospores, (d) Hypnospores, (e) Akinetes, (f) Autospores (g) Auxospores, (h) Endospores, and (i) Cysts.

(a) Zoospores. These are formed during favourable conditions and they resemble exactly similar to the adult form or parent cell except for size, and are liberated by breaking up of the cell wall. Normally they are biflagellated, motile, and they give rise to adult plants after resting, settling and shedding of cilia. In *Ulothrix* the zoospores are quadriflagellated and in *Oedogonium* and *Vaucheria* they are multiflagellated.

(b) Palmella stage. If the zoospores are surrounded by a mucilaginous sheath and are not liberated which give the appearance of a parent colony. This temporary association is called Palmella stage and after sometime they are liberated and form new colonies as in *Chlamydomonas*.

(c) Aplanospores. During unfavourable conditions aplanospores are formed which are non-motile with a thick wall and they germinate into new plants on the onset of favourable conditions as in *Chlorella* and *Draparnaldia*.

(d) Hypnospores. Hypnospores are thick walled long resting spores and are formed in *Pediastrum* and *Sphaerella*.

(e) Akinetes. Akinetes are characteristic of *Cladophora* and *Pithophora* where the entire cell gives rise to a new plant and grows in size with a thick wall.

(f) Autospores. In some Scenedesmaceae and Chlorococcales the resting spores develop structures exactly like parent cell except in size and these are called autospores which give rise to adult forms.

(g) Auxopore. It is a specialized structure found in all Diatoms. During cell division some of the small cells get enveloped by mucilage which push the valves apart liberating the protoplast. This protoplast later develops into a parent cell.

(h) Endospores. Endospores are formed in Rhodophyceae which are also called tetraspores as in *Polysiphonia*.

(i) Cysts. Cysts are common in *Vaucheria* which are formed during unfavourable conditions and they are many layered.

ALGAL ECOLOGY

Planktonic Algae

The term 'planktonic algae' refers to the forms found floating or freely swimming in water. Among the freshwater planktonic algae, forms such as *Chlorella, Scenedesmus, Hydrodictyon, Chlamydomonas, Volvox* and *Eudorina* of Chlorophyceae, *Euglena* and *Phacus* of Eugleninae; *Microcystis, Anabaena, Aphanotheca, Spirulina, Arthrospira, Anabaenopsis* and *Oscillatoria* of Myxophyceae and *Melosira, Cyclotella, Pinnularia, Navicula, Fragilaria* and *Asterionella* of Bacillariophyceae are common while among marine planktonic forms *Phalacroma, Dinophysis, Exuviaella* and *Prorocentrum* of Desmophyceae; *Gymnodinium, Peridinium, Gonyaulax* and *Ceratium* of Dinophyceae; *Sceletonema, Cyclotella, Planktoniella, Eucampia, Hemidiscus, Chaetoceros, Biddulphia, Fragilaria, Asterionella* and *Nitzschia* of Bacillariophyceae; *Trichodesmium, Anabaena Oscillatoria* and *Aphanothece* of Myxophyceae and *Chlamydomonas* of Chlorophyceae are well-known.

Benthic Algae

The term 'benthic algae' refers to aquatic algae found attached to one or the other substratum. Among the freshwater forms, *Cladophora, Pithophora, Chara, Nitella* etc., and among marine forms most members of Phaeophyceae and Rhodophyceae are the common examples. *Cladophora, Enteromorpha, Porphyra, Polysiphonia, Sargassum, Laminaria, Chondrus, Ulva, Ectocarpus, Sphacelaria,* & *Acetabularia.* The prolific growth of benthic forms are found on rockycoasts such as in the areas of Rameshwaram in South India and Dwarka in Gujarat.

Thermal Algae

Some algae withstand or tolerate a very high temperature and these are often called thermal algae. Such forms are known to grow upto 85°C, nearly boiling water.

Majority of thermal algae belong to Myxophyceae, e.g., *Synechococcus elongatus, Mastigocladus laminosus, Phormidium tennue, Conferva thermalis* etc. A few forms belong to Chlorophyceae (Zygnematales) and Bacillariophyceae. Thermal algae reproduce by means of cell division and fragmentation and very rarely by spores.

Soil Algae

Such forms of algae that grow on or in soil are called soil or terrestrial algae or edaphophytes. *Vaucheria, Botrydium, Zygnema, Oedogonium, Microcoleus, Nostoc, Oscillatoria* etc. occur on soils.

Crybophytes

Certain algae are found growing on snow covered peaks of high mountains imparting attractive colours to snow. Common examples are—*Haematococcus nivalis, Chlamydornonas yellowstonensis, Raphidonema, Cylindrocystis, Protoderma, Scotiella. Ancyclonema nordenskioldii* imparts brownish to purple colour to snow.

Lithophytes

The algae growing attached to stones and rocky surfaces are called lithophytes. These may be of two types:
 (i) **Epilithic.** These include algae living on surface of rocks, e.g., *Calothrix, Rivularia, Gloeocapsa, Pleurocapsa, Ectocarpus, Polysiphonia* etc.
 (ii) **Endolithic.** These include algae which live inside the rocks, e.g., *Dalmatella* and *Podocapsa*.

Epiphytes

Some algae grow attached on the other plants and are called epiphytes. Such algae do not obtain the food from the plants on which they grow rather require support only. *Bulbochaete, Oedogonium, Ulothrix* etc., grow on other larger algae, besides, *Coleochaete* in association with *Chara* and *Nitella, Chaetophora* on leaves of *Vallisnaria* and *Nelumbo* and *Oedogonium* on *Hydrilla*.

Halophytes

Certain algae inhabit in water with high percentage of salt, as *Dunaliella* and *Stephanophora*. However, *Chlamydomonas ehrenbergii* and *Ulothrix flacca* have also been reported to grow in salt water.

Symbionts

A pretty large number of algae live in association with dissimilar organisms for their mutual advantage and are called symbiotic algae. *Nostoc* in *Anthoceros, Anabaena cycadae* in the coralloid root of *Cycas, Anabaena* species in *Azolla* etc.

However, lichens are the best examples of symbiosis where the association lies in between algae and fungi. *Trebauxia, Calothrix, Chlorella, Gloeocapsa, Nostoc* etc.

Endozoic Algae

Endozoic algae inhabit the protoplasm of other organisms, e.g., *Euglenomorpha, Zoochlorellae, Zooxanthellae, Carteria* etc. *Chlorella* like algae are found living within *Paramecium, Hydra* and certain molluscs and sponges (Cooke, 1975). *Zooxanthellae* live in intimate association with coral community.

Parasitic Algae

Some algae, for their food, are dependent on other plants and are termed as parasitic forms. The common intercellular parasite *Cephaleuros* (Chlorophyceae) grows on the leaves of angiosperms like *Magnolia, Rhododendron. Polysiphonia fastigata* is a semiparasite occurring on another algae

Ascophyllum nodosum (Phaeophyceae). Some blue green algae *Anabaeniolum, Oscillatoria* and *Simonosiella* are found as parasite on man and in the intestines of animals.

Algae of East Coast of India

Chlorophyceae. *Ulva, Cladophora, Bryopsis, Acetabularia, Neomeris, Udotea, Dictyosphaeria. Boodlea, Halimeda Caulerpa, Chaetomorpha littorea, Rhizoclonium kerneri, Caulerpa fergusonii, C. freycinetti.*

Phaeophyceae. *Ectocarpus, Giffordia, Streblonema, Hecatonema, Chnoospora, Colpomenia, Hydroclathrus, Iyengaria* and *Rosenvingea. Sphacelaria, Dictyota, Dictyopteris, Padina, Spatoglossum, Stoechospermum, Zonaria* and *Procockiella*.

The common species of various forms are—*Ectocarpus breviarticulatus, E. filifer, E. enhali, E. geminifructus, Chnoospora minima, Sphacelaria tribuloides, S. furcigera, Dictyota dichotoma, Padina gymnospora, P. tetrastromatica, Turbenaria conoides, Zonaria latisima, Z. crenata, Dictyopteris delicatula, D. muelleri, Sargassum wightii, S. cristaefolium* etc.

Rhodophyceae. *Chondria armata, C. transversalis, Acanthophora muscoides, Polysiphonia unguiformis, P. tuticorinensis, Rhodymenia dissecta, Liagora erecta, Porphyra tenera, Martensia fragillis, Gracilaria disticha, G. lichenoides, Ceramium gracillimum* etc.

ALGAL FLORA OF VISAKHAPATNAM (ANDHRA PRADESH) COAST.
(UMAMAHESWARA RAO & SRIRAMULU, 1970)

	Chlorophyceae	Phaeophyceae	Rhodophyceae	Total
Genera	11	13	31	55
Species	20	16	44	80

Class: Chlorophyceae

Family: Ulvaceae
    Enteromorpha compressa (L.) Greville   *Ulva fasciata* Delile (—)

Family: Cladophoraceae
    Chaetomorpha antennina (Bory) Kutzing   *Chaetomorpha brachygona* Harvey
    Chaetomorpha linoides Kutzing   *Chaetomorpha torta* (Farlow) McClatchie
    Cladophora   *Cladophora colabense* Borgesen
    Cladophora patentiramea (Mont.) Kutz.   *Cladophora utriculosa* Kutzing
    forma *longiarticulata* Reinbold.
    Spongomorpha   *Spongomorphia indica* Thivy et Visalakshmi

Family: Valoniaceae
    Boodlea struveiodes Howe

Family: Derbesiaceae
    Derbesia turbinata Howe et Hoyt

Family: Bryopsidaceae
    Bryopsis pennata Lamouroux   *Pseudobryopsis mucronata* Borgesen

Family: Caulerpaceae
    Caulerpa fastigata Montagne   *Caulerpa racemosa* (Forsk.) J. Agardh var.
                                                   macrophysa (Kutzing) Taylor
    Caulerpa sertularioides (Gmelin) Howe   *Caulerpa taxifolia* (Vahl) C. Agardh

Family: Codiaceae
    Codium iyengarii Borgensen

Class: Phaeophyceae

Family: (Ectocarpaceae)
    Bachelotia antillarum (Grunow) Gerloff   *Feldmannia irregularis* (Kutzing) Hamel
    Giffordia mitchellae (Harvey) Hamel   *Strebolnema turmale* Borgesen
Family: Sphacelariaceae:
    Sphacelaria furcigera Kutzing   *Sphacelaria tribuloides* Meneghini
Family: Dictyotaceae
    Dictyota dichotamoa (Hudson) Lamouroux   *Padina tetrastromatica* Hauck
    Pocockiella variegata (Lamouroux) Papenfuss
Family: Myrionemataceae
    Myrionema sp.
Family: Ralfsiaceae
    Ralfsia expansa J. Agardh
Family: Punctariaceae
    Rosenvingea nhatrangensis Dawson   *Chnoospora minima* (Hering) Papenfuss
Family: Sargassaceae
    Sargassum illicifolium (Turner) C. Agardh   *Sargassum tenerrimum* J. Agardh
    Sargassum vulgare C. Agardh

Class: Rhodophyceae

Family: (Bangioideae or Bangiaceae)
    Bangiopsis subsimplex (Montagne) Schmitz   *Erythrocladia subintegra* Rosenvinge
    Erythrotrichia obscura Berthold   *Porphyra vietnamensis* Tanaka et Ho

Sub-class: Florideae

Family: (Acrochaetiaceae or Acrochaetium Naegeli)
    Acrochaetium iyengarii Borgesen   *Acrochaetium krusadii* Borgesen
    Acrochaetium sargassicola Borgesen
Family: Helminthocladiaceae
    Liagora erecta Zeh   *Liagora visakhapatnamensis* Umamaheswararao
Family: Chaetangiaceae
    Scinaia bengalica Borgesen
Family: Gelidiaceae
    Gelidiella myriocladia (Borgesen) Feldmann et Hamel   *Gelidiopsis variabilis* (Greville) Schmitz
    Gelidium heteroplatos Borgesen   *Gelidium pusillum* (Stackhouse) LeJolis
Family: Squamariaceae
    Hildenbrandia prototypus Nardo   *Peyssonnelia conchicola* Piccone et Grunow
    Peyssonnelia obscura Weber-Van Bosse var. *bombayensis* Borgesen

Family: Corallinaceae (Melobesieae)
    Fosliella farinosa (Lamouroux) Howe   *Fosliella minutula* (Foslie) Ganesan
    Dermatolithon ascripticium
    (Foslie) Setchell et Mason
Family: Corallinaceae (Corallineae)
    Amphiroa fragilissima (L) Lamouroux   *Fania rubens* (L.) Lamouroux
Family: Grateloupiaceae
    Grateloupia filicina (Wulfen) C. Agardh   *Grateloupia lithophila* Borgesen
Family: Gracilariaceae
    Gracilaria corticata J. Agardh   *Gracilaria textorii* (Sur.) J. Agardh
    Gracilariopsis sjoestedtii (Kylin) Dawson
Family: Hypneaceae
    Hypnea musciformis (Wulfen) Lamouroux
Family: Gigartinaceae
    Gigartina acicularis (Wulfen) Lamouroux
Family: Ceramiaceae
    Wrangelia argus Montagne   *Aglaothamnion cordatum* (Borgesen) Feldmann-Mazoyer
    Spermothamnion speluncarum   *Spermothamnion* sp.
    (Collins et Hervey) Howe
    Ceramium cruciatum Collins et Hervey   *Ceramium fimbriatum* Setchell et Gardner
    Ceramium gracillimum (Kitz.) Griff. et Harv.   *Centraceras clavulatum* (C. Agardh) Montagne
    var. *byssoideum* (Harvey) Mazoyer
Family: Rhodomelaceae
    Polysiphonia ferulaceae Subr   *Polysiphonia plactycarpa* Borgesen
    Bryocladia thwaitesii (Harvey) De Toni   *Herposiphonia secunda* (C. Agardh) Ambronn
    Herposiphonia tenella (C. Agardh) Ambronn   *Chondria cornuta* Borgesen
    Acanthophora spicifera (Vahl) Borgesen

Algae of West Coast of India

Chlorophyceae. West coast algal flora is enriched with 28 genera and 72 species of Chlorophyceae. A few characteristic species are *Enteromorpha tublosa, Ulva reticulata, Bryopsis hynoides, Acetabularia moebii, Struvea anastomosans. Caulerpa* with its several species is quite common throughout the coast.

Phaeophyceae. The common forms are *Ectocarpus arabicus, E. enhali, Giffordia mitchellae, Colpomenia sinuosa, Iyengaria stellata, Rosenvingea intilicata, Sphacelaria tribuloides, S. furcigera, Dictyota dichotoma, D. divaricata, D. cervicornis, Dictyopteris australis, Padina gymnospora, P. tetrastromatica, Spatoglossum variable, Cystophyllum muricatum, Sargassum tenerrimum, S. cinereum, Turbenaria decurrens* etc.

Rhodophyceae. West Coast is quite rich as regards to Rhodophyceae being represented by 89 genera and 175 species. A few most characteristic forms are *Scinaia hatei, Asparagopsis taxiformis, Nitophyllum punctatum, Rhodymenia australis, Hypoglossum spathulatum* etc.

Role of Algae in Nitrogen Fixation
List of Algae Capable to fix Atmospheric Nitrogen

Order	Nitrogen fixing forms
1. Chroococcales	*Chlorogloea fritschii, Gloecocapsa,* sp.
2. Nostocales	*Oscillatoria princeps. Anabaena ambigua A. solicola, A. cylindrica, A. doliolum, A. fertilissima, A. hemicola, A. naviculides, A. oryzoe, A. variabilis, A. ozillii. Anabaenopsis cirularis, Cylindospermum majus, C. gorakhpurense, C. licheniforme, C. sphaerica, C. muscicola, Nostoc commune, N. sphaericum, N. paludosum, N. calcicola, N. punctiforme, N. entophytum, N. muscorum, Aulosira fertilissima, A. prolifica, A. ambigua, Calothrix brevissima, C. elenkinii, C. parietina, Nodularia harveyana, Fischerella mucicola.*
3. Stigonematales	*Tolypothrix tenuis, Scytonema oscillatum, S. hofmanii. S. bohneri. S. arcangelii, Stigonema dendroideum, Mastigocladus laminosus, Westeilopsis prolifica.*

Mechanism of Nitrogen Fixation

A fairly large amount of work has been done to establish the structural and functional linkages between N_2-fixation and photosynthesis and it is now a well-known fact that nitrogen fixing enzyme (nitrogenase) in blue green algae has a definite requirement of light for providing ATP through photophosphorylation. In this process, ferredoxin as a reducing agent is also required which is generated in the light by action of photosystem I of photosynthesis.

The recent findings are still more interesting establishing the fact that nitrogen fixation is controlled by genes. A block of genetic set consisting of nitrogenase complex and heterocystulon complex which contain nitrogen fixing genes *(nif)*. Each complex has one operator gene and a few structural genes or *nif* but both are regulated by common regulator gene.

Heterocyst as Nitrogen Fixing Organ

Fogg (1949) for the first time suggested the role of blue green algae in nitrogen fixation.

Stewart, Haystead and Pearson (1969) have provided further information about the role of heterocysts in N_2-fixation. Nitrogen fixation is a reductive process and is inhibited in blue green algae by high oxygen levels (Stewart and Pearson, 1970). The heterocysts are unlikely to evolve oxygen and, therefore, it seems reasonable to suppose that nitrogenase should be active in non-oxygen evolving heteroysts rather than in oxygen evolving vegetative cells. They observed that heterocysts were the sites of reducing activity associated with nitrogen fixation.

CYANOPHAGES OR PHYCOPHAGES

In some recent years a few phages (viruses) have been found infecting blue-green algae. These phages are called cyanophages. The first known isolated cynophage is LPP-I. The name LPP-I refers to its hosts, *Lyngbya, Plectonema* and *Phormidium*.

EXTRACELLULAR PRODUCTS OF ALGAE

A variety of kinds of organic substances are liberated in extracellular form algae. The amount may be quite high sometimes equal to that of intracellular materials. These extracellular products are liberated from healthy cells of algae belonging to several different groups, *viz.*

Extracellular products	Name of algae
Carbohydrates	*Chlamydomonas, Scenedesmus, Chlorella, Oocystis, Trebauxia, Zooxanthellae, Nitzschia, Synendra, Strephonodiscus, Gymnodinium, Fucus, Laminaria Ascophyllum, Chondrus, Anabaena, Nostoc, Scytonema, Calothrix.*
Nitrogenous substances	*Chlamydomonas, Scenedesmus, Chlorella, Volvox, Ochromonas, Euglena, Cyclotella, Synendra, Stephanodiscus, Gymnodinium, Ectocarpus, Fucus, Laminaria, Polysiphonia, Chondrus, Anabaena, Lyngbya, Nostoc, Microcystis, Oscillatoria.*
Organic acids	*Chlamydomonas, Chlorella, Scenedesmus, Chlorococcum, Ankistrodesmus, Coccoliths, Gymnodinium, Peridinium, Cyclotella, Stephanodiscus, Oscillatoria, Anacystis.*
Vitamins	*Dunaliella, Gonyaulax, Cyclotella, Phaeodactylum.*
Phenolic substances	*Fucus, Ascophyllum, Laminaria.*
Sex factors	*Chlamydomonas, Volvox, Ectocarpus.*
Volatile substances	*Cryptomonas, Chlamydomonas, Synura.*

Ecological Implications

The liberation of extracellular products by algae is of quite ecological importance in many ways. In most simple cases, the nutrient cycle is short circuited, i.e., photosynthetic products of algae liberated by healthy cells are used directly by bacteria and some animals as food material. Extracellular products such as organic acids and polysaccharides are utilised quickly. Many algae and bacteria are able to use glycolate as a carbon source.

The phenomenon of extracellular release of enzymes by algae is of another importance. In this light, the growing use of algae in sewage disposal deserves further investigation.

FOSSIL ALGAE

Majority of algae are not easily preserved because of delicate and readily decomposable composition. However, fossil algae include members belonging to Chlorophyceae, Rhodohyceae, Cocoliths, Bacillariophyceae, Silicoflagellates and Myxophyceae. Some of them are calcium or magnesium deposits (Chlorophyceae, Rhodophyceae, Cocoliths) and others are siliceous deposits. Fossil algae have been recorded from the rocks of all ages as far back as pre-Cambrian (about 1200 million years ago) and have been found preserved as impressions, carbon films, molds and casts and petrifactions.

Halimeda, Rhabdoporella, Vermiporella, Cenolosphaeridium, Primicorallina, Mizzea etc.

Diplopora, Cyclocrinus, the fossil members of Charales are found as hollow calcareous bodies with a basal aperture.

Chlamydomonas-like alga, *Gloeocystis* has been reported from Upper Jurassic and flagellated form *Phacotus* from Tertiary.

Silicoflagellates, diatoms and members of Chrysophyceae readily fossilize. Most of them occur as thick deposits. The fossil diatoms date back to Jurassic and majority of them belong to the centric groups. The important diatom fossil forms are *Cyclotella iyengarii, Stephanodiscus aestreal, Cymbella affinis,* etc., *Cymbella affinis, Gyrosigma accuminatum, Fragilaria kasmeriensis* are few important Indian Pleistocene diatoms.

Few fossil blue-green algae have also been found. *Marpolia, Gloeocapsomorpha, Archothrix oscillatorifomis* (cellular dimensions resembling to Oscillatoria). *Anabaenidium, Palaeomicrocoleus* and *Palaeonostoc* etc. *Palaeonostoc indica* has been recently reported from India.

The important fossil forms belonging to Pyrrophyceae are *Eodinia, Lithodinia, Biecherella, Cacisphaerella,* Phaeophyceae and *Lithothamnion, Lilkophyllum Melobasia* etc.

Myxophyceae—Paleozoic (Archaeozoic)
Chlorophyceae—Paleozoic (Ordovician)
Xanthophyceae—Paleozoic (Mississippian)
Bacillariophyceae—Mesozoic (Triassic)
Dinophyceae—Mesozoic (Triassic)
Phaeophyceae—Paleozoic (Silurian)
Rhodophyceae—Paleozoic (Ordovician)
Charophyceae—Paleozoic (Ordovician)

Examples. *Gloeocapsa, Gloeotheca, Lithothamnium glaciale, Halimeda, Chlorellopsis, Chara, Gloeocapsamorpha, Epiphyton, Palaeoporella, Cyclorinus, Bornetella, Dimorphosiphon, Sphaerocodium, Solenospora, Botryococcus, Pyxidicula, Conscinodiscus, Actinoclava.*

3 Commercial Cultivation of Algae

INTRODUCTION

Commercial cultivation of seaweeds involves the large-scale raising of macroscopic marine algae for commercial purposes. It can be viewed as a type of agricultural practice carried out in the sea, to which the term 'marine agronomy' has been applied (Doty 1977). The following account treats this subject by focusing on four major marine crop plants: *Porphyra*, *Eucheuma*, *Laminaria japonica* and *Undaria pinnatifida*. It will examine the scientific principles that affect this practice as well as the actual techniques that have been developed in this relatively new technology, in China, Japan and the Philippines, where the large-scale farming or valuable seaweeds is underway. Some references to work in other countries are also included, but in many countries this science is often still in an embryonic stage.

Economic Seaweeds Under Commercial Cultivation and their Economics

According to Bonotto (1976) there are five genera of green seaweeds including 27 species, 40 genera of brown seaweeds including 88 species, and 56 genera of red seaweeds including 344 species, totalling 101 genera and 459 species of seaweeds which are of economic value as human food, as animal food, as manure, for medical or pharmaceutical purposes, or for industrial use. Bonotto's list may be reduced to 99 genera and 457 species by realizing that a freshwater species of *Prasiola* was included and that *Gracilariopsis* and *Gracilaria confervoides* can be regarded as synonymous with *Gracilaria* and *Gracilaria verrucosa*, respectively. Bonotto had overlooked the manual of economic seaweeds of China in which 46 genera including 100 species were described. Eight genera with 36 species were not contained in Bonotto's list. So the present total of economic seaweeds is 107 genera and 493 species. It is noteworthy that two species of blue-green algae are included, namely, *Brachytrichia quoyi* and *Calothrix crustacea*.

Among the seaweeds utilized by man for various purposes, only 11 genera with less than 20 species are commercially cultivated to any significant extent. These eleven genera are *Porphyra*, *Gelidiella*, *Gloiopeltis*, *Eucheuma* and *Gracilaria* of the red algae, *Laminaria*, *Undaria*, and *Macrocystis* of the brown algae, and *Monostroma*, *Enteromorpha* and *Caulerpa* of the green algae. Most of these algae, however, are not actually domesticated in the sense of the crop plants we know in agriculture. Perhaps only four genera, namely, *Porphyra*, *Eucheuma*, *Laminaria* and *Undaria*, for which the amount harvested from cultivated sources exceeds that taken from wild populations, can be regarded as truly marine crop plants.

The Marine Crop Plant *Porphyra* (Figs. 3.1 to 3.3)

Species of *Porphyra* are popularly known in China as *zicai* (meaning purple vegetable), in Japan as *nori*, and in the western world as 'purple laver', and have long been appreciated by coastal people as a delicious food, especially by the Japanese and Chinese. Both China and Japan have a long history of the cultivation and utilization of *Porphyra*. This is a very widely distributed genus, primarily temperate but also occurring in many subtropic and subarctic regions. The purple lavers are in general littoral seaweeds, growing on rocks and only in few cases growing epiphytically or in the sublittoral zone. In China and Japan seven species are used in commercial cultivation: *P. yezoensis, P. tenera* (China and Japan), *P. haitanensis* (China) and *P. pseudolinearis. P. kuniedai, P. arasaki* and *P. seriata* (Japan), of which the first three species are the most important, perhaps accounting for more than 90% of the total production. In Santou, Guangdong Province, *Porphyra guangdongensis* and in Penghu, Taiwan Province, *Porphyra* sp. are also cultivated but on a very small scale.

In China the earliest records of the occurrence of *Porphyra* and its food value appeared in books published in the years 533–544. In the Sung Dynasty (A.D. 960–1279) *Porphyra* had already become a local good from Haitan Island of the Fujian Province selected for presenting to the emperor annually. This means that about one thousand years ago the Chinese had already come to regard *Porphyra* as a delicacy. More than 200 years ago the simple 'rock clearing' method of cultivation was devised by mechanically clearing various seaweeds and barnacles from the rocks in early autumn. This was done just before the mass liberation of the *Porphyra* spores so that the liberated spores had the required space for attachment and growth. Modern methods of commercial cultivation involving artificial collection of spores were initiated in the early sixties on the basis of the research of the previous decade, which in turn had received impetus from Drew's discovery of the life history of *Porphyra* in 1949. These methods are now routinely used in the *Porphyra* cultivation industry in China. Two principal species are involved: *P. yezoensis* north of the Changjiang (Yangtze) River and *P. haitanensis* south of it. About 3400 hectares of the sea surface are presently devoted to *Porphyra* farms, although the actual area of the cultivation raft is only about 27% of that area. In 1979 China's production was about 7200 tonnes in dry weight, averaging 2.1 tonnes per hectare, and worth about U.S.$30,000,000. The product is in the form of paper-like sheets, as in the case of the Japanese product.

Commercial cultivation of *nori* in Japan was initiated more than 300 years ago by the primitive method of inserting bundles of bamboo twigs, called *hibi*, for the 'seeding' process, i.e. for collecting spores. This method was much more efficient than the primitive 'rock cleaning' method used in China. The two methods are similar, however, in that the number of spores attaching to the substrates, hence the resulting production, is entirely at the mercy of nature, and the annual production fluctuated greatly. The net cultivation method was introduced in the 1920's, and some increases in productivity resulted. It was not until the 1960's that, with the introduction of the artificial collecting of spores, the commercial cultivation methods became truly modernized, as is true also in China. This progress resulted from the research of several workers. With the innovation of the cold-storage net and the use of floating nets, *nori* production increased steadily. In 1978, according to Dr. Miura, 612.8 square kilometres of the sea surface were devoted to the cultivation of *nori*, involving more than 30,000 production teams and 3 75 8000 *satsus**. A total of 21 150 tonnes (averaging 345 kg/hectare) of dry *nori* was produced with a value of 129 700 million Japanese yen, or about U.S. $540 000 000. It is undoubtedly the most economically important food seaweed industry in the world. *Porphyra tenera* and *P. yezoensis* are the two principal species cultivated, although six species are grown commercially.

---

* Satsu is the Japanese unit of production area, equivalent 21.6 m^2 nettings.

The Marine Crop Plant *Eucheuma* (Figs. 3.4 to 3.6)

Eucheuma is a genus of tropical red seaweed which grows generally on limestone rich substrates, especially coral reefs, and usually is most abundant in the zone from the lowest tide level to the upper 1 m level of the sublittoral region. Appreciated as a food in China and Malaysia and other southeastern Asian countries, *Eucheuma* is known as *gilin cai* (meaning unicorn vegetable) in China and as *agar-agar* in Malaysia—different in meaning, however, from the commercial agar-agar. *Eucheuma* has served more recently as the raw material from which the useful phycocolloid carrageenan is extracted, a key substance in today's food and dairy industries. This genus can be divided into two natural groups: the axiferous *E. denticulatum* (= *E. muricatum*) group, characterized by a dense central axis of slender cells in the centre of each branch and producing iota carrageenan, and the anaxiferous *E. striatum* group, characterized by kappa carrageenan.

The traditional raw material for the carrageenan industry is *Chondrus crispus*, or Irish moss, also a red seaweed. With the increasing demand for this seaweed colloid, the production of *Chondrus* has not met the demand, and the carrageenan manufacturers have had to turn their attention to other sources, particularly *Eucheuma* and other genera in the order Gigartinales. Some years ago, 3–4000 tonnes of *Eucheuma* in dry weight reached the world market annually. Over-harvesting resulted in the decrease in production of wild *Eucheuma*, and cultivation experiments were started in the Philippines, where some innovative techniques were devised. By mid-1974 there were more than 1000 farms engaged in the production of *Eucheuma*, exporting over 600 tonnes in dry weight per month, thus producing about twice as much *Eucheuma* as the original total world supply. The present annual production is more than 10 000 tonnes in dry weight.

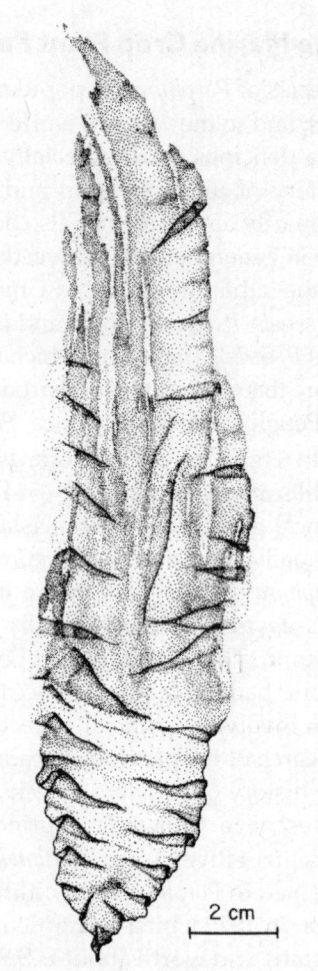

Fig. 3.1 Habit sketch of *Porphyra yezoensis* from Qingdao, China.

In China *Eucheuma gelatinae* has been commercially cultivated in the Qionghai and the Wenchang districts of Hainan Island since 1960. At first, *Eucheuma* was cultivated by sending divers to insert cuttings of the seaweed in the sublittoral reefs. Since 1974, new cultivation methods have been devised, and the cuttings are fastened to coral branches with rubber rings or by some other means, thrown into the sublittoral reefs, and then finally rearranged in order by divers. At present the annual production is about 300 tonnes in dry weight, valued at about U.S.$450 000.

The Marine Crop Plant *Laminaria japonica* (Figs. 3.7 to 3.9)

The largest single-species seaweed cultivation industry, as far as the quantity of production is concerned, is undoubtedly that of the brown seaweed *Laminaria japonica*, as is now practised in

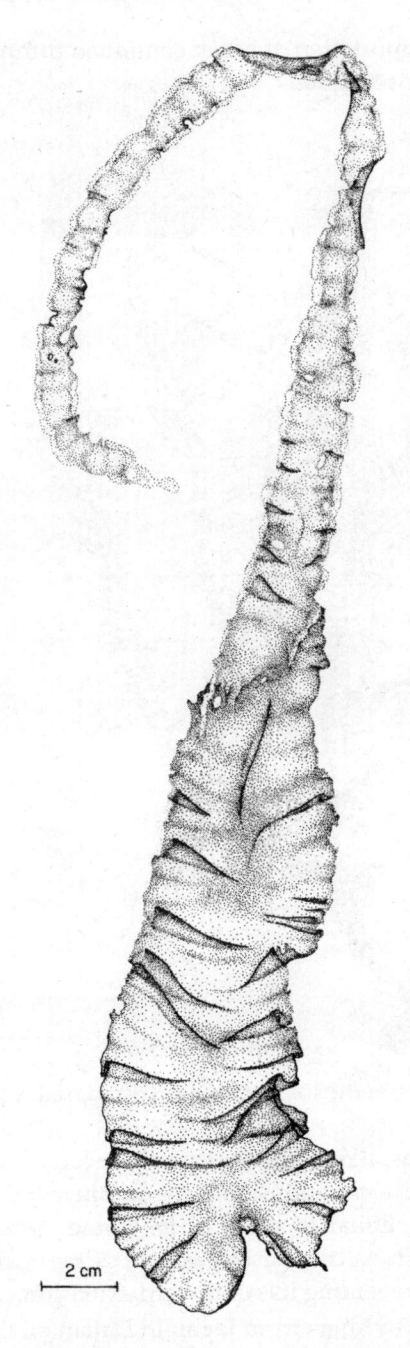

Fig. 3.2 Habit sketch of *Porphyra haitanensis* from Haitan Island, Fujian Province, China.

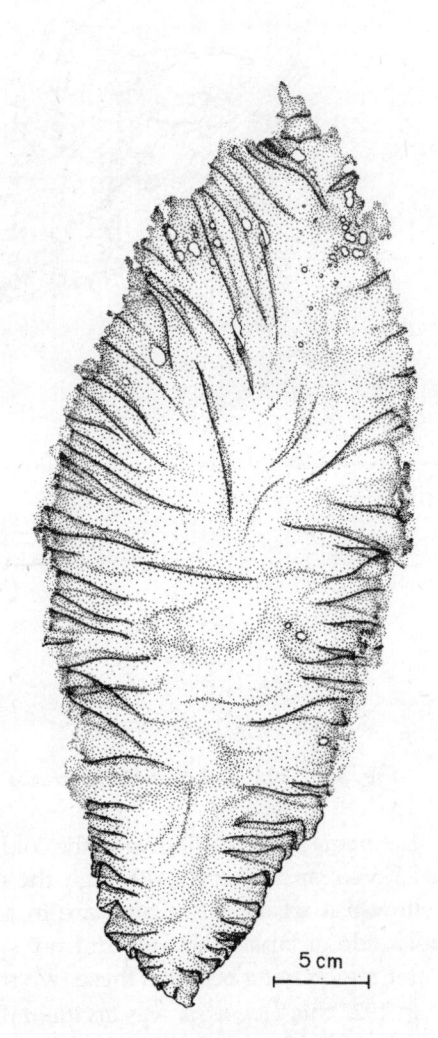

Fig. 3.3 Habit sketch of *Porphyra tenera* from Qingdao.

China and Japan. In China this *Laminaria*, commonly known as *haidai*, has served as a food for more than 1000 years. The popularity of eating this kelp, with its very high iodine content, has helped greatly in the prevention and treatment of goiter in many places in China. Indeed, it is hard to believe that such a widely accepted food material is not native to China but had been imported in dry form from Japan and Korea in great quantity. Importation of *haidai* continued during the first few years after the founding of the People's Republic of China.

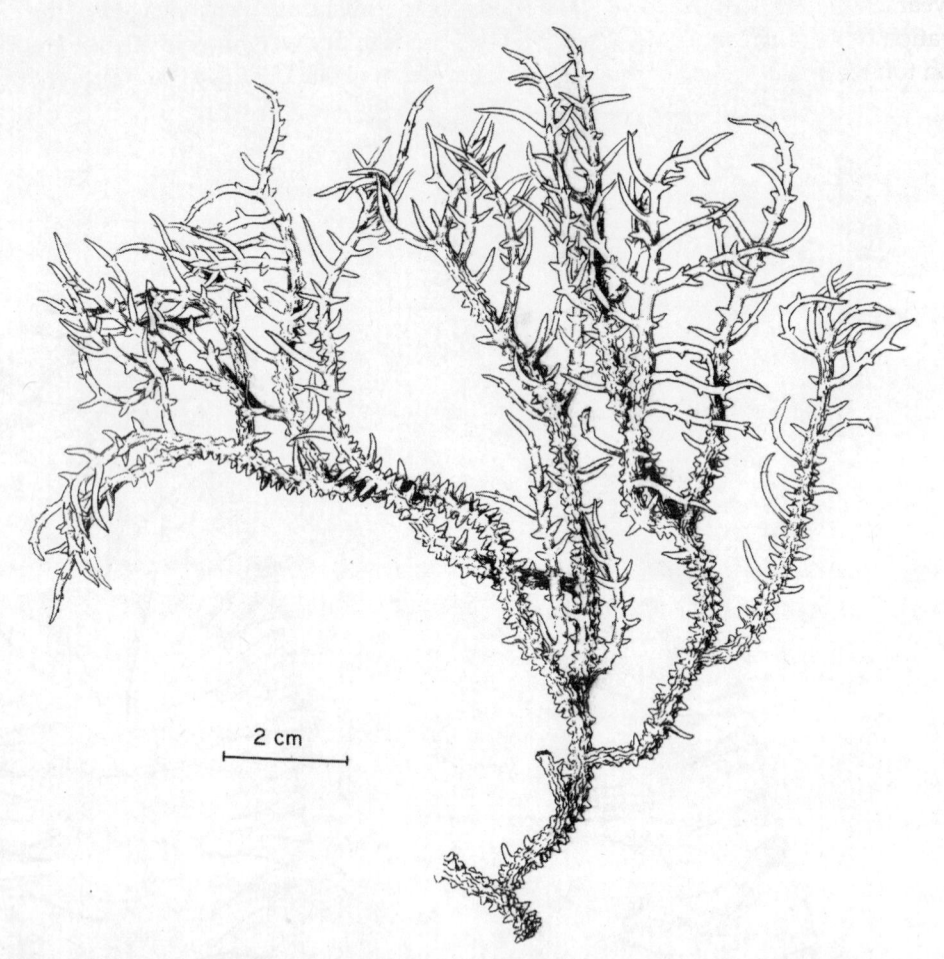

Fig. 3.4 Habit sketch of *Eucheuma denticulatum* from the local market in Shanghai, China.

Laminaria japonica is native to the cold temperate coastal regions along northern Japan, northeastern Korea, and Siberia. Although the oceanographic conditions along the northern coast of the Yellow Sea adjacent to China are in some respects similar to those of its 'home' near western Hokkaido of Japan, this kelp did not spread to the China coast itself because of the wide warm-water region lying between these two regions, thus preventing its southward extension.

In 1927 this *Laminaria* was accidentally introduced by ships from Japan to Dalian on the northern coast of the north Yellow Sea and settled in the vicinity of the harbour. After a few more trials in transplanting this kelp from Japan, *haidai* finally became a permanent component of the seaweed flora of China. In the forties *haidai* was transplanted to Yantai on the southern coast of the north Yellow Sea and in 1950–1951, after the founding of the People's Republic of China, it was further successfully transplanted to Qingdao (Tsingtao) on the southern coast of the Shandong Peninsula. In 1952 the newly devised floating raft cultivation method was employed with success. Since then, new methods for enhancing the production and improving the quality of the kelp have been successfully devised one after the other. Now, commercial cultivation of *haidai* is being practised on the entire China coast of the East China Sea (including the Yellow Sea), from Dalian of the Liaoning Province in the north to Xiamen (Amoy) and Dongshan of the Fujian Province in the south. In the kelp year 1978–1979, more than 18 000 hectares of marine farms were engaged in the commercial cultivation of this *Laminaria*, and about 275 000 tonnes in dry weight (equivalent to more than one million tonnes in wet weigt) of this seaweed, valued at about U.S.$300 000 000, were produced.

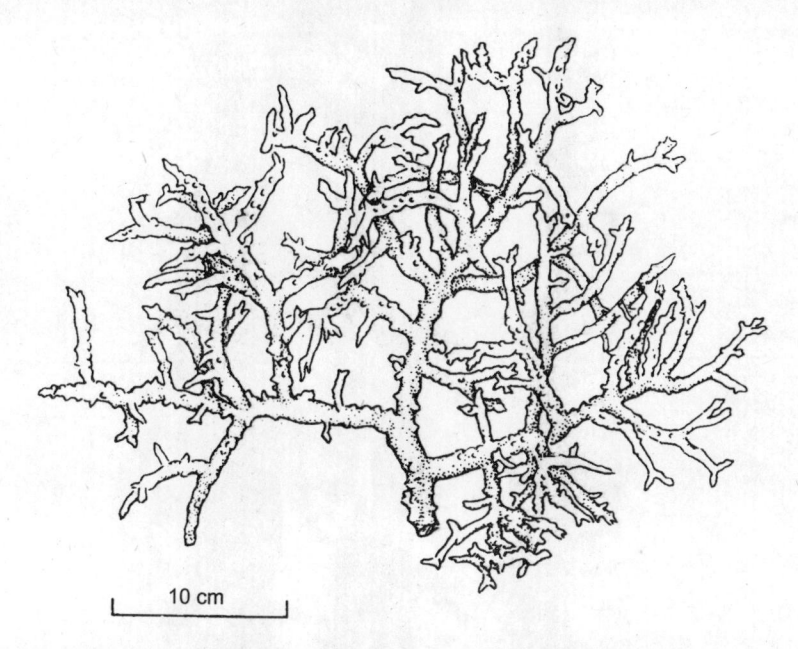

Fig. 3.5 Habit sketch of *Euchetum striatum* from the Philippines (after Doty 1973).

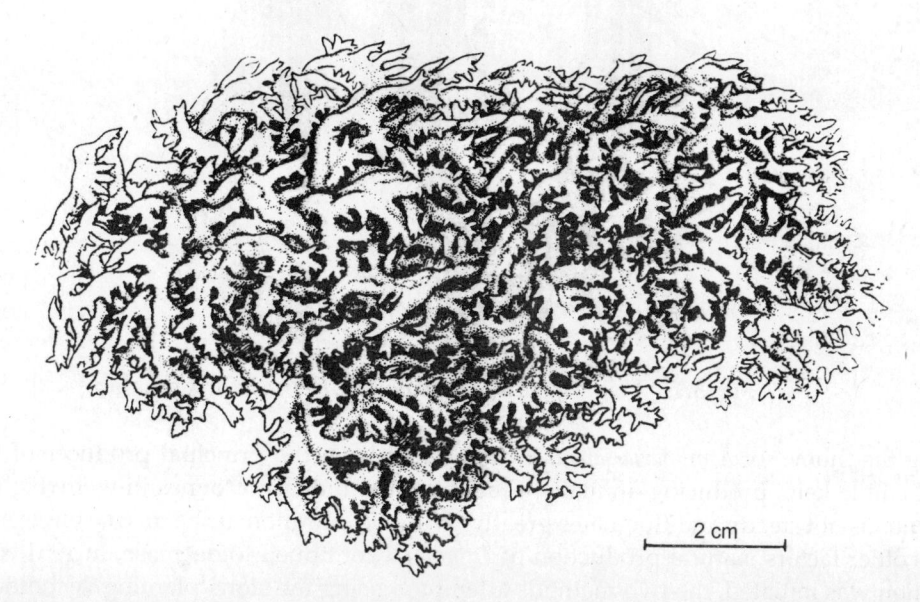

Fig. 3.6 Habit sketch of *Eucheuma gelatinae* from Qionghai, Hainan Island, China.

38 *A Textbook of Algae*

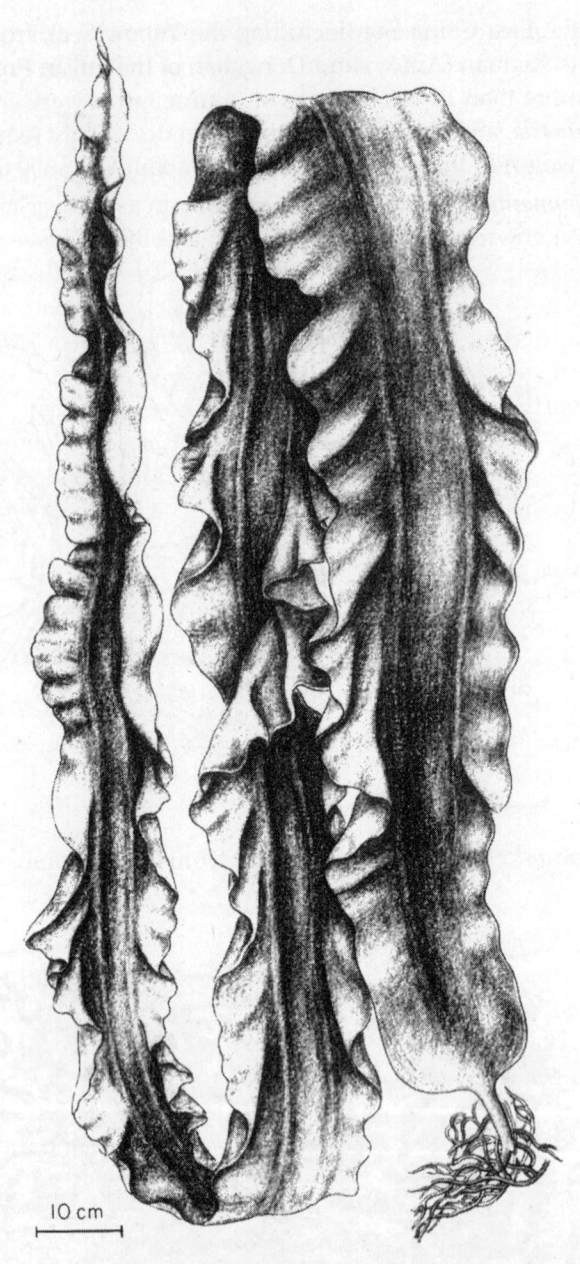

10 cm

Fig. 3.7 Habit sketch of *Laminaria japonica* from Qingdao.

Being the 'home' or *Laminaria japonica*, Japan had been the principal producer of this much desired edible kelp, producing annually a few hundred thousand tonnes in wet weight. Natural production is not steady and fluctuates greatly. With the pollution of the marine environment and various other factors, natural production of *Laminaria* continued to decrease. In 1951 commercial cultivation was initiated, the two methods attempted being the stone-planting, technique and the blasting or reefs. Improvement in production was, however, not promising, and production was unable to meet the demands of the people. Since 1968, the so-called forced cultivation method has been employed, and the output has increased from 30 tonnes in 1969 to over 7500 tonnes in wet weight in 1974, worth about U.S.$1 000 000, amounting to 16% of the natural production on reefs.

The Marine Crop Plant *Undaria pinnatifida*

The other brown seaweed now under commercial cultivation and qualified to be called a marine crop plant is *Undaria pinnatifida*, popularly known in Japan as *wakame* and very much prized as a food alga. In China it is known as *qundai-cai*, also serving as food. This seaweed is a warm temperature species, widely distributed in Japan except for the eastern and northern coasts of Hokkaido, which receive the cold Kurile Current, and the eastern coasts of Kyushu and Shikoku, which receive the warm Kuroshio. It grows on rocks and reefs at a depth of 1–8 m below the low tide level in open seas or within bays near the open sea.

In Japan *Undaria pinnatifida* is a more important marine crop than *Laminaria japonica* both in value and in production. For years, maintenance and increase of the output of its natural production had been accomplished by various modes of propagation, such as depositing stones on the sea bottom and blasting rocky reefs to increase the area for its attachment. In this way the annual product of *Undaria* from its natural habitat had been maintained at 40 000–60 000 tonnes in fresh weight in the years 1960–1969. Since 1955, mass production of *Undaria* sporelings has been carried out by artificial spore-sowing on ropes of hemp-palm and later on synthetic ropes. In 1962 production in this way, in new cultivation areas, reached 21 000 tonnes in fresh weight, and in 1968 annual production by this method reached a peak production of 69 680 tonnes in fresh weight, surpassing the production of 48 300 tonnes in fresh weight in its natural habitat by a broad margin.

In China *Undaria pinnatifida* has been collected for food from the natural habitat for centuries in the Zhoushan Island, Zhejiang Province on the East China Sea coast. *Undaria* now growing in Qingdao and Dalian on the Yellow Sea coast was, however, transplanted from Korea and perhaps also from Japan.

The first transplantation of *Undaria* from southern Korea to Qingdao took place in 1935 by a Korean who established a mariculture company to grow *Undaria* and some other marine products in Qingdao. His method of cultivation, however, was limited to transplantation and natural propagation, effected entirely on natural substrates, i.e. rocks. In the late fifties, an artificial method of sowing spores on ropes of hemp or palm fibres and later on ropes of synthetic fibres was employed, like that of the commercial cultivation of *Laminaria*. *Undaria* is often mixed-planted with *Laminaria* on the *Laminaria* floating raft. Since *Undaria* has an earlier short-growing season, maturing much earlier than *Laminaria*, and is harvested in March, it does not interfere with the growth and maturation of *Laminaria*, which is harvested in June. *Undaria* in China is not as popular as *Laminaria*, and the manipulation during the drying process is very tedious. Hence, maricultural firms are not very enthusiastic about its cultivation, and the annual production is much smaller than that of *Laminaria*, amounting to only a few hundred tonnes in dry weight.

For a better idea of the seaweed mariculture industry in China, the accompanying map of China indicates the principal areas for the cultivation of these marine crops.

Other Commercially Cultivated Seaweeds

A few other species are commercially cultivated but on a much smaller scale, and this production does not compare to that from the natural habitat. Thus, they are not regarded as 'marine crop plants'. Some of these algae are discussed below.

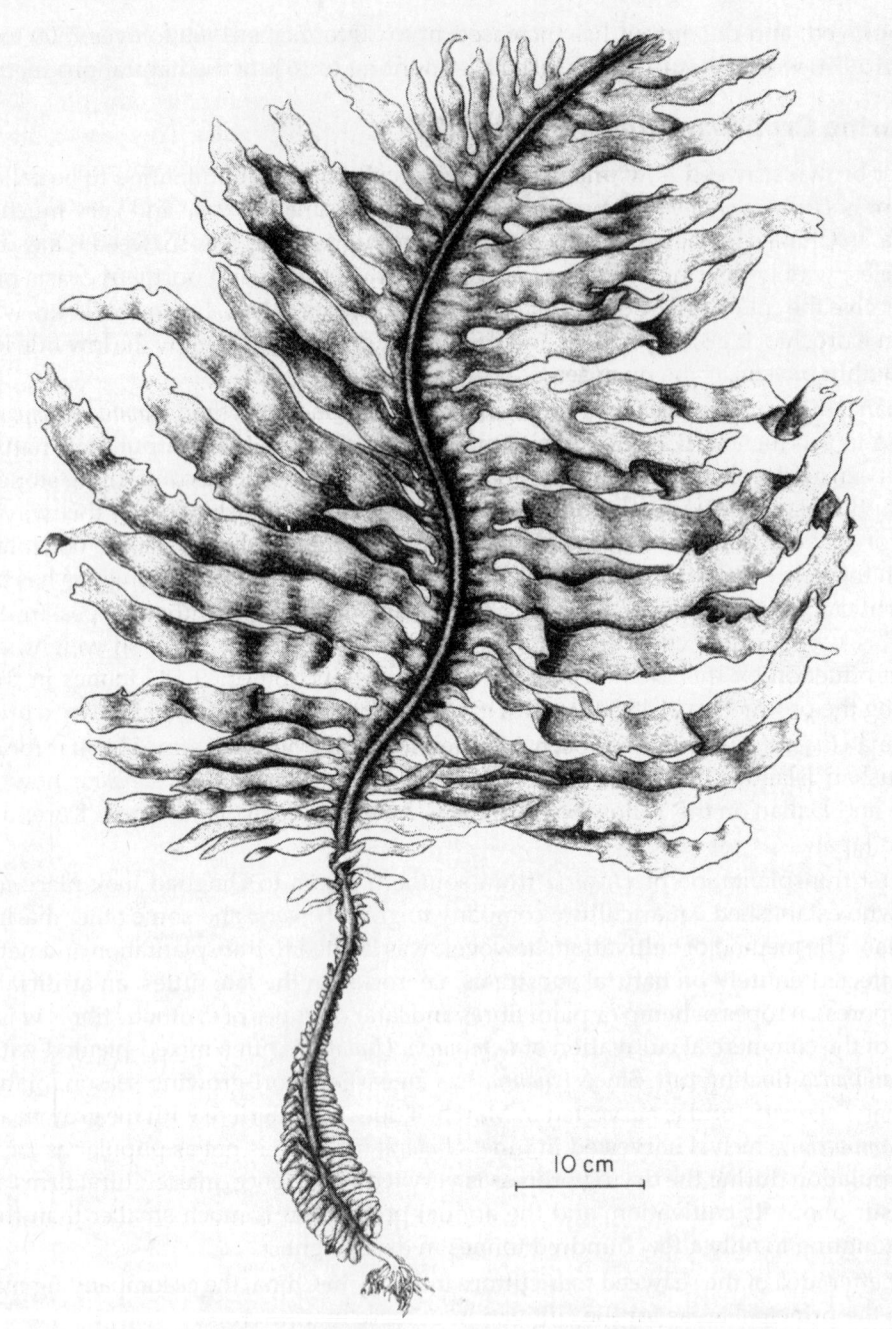

Fig. 3.8 Habit sketch of *Undaria pinnatifida* from Qingdao.

Red Seaweeds *Gracilaria* and *Gloiopeltis*

Gracilaria, known as *jiangli* in China, has been appreciated as a food and as a binding material in the preparation of lime for painting walls. In recent years these seaweeds are better known for their agar content and are now standard agarophytes used as raw material for agar production in China as well as in Australia, Italy, Japan, New Zealand, South Africa, and the United States. Production from the natural habitat, however, is not sufficient to meet the increasing demand for raw material by the agar industry. Hence, commercial cultivation has been practised in various places, the production being still on a small scale and not comparable to its natural production. The principal species under cultivation is the widely distributed *Gracilaria verrucosa*. This seaweed grows abundantly on rocks, stones, gravel, and in fact any solid substrate in the littoral zone.

In the late fifties to early sixties, commercial cultivation was carried out in Guangdong Province, China, by inserting branches of the seaweed into the splits of bamboo sticks about 15 cm long, which were then planted in littoral

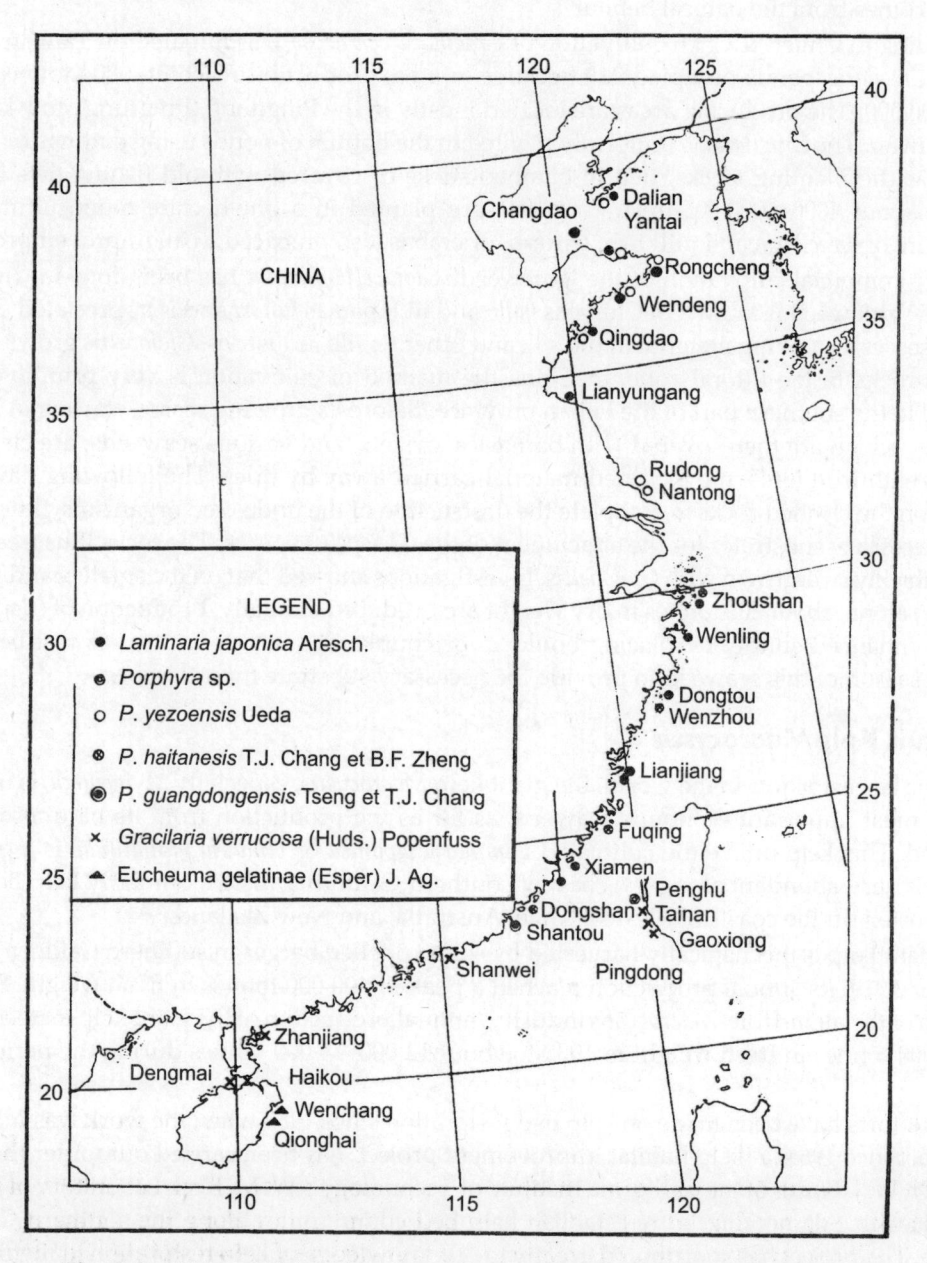

Fig. 3.9 Map of China showing principal areas of maricultural activities.

farms. Later, cultivation of the seaweed was carried out by scattering juvenil *Gracilaria* of more than 5–6 cm length, either gathered from its natural habitat or cultured from spores on stones and pebbles in littoral farms sheltered from strong wave action. In recent years the net cultivation method has been employed, in which branches of *Gracilaria* are inserted in the intersections of the net as in *Eucheuma* cultivation. The floating raft method in *Laminaria* cultivation has also been employed with some success. In a cultivation season or about three months, the production may reach 2–3 tonnes in dry weight per hectare. To date, however, the commercial cultivation in mainland China is on a small scale, and the principal source of *Gracilaria* as a raw material for the agar industry comes from the natural habitat.

According to Chueh & Chen cultivation of *Gracilaria verrucosa* was initiated on Taiwan in 1967. By 1977 *Gracilaria* ponds occupied 223 hectares in area, production reaching 6804 tonnes, worth U.S.$2 700 000. The producing areas are located mostly in the Pingdong (Pingtung) and Tainan in South Taiwan. The *Gracilaria* is uniformly planted in the bottom of ponds using cuttings or the torn sections as the planting stocks, tied to bamboo sticks or covered with old fishing nets to avoid drifting. About 4000–6000 Kg fresh *Gracilaria* are planted in a one-hectare monoculture pond. Polyculture of *Gracilaria* and milkfish, shrimp, or crab is also practiced, with improved profits.

Some commercial cultivation of the 'glueweed' *Gloiopeltis furcata* has been done in China and Japan. The glueweed is known in China as *hailo* and in Japan as *funori* and is appreciated as a food but also serves as a sizing material in the silk and other textile industries. *Gloiopeltis* grows gregariously on rocks in the littoral zone. In China the method of cultivation is very primitive and is practiced in the southern part of the Fujian province. Before its growing season starting in November, rocks, which are then covered with barnacles, oysters, and various seaweeds, are cleaned by scraping with iron tools, the loosened material carried away by tides. The following day lime is sprayed on the cleaned rocks to complete the destruction of the undesired organisms, thus preparing the necessary substrates for the attachment of the *Gloiopeltis* spores. The rocks thus treated will become thickly overgrown with *Gloiopeltis*. Investigations showed that on the small island Jinmeng (Quimoy) alone, about 200 tonnes in dry weight are produced annually. Production of *Gloiopeltis* in Japan is enhanced simply by placing boulders or chunks of concrete in the sea just before the growing season of this seaweed to provide the necessary substrate for it to grow.

The Giant Kelp *Macrocystis*

One of the largest plants in the world, the giant kelp *Macrocystis*, especially *M. pyrifera*, is undoubtedly the most important economic seaweed, as far as the production from its natural habitat is concerned. This kelp unlike the cultivated *Laminaria japonica* or *Undaria pinnatifida*, is a perennial plant. It is very abundant along the coast of southern California, U.S.A. Similarly kelp beds have been reported off the coasts of Mexico, Chile, Australia, and New Zealand.

The giant kelp is mechanically harvested by self-propelled barges in southern California. In the year 1916–1919, its annual production reached a peak of 400 000 tonnes in fresh weight. Since the initiation of the algin-from-*Macrocystis* industry, annual production of the giant kelp increased from about 1000 tonnes in fresh weight in 1932 to about 12 000–14 000 tonnes during the period 1950–1970.

Great efforts have been made on kelp bed restoration since 1963 when the work was initiated at Pt. Loma. Since 1964, a 'kelp habitat improvement project' has been carried out under the leadership of Dr. W. J. North of the California Institute of Technology's W. M. Keck Laboratory of Environmental Health Engineering, to re-establish kelp bed communities along the southern California coastline. The project has contributed greatly to our knowledge of kelp restoration in methodology and has recently switched to investigating possibilities for increasing organic productivity in kelp beds through fertilizing operations.

To date, however, *Macrocystis* is still a wild plant and cannot be regarded as a marine crop plant. For a perennial plant such as *Macrocystis* growing in the open sea to be subjected to commercial cultivation and to become a genuine crop plant, it is a very difficult matter, and many biological and engineering problems have yet to be solved. Efforts to construct support structures to grow the giant kelp have been futile thus far. Initial trials have met with failure, but the American Gas Association is said to be continuing work with a quarter-acre module to be set in water about 300 m deep. Undoubtedly, the success of such a venture will eventually revolutionize the whole structure of this industry.

Green Seaweeds *Monostroma Enteromorpha* and *Caulerpa*

In a broad sense, *Monostroma*, *Enteromorpha* and *Caulerpa* are under commercial cultivation and are appreciated as foods in China, Japan, and other oriental countries. *Monostroma* and *Enteromorpha* are common seaweeds growing on stones, pebbles, etc. in the littoral zone. In Japan *Monostroma* and *Enteromorpha* grow on *Porphyra* cultivation nets in spring usually after the growing season of *Porphyra* comes to an end and are later harvested and used to manufacture the much appreciated *hoshi-nori*. Its annual production amounts to 2–10% of that of *Porphyra*.

Caulerpa is dull greenish in colour, its plant body being differentiated into a prostrate stolon with rhizoids and erect branches. The common edible species is *C. racemosa*, the stolon of which may reach 1 m in length. On Taiwan, China.

Caulerpa is grown in brackish water ponds by seeding with chopped pieces of the mature thallus. It is later shipped to the Philippines for the fresh vegetable market.

Scientific Bases for Seaweed Mariculture

The commercial cultivation of seaweeds may be distinguished from agronomy of land crop plants not only in regard to growth media–water in the former and soil and air in the latter–but also in that the former involves unicellular spores as the reproductive units and rhizoids as the attachment organs, whereas the latter involves very complex, multicellular seeds and structurally complicated roots as the attachment and absorption organs. These differences naturally lead to different cultivation methods. These facts constitute reasons for also distinguishing the cultivation of seaweeds from that of aquatic seed plants, even though both are aspects of aquaculture.

THE SUBSTRATE

The rhizoids of seaweeds are not responsible for the absorption of water or nutrients, a function shared by all parts of the seaweed immersed in the aquatic medium. The function of the substrate is merely to provide a suitable attachment for the seaweed. Yet the selection of the substrate is important in some other ways, such as convenience, availability, durability and economics. Algal spores are very delicate unicellular structures and die as soon as they are removed from the medium. Therefore, it is imperative that in the 'seeding' process, i.e. the collecting of spores, the substrates must be brought close enough to the sporeliberating frond so that the spores will have the best chance to adhere to this substrate. In general, the spores of a seaweed will germinate only after adhering to the substrate.

The growth medium for seaweed cultivation is seawater, which contains many kinds of salts, some of which are quite corrosive. Therefore, the substrate employed should be able to resist this corrosive action of the seawater. The sea is in constant motion because of current, waves, and tide. Waves can be very destructive, and the positioning of the substrate has to be strong enough to withstand this wave action. Because of the constantly changing water level due to the tide, in many places amounting to several metres for the maximum tidal amplitude, the positioning of the substrate should be able to adapt to this tidal change.

It is evident that the selection of substrate and its positioning are of primary importance even though the seaweed does not derive any nutrients from the substrate. In general, the substrates may be natural, such as rocks and reefs, or artificial, such as ropes made of hemp, palm, or synthetic fibres. In the early stages of seaweed cultivation, natural substrates were employed, but with the advance of this science an artificial substrate in the form of a raft was adopted.

For years, both in Japan and China, the cultivation of *Laminaria* was done on rocks and reefs in its natural habitat by depositing stones at the desired depths, all for the purpose of increasing surface area for the attachment of *Laminaria* spores. Later, mature *Laminaria* plants were tied to sublittoral substrates by divers to provide more spores for those natural substrates. For some years there had been much controversy as to which method was better : propagation on sublittoral rocks or cultivation on rafts. Production practice shows the definite superiority of the artificial substrate (the floating raft), as shown by the production data of two mariculture stations, one at Qingdao (Shandong Mariculture Station) and the other at Dalian (Luda Mariculture Station), in 1952–1958. *Laminaria* was introduced to Qingdao (36° N Lat.) in 1950–1951. In 1952 when the floating raft (of hemp ropes) method was devised, Qingdao produced 62 tonnes of kelp in fresh weight. The production rose steadily to 13 207 tonnes in 1958, far ahead of the natural habitat production from the sublittoral rock of 1378 tonnes. Dalian (40° N Lat.) has a lower water temperature, more favourable for the growth of *Laminaria*, but, because of the continued use of the natural substrate by the Luda Station, Dalian produced less kelp than Qingdao in 1955 and 1956. In 1957 the Luda Station finally adopted the floating raft method, and in the same year production slightly surpassed that of Qingdao; by 1958 production (22,936 tonnes) surpassed that of Qingdao (14,585 tonnes) by a broad margin of 157%.

Chinese mariculturists for many years used ropes made of hemp or palm fibres for artificial substrates. These had to be soaked in water for a few months in order to leach out undesirable substances in the fibre. In the course of the development of the kelp cultivation industry, it was noted that raw materials for the raft (the hemp and palm fibres) could not keep up with the demand, and so in the early sixties synthetic fibres were used with success. At present, practically all of the ropes used for growing kelp are made of synthetic fibres. As to the floating raft itself, there is a tendency to abandon the double-line raft and adopt the single-line raft, which appears to be better adapted for places with stronger currents and heavier wave action.

The Temperature Factor

Temperature is one of the most important factors controlling the activities of seaweeds. Every species has its maximal, optimal, and minimal temperature for growth and development, which may differ in the same species depending on the phase of the life history and also on the different stage of growth in the same phase. Differences in the various phases are particularly conspicuous; in some cases the sporophytic phase has a higher and the gametophytic phase a lower optimal growth temperature, whereas in other cases the reverse situation is true.

The Light Factor

Light, of course, is another basic environmental factor in the normal activities of seaweeds. The quality of light changes at different levels of the sea because of the differential absorption of light by seawater. However, with different levels, light intensity plays an even more important role, and hence the quality of light is generally not separately considered.

Light Intensity

Each species of seaweed has an optimal intensity for its vegetative growth. Obviously, a knowledge of this characteristic is important for successful seaweed cultivation. For instance, the best growth of the sporophytic phase (i.e. the *Conchocelis*) of *Porphyra yezoensis* occurs at about 3000 lux and that of the gametophytic phase of *Laminaria japonica* at 3000–4000 lux, both being cultivated indoors. The leafy thallus (gametophytic phase) of *Porphyra* is an intertidal plant and is better adapted to high light intensity. Therefore, in the floating raft cultivation method, it is necessary to keep the raft as close to the water surface as possible. On the other hand, the blade of *Laminaria japonica*, representing the sporophytic phase, is a sublittoral plant and in its native region of Japan grows well even at a depth of 10 m. On the East China Sea coast, however, the seawater is rather turbid and the transparency of the water quite low. In one experiment at Qingdao it was observed that when the sporelings were only a few cells to about 50 cells high, the best growth occurred in the uppermost layer of the raft near the surface (17 cm). But then as the fronds grew, the depth for best growth was successively lower such that after two and a half months the best growth was at the 3 m level.

In another experiment conducted under controlled conditions, young *Laminaria japonica* sporelings smaller than *c.* 800 µm high grew best under 3000 and 4000 lux illumination, but soon afterwards they became unhealthy and died; those under 2000 lux illumination grew best. Therefore, in the greenhouse cultivation of the summer sporelings under low temperature, the light intensity has to be carefully controlled at about 2000 lux to obtain the best results.

Regulation of Light Intensity in the Field

The transparency of seawater is different depending on the location and the particular season. For example, on the Zhejiang coast the seawater is quite turbid and the transparency low, especially during winter, the best growing season for kelp. During this period the transparency scarcely exceeds 1 m, and the level for the kelp under cultivation must be raised almost to the surface to avoid the so-called 'green rot' caused by insufficient light. Between late March and early April, the water suddenly becomes clearer because of the incoming warm current from the south. Under such circumstances the level of the cultivated kelp must be lowered 1–2 m or else the kelp will turn pale, eventually suffering from 'white rot', caused by too strong light intensity.

Light Period

Photoperiodism is especially important in the development of reproductive structures. Diurnal light rhythms may also trigger spore release.

LIGHT AND DEVELOPMENT OF *LAMINARIA* GAMETOPHYTES

Similar to the sporophyte of *Porphyra* (the *Conchocelis*), the gametophyte of *Laminaria* is the summer phase and grows only vegetatively, unable to carry out sexual reproduction or to develop sporophytes under 24 h of continuous illumination. Unlike the *Porphyro* sporophyte, however, the number of days required for the gametophyte to effect sexual reproduction and develop sporophytes is inversely proportional to the number of hours of illumination per day. For example, when the illumination is 2 h per day, it takes 44 days for sporophytes to appear and 59 days for the sporophytes to reach 50% of the total number of gametophytes. When the illumination is 19 h per day, it takes only 8 days to develop sporophytes and 11 days for the sporophytes to reach 50% of the total number of gametophytes.

It was also discovered that a certain dark period is necessary for the discharge of eggs and sperm in *Laminaria* gametophytes. In general, about an hour of darkness is required for this discharge, which may explain why 24 h of continuous illumination prevents the production of sporophytes. The period of darkness required depends upon the previous light treatment. Thus, when the culture is exposed to 24 h continuous illumination, only 20 min are required for initiating the discharge of eggs and sperm, whereas if previous lighting lasted only for 1 h, then 3 h of darkness are necessary for bringing about the discharge. It seems that there is some material synthesized in the light and in the darkness which is necessary for egg discharge. As shown in the experiment, the sperm are discharged only after the discharge of eggs and are attracted to the eggs by certain substances.

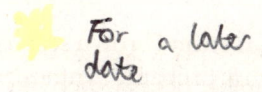

4 Chemical Constituents of Seaweeds

PHOTOSYNTHETIC PIGMENTS (Figs. 1 to 113)

Chlorophylls

Chlorophyll *a* is the most abundant chlorophyll, and the primary photosynthetic pigment, of seaweeds. As a specialized chlorophyll-protein complex (P700), it is directly involved in the Photosystem I light reaction. Seaweeds probably synthesize chlorophyll *a* from 5-aminolevulinic acid, as do other plants; however, little is known about the biosynthetic origin of 5-ALA itself in seaweeds. In certain unicellular protists (*Cyanidium caldarium*, *Porphyridium purpureum*, *Chlorella vulgaris*, *Euglena gracilis*) cyanobacteria and higher plants, 5-ALA is produced from α-ketoglutaric acid via 4,5-dioxovaleric acid.

(a) $R_1 = CH_3$ $R_2 = CH_3$
(b) $R_1 = CH_3$ $R_2 = CHO$

(a) Chloropyll *a*. (b) Chlorophyll *b*.

(a) $R = CH_2CH_3$
(b) $R = CH = CH_2$

(a) Chlorphyll c_1. (b) Chlorophyll c_2.

Green and brown seaweeds produce accessory chlorophylls in addition to chlorophyll a, Chlorophyll b, probably synthesized directly from chlorophyll a, is found in green algae. Chlorophyll(ide)s c_1 and c_2, present in brown seaweeds, may be derived from magnesium 2,4-divinyl phaeoporphyrin a_5, an intermediate in the biosynthesis of chlorophyll a. Red seaweeds do not produce chlorophylls b or c; a 'chlorophyll d', found in extracts of some red seaweeds, has not been detected *in vivo*.

Phycobiliproteins

Phycobiliproteins are open-chain ('linear') tetrapyrrole pigments found in Rhodophyceae, Cyanobacteria and Cryptophyceae. The bilin unit is derived from 5-aminolevulinic acid via protoporphyrin IX. Two types of bilins have been identified in seaweeds: phycocyanobilin (Fig. 1) and phycoerythrobilin (Fig. 2). A third phycobilin, phycourobilin, may or may not be a structurally distinct entity. These bilins are covalently attached to polypeptides to form the major classes (α, β, γ) of subunit component proteins. Phycobiliproteins are then assembled from these subunits.

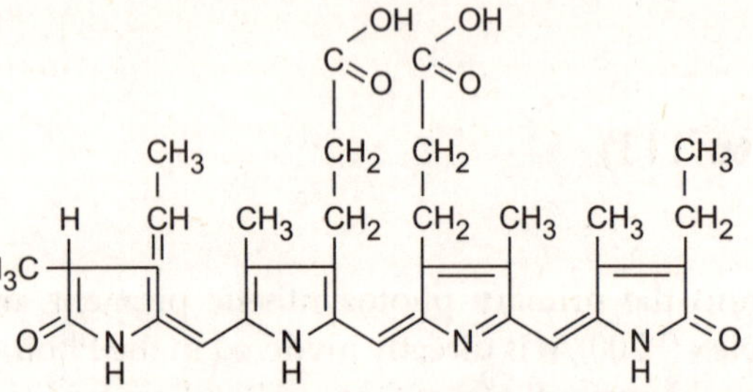

Fig. 1 *Phycocynobilin.*

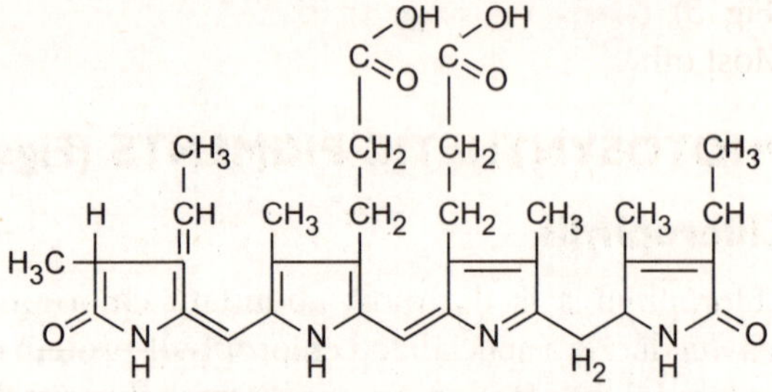

Fig. 2 *Phycoerythrobin.*

Phycobiliproteins were originally grouped on the basis of their spectral characteristics and their distributions in the different algal taxa; for example, R-phycocyanin was the blue pigment present in Rhodophyceae. Subsequent research has in many cases revealed more widespread taxonomic distributions, and the present nomenclature refers only to spectral characteristics, not to taxonomic distribution. Even this is an oversimplification, as variation exists within the present groups; a more natural classification should incorporate information from amino acid sequences of the apoprotein moieties.

Carotenoids

Carotenoids are isoprenoid polyene pigments present in all photosynthetic (and many non-photosynthetic) organisms. Biosynthetically, carotenoids are derived from mevalonic acid via isopentenylpyrophosphate (Fig. 3) and lycopene (Fig. 4) and/or its isomer β-zeacarotene. β,β-Carotene ('β-carotene', Fig. 5) and β,ε-carotene ('α-carotene', Fig. 6) are formed by enzymatic cyclization; further biosynthetic alterations (hydroxylation, epoxidation, in-chain oxidation, etc.) subsequently give rise to the variety of carotenoids and xanthophylls found in seaweeds.

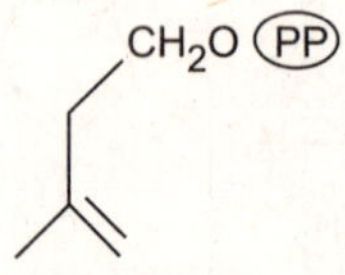

Fig. 3 *Isopentenylpyrophospate.*

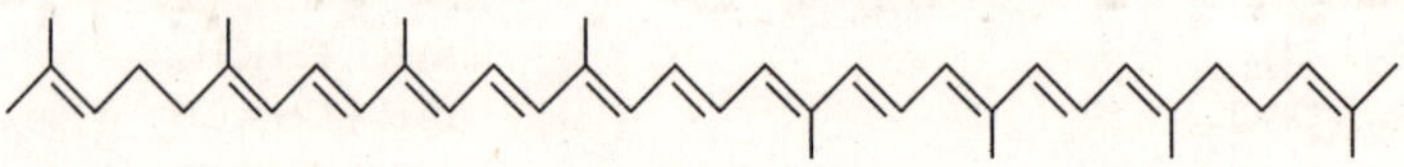

Fig. 4 *Lycopene.*

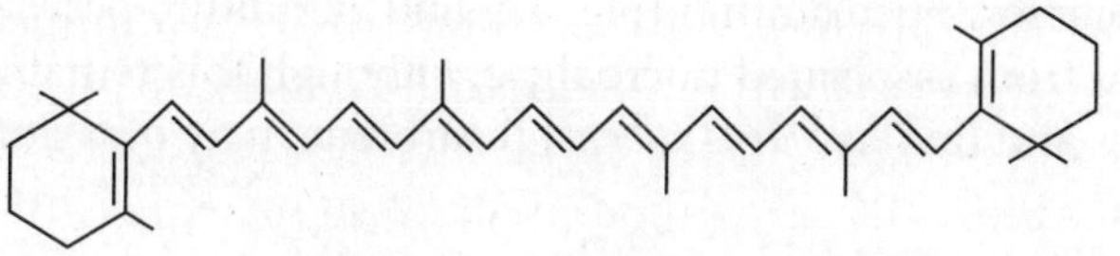

Fig. 5 β-*Carotene.*

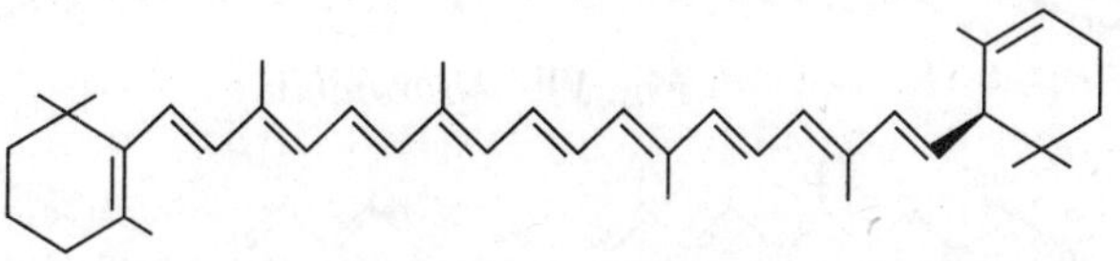

Fig. 6 α-*Carotene.*

Most red seaweeds exhibit a rather simple carotenoid pattern, accumulating mainly β-carotene (Fig. 5), α-carotene (Fig. 6), and their dihydroxy derivatives zeaxanthin (Fig. 7) and lutein (Fig. 8). Most other carotenoids occasionally reported to be minor constituents of red seaweeds are closely

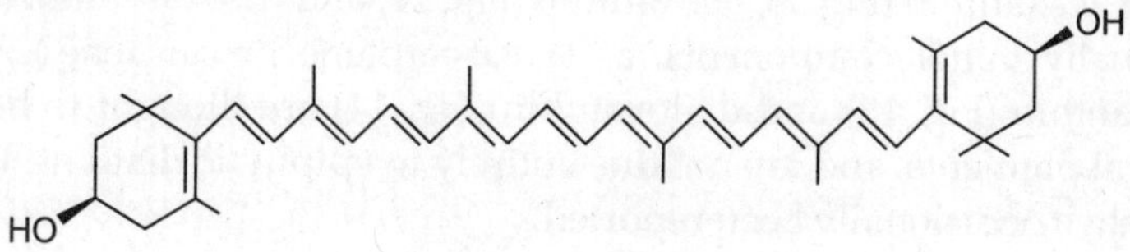

Fig. 7 *Zeaxanthin.*

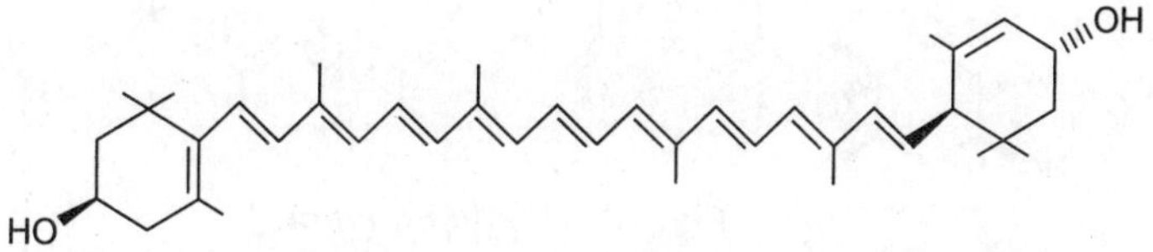

Fig. 8 *Lutein.*

related to these four major pigments: the corresponding monohydroxides β- and α-cryptoxanthin, lutein epoxide ('taraxanthin'), zeaxanthin mono- (antheraxanthin, Fig. 9) and di- (violaxanthin, Fig. 10) epoxides, and auroxanthin (violaxanthin furan oxide). Neoxanthin (Fig. 11), reported from *Antithamnion plumula* and *Nemalion helminthoides* (=*N. multifidum*), is presumably derived from a

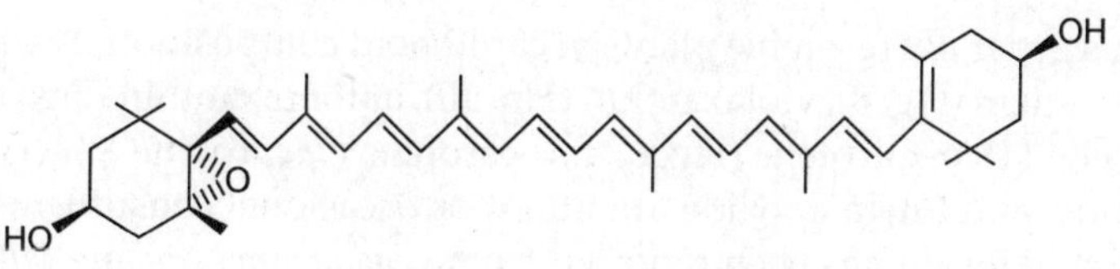

Fig. 9 *Antheraxanthin.*

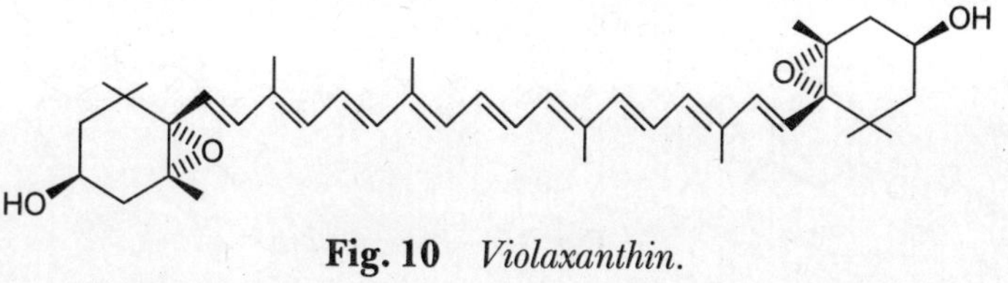

Fig. 10 *Violaxanthin.*

violaxanthin-like precursor. Fucoxanthin (Fig. 12) and peridinin, occasionally reported from red seaweeds, are thought from associated microalgae, although this remains to be conclusively demonstrated. α-Carotene, and (in *Phycodrys rubens*) β-carotene, may occasionally be absent.

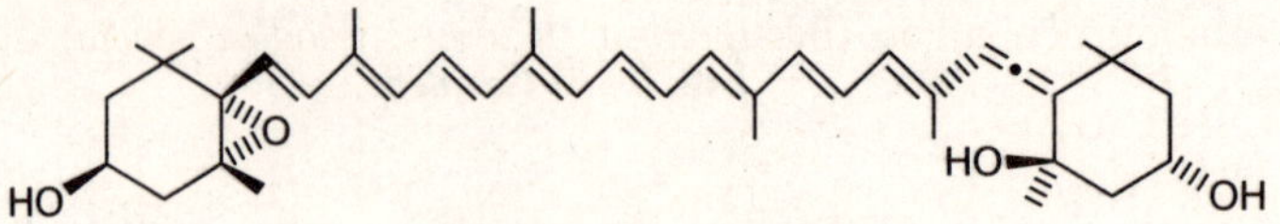

Fig. 11 *Neoxanthin.*

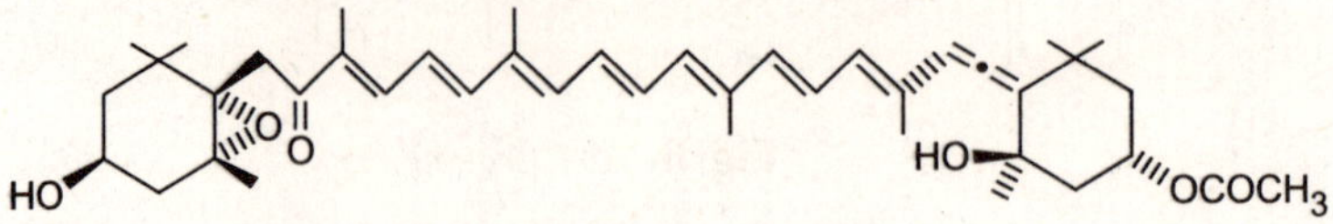

Fig. 12 *Fucoxanthin.*

Brown seaweeds accumulate β-carotene (Fig. 5), fucoxanthin (Fig. 12), and violaxanthin (Fig. 10); the closely related zeaxanthin (Fig. 7), neoxanthin (Fig. 11) and fucoxanthinol (desacetoxy-fucoxanthin) are occasionally minor components, as is ε,ε-carotene ('e-carotene'). The acetylenic carotenoids diadinoxanthin (Fig. 13) and diatoxanthin (Fig. 14) are thought to be endogenous, albeit minor, brown algal pigments, and are not due entirely to epiphytic diatoms. Lutein (Fig. 8) and α-carotene (Fig. 6) have occasionally been reported.

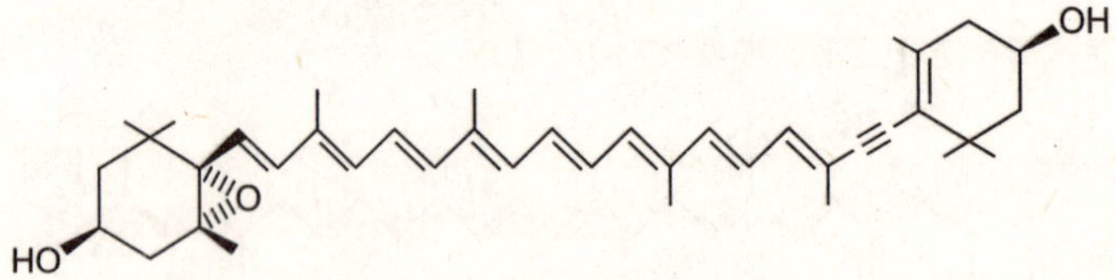

Fig. 13 *Diadinoxanthin.*

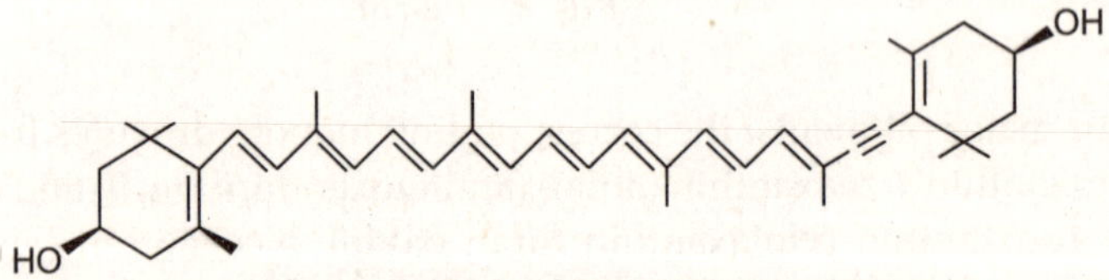

Fig. 14 *Diatoxanthin.*

Green seaweeds generally resemble plants in carotenoid composition. The major carotenoids are β-carotene (Fig. 5), lutein (Fig. 8), violaxanthin (Fig. 10), antheraxanthin (Fig. 9), zeaxanthin (Fig. 7), and neoxanthin (Fig. 11). α-Carotene (Fig. 6), β,ψ-carotene ('γ-carotene'), β-cryptoxanthin, cryptoxanthin-5′,6′-epoxide, and lutein epoxide are minor or occasional constituents. The α-carotene derivative loroxanthin (Fig. 15) has been reported from *Cladophora* spp. and *Ulva rigida*.

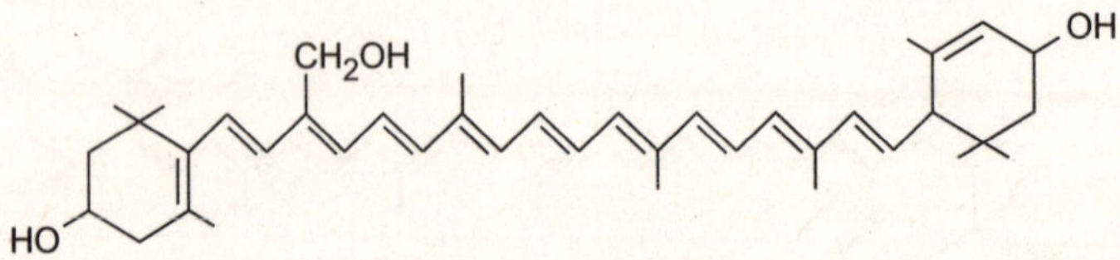

Fig. 15 *Loroxanthin.*

Most siphoneous green seaweeds (*Avrainvillea, Bryopsis, Caulerpa, Codium, Halimeda, Pseudobryopsis, Udotea*), as well as members of the Derbesiaceae (*Derbesia, Halicystis*), also accumulate the α-carotene derivative siphonaxanthin (Fig. 16) and its fatty acid ester(s), siphonein. The latter is also accumulated in Dichotomosiphon sp., and one or both of these compounds are occasionally found in other green algae (*Blastophysa rhizopus; Boodlea coacta; Cladophora* spp.; *Ulva japonica; Valonia* spp.).

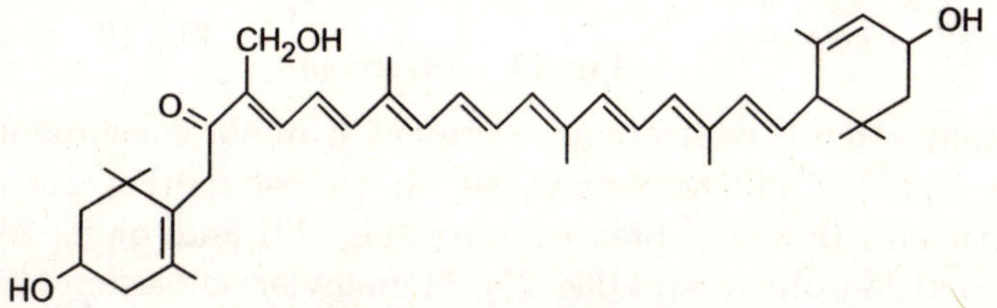

Fig. 16 *Siphonaxanthin*.

Environmental conditions may significantly influence the carotenoid composition of seaweeds. An example is the light-mediated de-epoxidation of violaxanthin to antheraxanthin to zeaxanthin other 'xanthophyll cycles' may operate with diadinoxanthin to diatoxanthin and lutein epoxide to lutein. The accumulation of siphonaxanthin may require low light levels: Yokohama *et al.* (1977) have shown that 'deep-water' green algae, including *Cladophora wrightiana*, accumulate siphonaxanthin, whereas intertidal green seaweeds, including another *Cladophora* sp., do not.

STEROIDS AND OTHER TRITERPENOIDS

Sterols and related compounds (4α-methyl- and 4,4-dimethyl sterols, ketosteriods, steryl esters and steryl glycosides) are biosynthesized from mevalonic acid via isopentenylpyrophosphate and squalene. Cyclization of squalene 2,3-epoxide produces cyclortenol (Fig. 17), thought to be the precursor of all algal steriods. Biosynthetic steps beyond cycoartenol are complex, and more than one pathway to an individual sterol may exist even within a single organism. Biosynthetic intermediates may or may not be accumulated in detectable quantities; for example, squalene is rarely encountered in seaweeds (*Fucus spiralis, F. vesiculosus, Bryocladia cuspidata* are exceptions), and cycloartenol has been reported infrequently (*Chondrus crispus, Hypnea japonica, Palmaria palmata* (= *Rhodymenia palmata*), *Plocamium telfairiae, Enteromorpha* spp., *Ulva lactuca, Fucus spiralis*). Sterol composition may vary seasonally within an individual species.

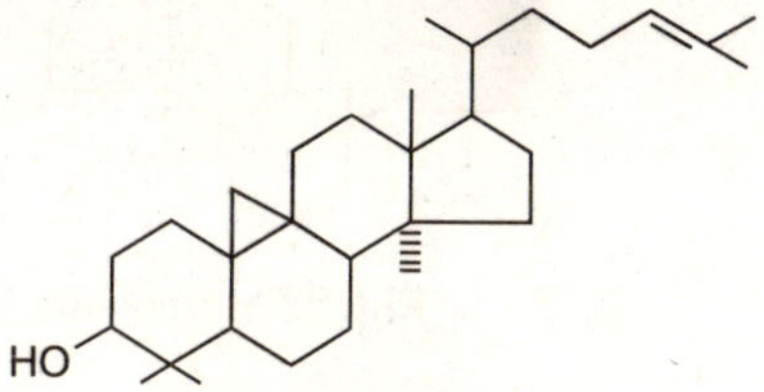

Fig. 17 *Cycloartenol*.

Although there is an extensive literature on seaweeds sterols, our understanding of their occurrence and distribution is far from complete as a result of all too frequently inadequate identification of isomeric sterols. Although positional isomers can be resolved by gas chromatography and identified by mass spectrometry, only with the recent application of ^{1}H- and ^{13}C-nuclear magnetic resonance spectroscopy to sterioid identification has direct identification of diastereomeric sterols become a routine matter. It is of great importance to bear this situation in mind when examining the relevant literature.

Sterols have been demonstrated in all red seaweeds investigated (approximately 120 species). C_{27} sterols such as cholesterol (Fig. 18), desmosterol (Fig. 19) and 22-dehydrocholestrol are almost always predominant; *Sphaerococcus coronopifolius* seems to be unique in accumulating no detectable

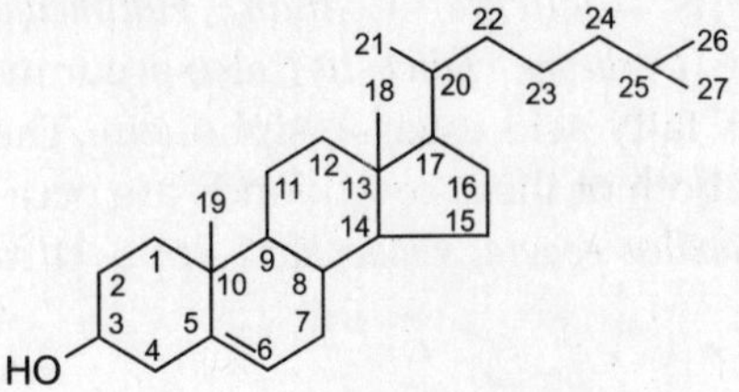

Fig. 18 *Cholesterol.*

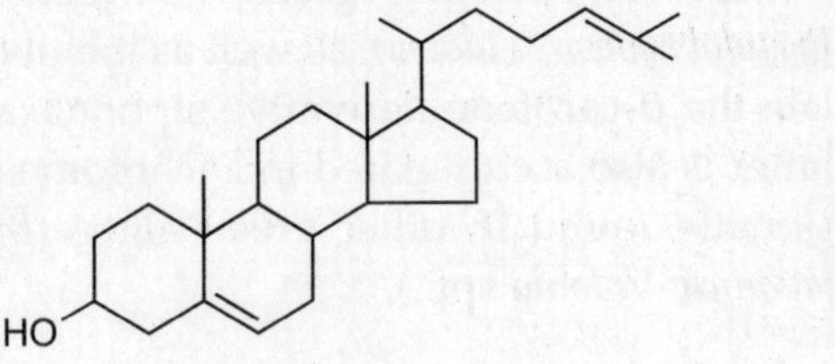

Fig. 19 *Desmosterol.*

cholesterol. Other phytosterols which may be present as minor constituents include campesterol (Fig. 20) and/or its isomer 22-dihydrobrassicasterol; β-sitosteral (Fig. 21) and/or its 24-S epimer clionasterol; stigmasterol (Fig. 22); brassicasterol (Fig. 23) and/or its 24-S epimer ergosterol; fucosterol (Fig. 24) and 28-isofucosterol (Fig. 25); 24-methylenecholesterol (Fig. 26); and liagosterol (Fig. 27) (*cis*- and/or trans-isomers). Δ^4-3-Ketosteroids have been reported in *Gracilaria textorii* and *Meristotheca papulosa*.

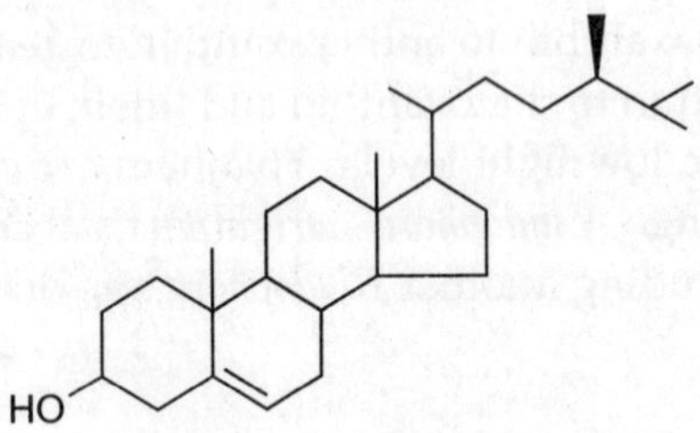

Fig. 20 *Campesterol.*

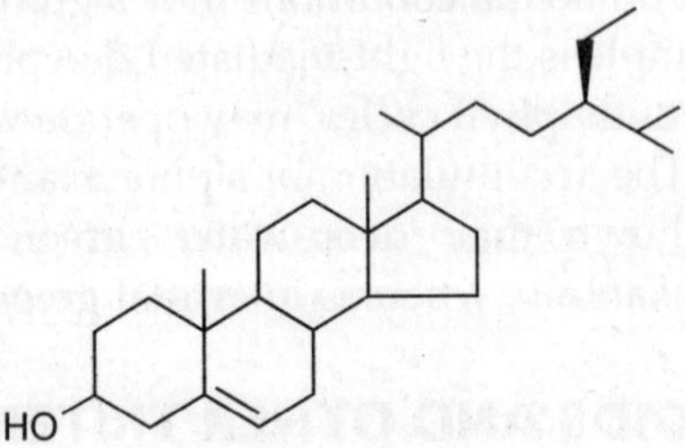

Fig. 21 *β-Sitosterol.*

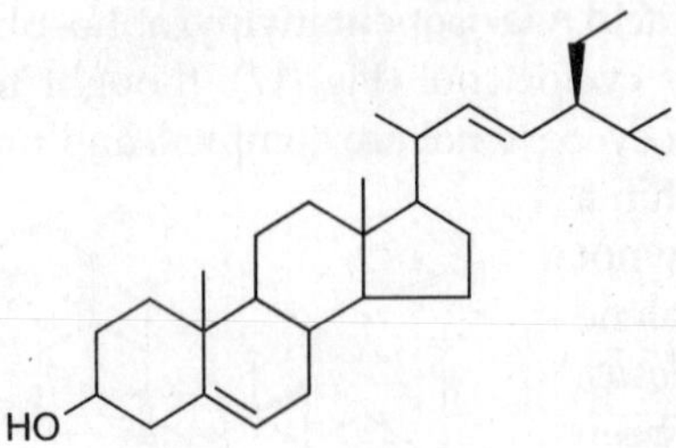

Fig. 22 *Stigmasterol.*

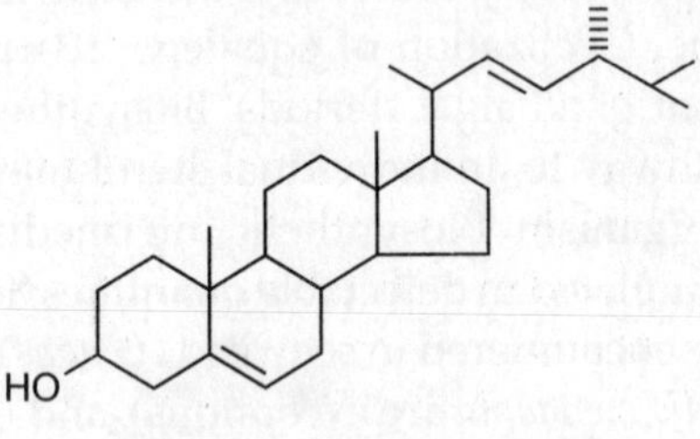

Fig. 23 *Brassicasterol.*

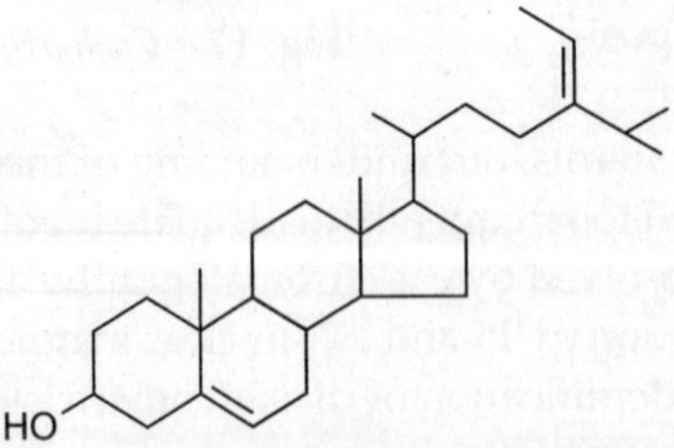

Fig. 24 *Fucosterol.*

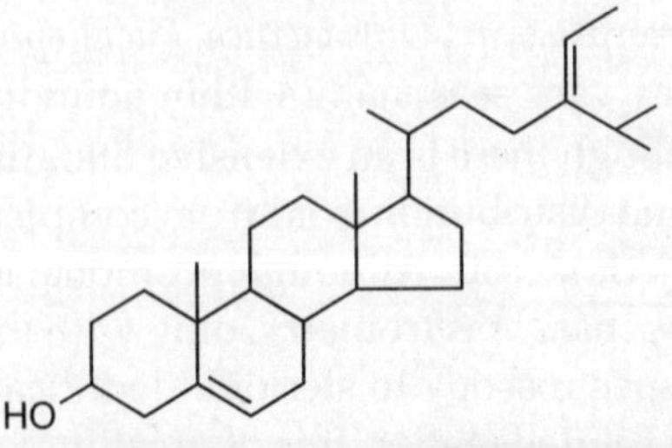

Fig. 25 *28-Isofucosterol.*

The predominant sterol of all investigated brown algae (some 60 species) is fucosterol, or in some cases possibly its $\Delta^{24[25]}$ isomer. Cholesterol, 24-methylenecholesterol, and 22-dehydrocholesterol are widespread among brown algae; brassicasterol or its 24-*S* epimer ergostenol, desmosterol, stig-

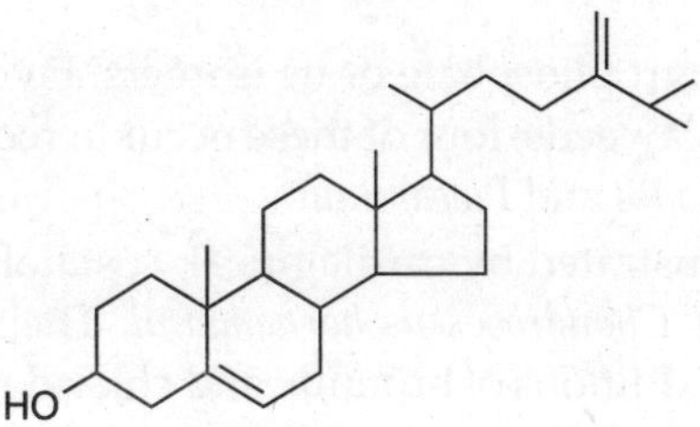

Fig. 26 *24-Methylenecholesterol.*

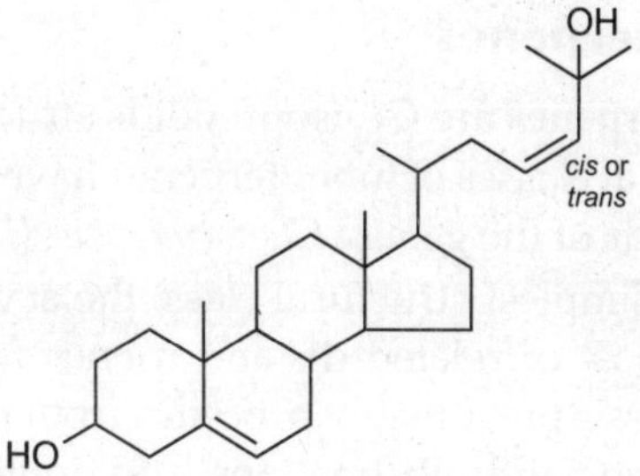

Fig. 27 *Liagosterol.*

masterol or its 24-*R* epimer poriferasterol, a C_{27} cyclopropanoid 'cystosterol' (Fig. 28), and the probable fucosterol oxidation product saringosterol (Fig. 29)

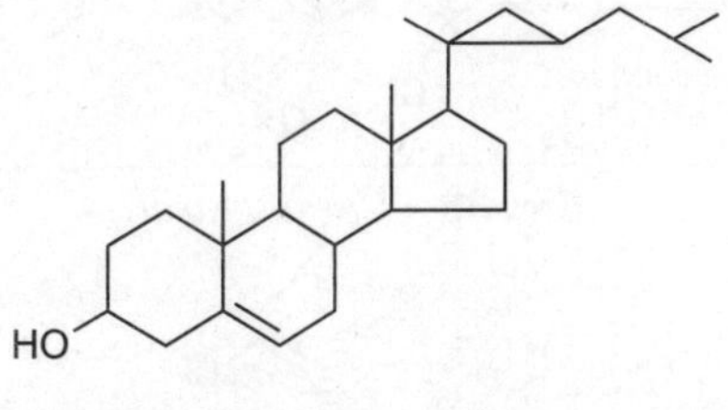

Fig. 28 *Cystosterol.*

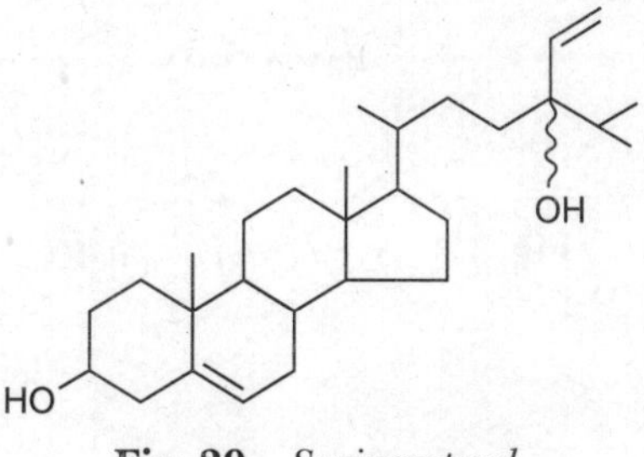

Fig. 29 *Saringosterol.*

are less frequently encountered. 'Sargasterol', a proposed C-20 isomer of fucosterol, may be an artifact.

Sterol profiles of green seaweeds are less predictable. Cholesterol, β-sitosterol, 24-methylenecholesterol and 28-isofucosterol appear to be present in most, although not all, green seaweeds. Campesterol and/or its C-20 isomer haliclionasterol, stigmasterol and/or its 24-*R* epimer poriferasterol, 22-dehydrocholesterol, desmosterol, fucosterol, 24-methylenecycloartenol, as well as 24-ethylidine and tentatively identified 24-methylenelophenols are less frequently reported. *Codium fragile* and *C. vermilara* differ considerably from other green seaweeds, accumulating only the unusual dienols clerosterol (Fig. 30) and codisterol (Fig. 31).

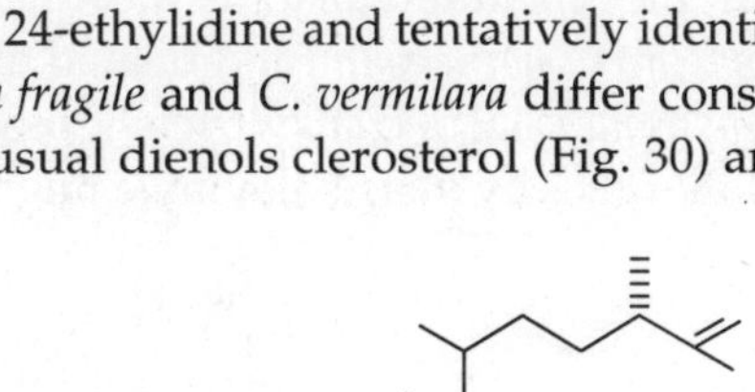

Fig. 30 *Clerosterol.*

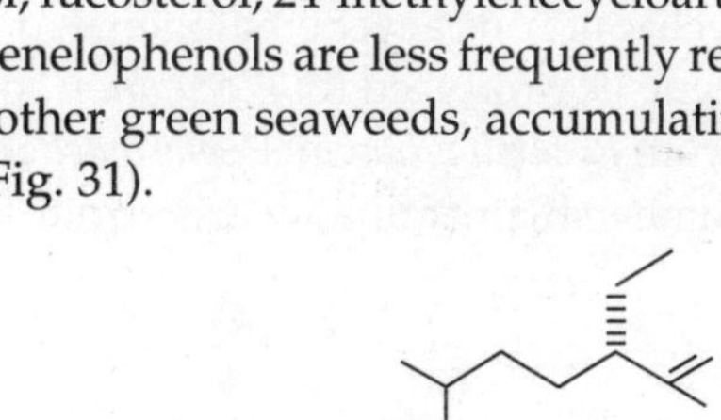

Fig. 31 *Codisterol.*

OTHER TERPENOIDS

There have been very few studies of the biosynthesis of algal terpenes. Thus classification of these metabolites must for the moment rely on a combination of comparative chemistry, classical terpene-biosynthetic proposals, and distributional data.

Monoterpenes

Monoterpenes are C_{10} isoprenoids structurally related to neryl pyrophosphate or its isomers. Five structural classes of monoterpenes have been described among seaweeds; four of these occur in red seaweeds of the genera *Chondrococcus* (*Desmia*), *Microcladia, Ochtodes* and *Plocamium*.

The simplest structural class, the acyclic monoterpenes, is illustrated by cartilagineal, costatol, and a host of related di- and trienes from *Plocamium* spp. and *Chondrococcus hornemanni*. Their biosythesis presumably procedes from neryl pyrophosphate by additions of bromide and chloride, hydrogen halide abstractions, and double bond isomerizations. Cyclization of a halogenated alcohol intermediate would lead to cyclic dihydropyran monoterpenes such as costatone and the related metabolites of *P. costatum*.

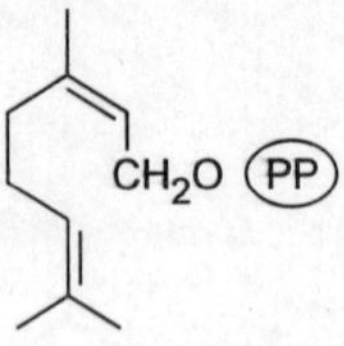

Fig. 32   *Neryl pyrophosphate.*

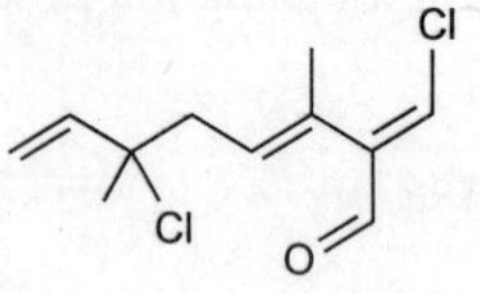

Fig. 33   *Cartilagineal.*

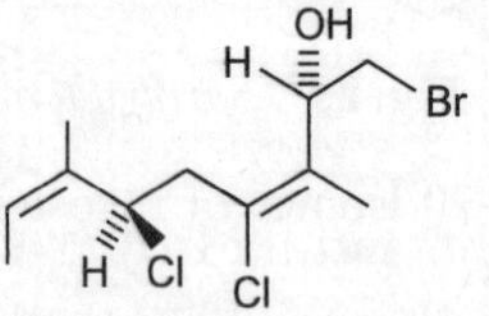

Fig. 34   *Costatol.*

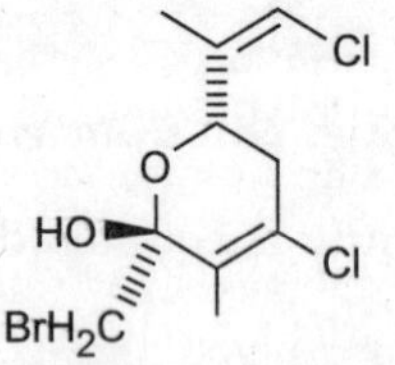

Fig. 35   *Costatone.*

Monoterpenes of a third structural class, the 1,1,5-trialkylcyclohexanes, arise from C-7 → C-2 carbocyclization of acyclic halogenated precursors. Examples include violacene and the related plocamene D, produced by *Plocamium* spp. chondrocoles from *C. hornemanni* and related monoterpenes from *P. cartilagineum, P. mertense* and *Ochtodes secundiramea*. 1,2,5-Trialkylcyclohexenes, such as plocamene B, presumably arise from 1,1,5-trialkylcyclohexanes by *trans*-halovinyle migration.

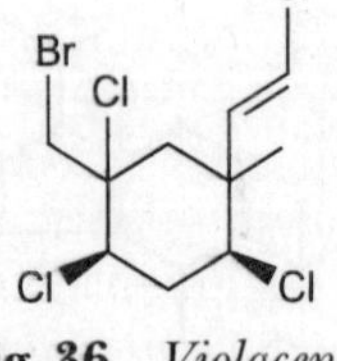

Fig. 36   *Violacene.*

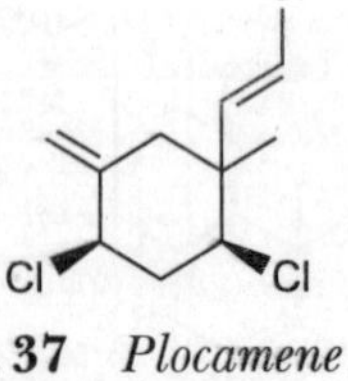

Fig. 37   *Plocamene D.*

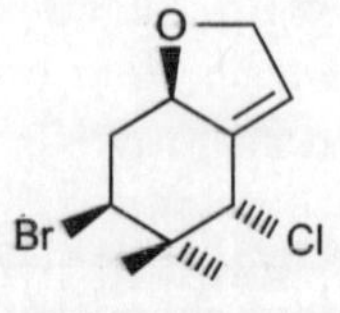

Fig. 38   *Chondrocola A.*

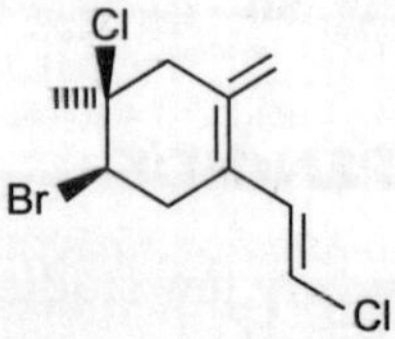

Fig. 39   *Plocamene B.*

A fifth structural class, monoterpene-substituted hydroquipones, is known only from the calcareous green seaweed *Cymopolia barbata*. The more complex meroterpenoids cyclocymopol and cymopochromenol presumably arise from cymopol by cyclization and rearrangement. Monoterpenes of other green seaweeds and of brown seaweeds have not closely examined.

Sesquiterpenes

Sesquiterpenes are C_{15} isoprenoids formally derived from *trans-cis*-farnesyl pyrophosphate or one of its isomers.

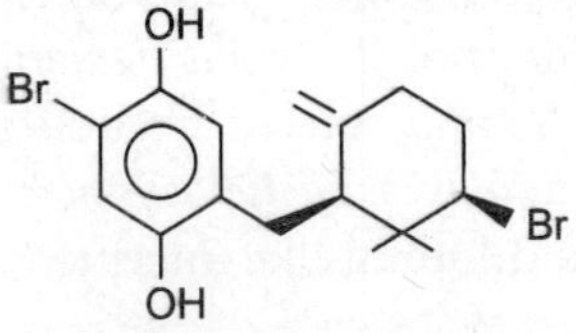

Fig. 40 *Cyclocymopol.*

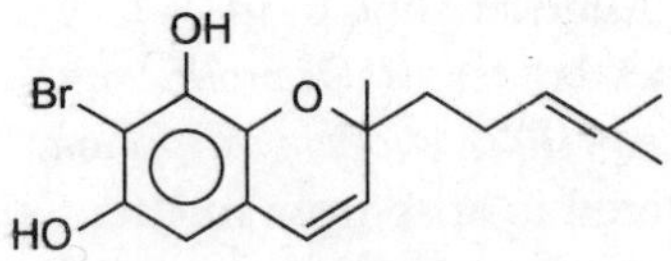

Fig. 41 *Cymopochromenol.*

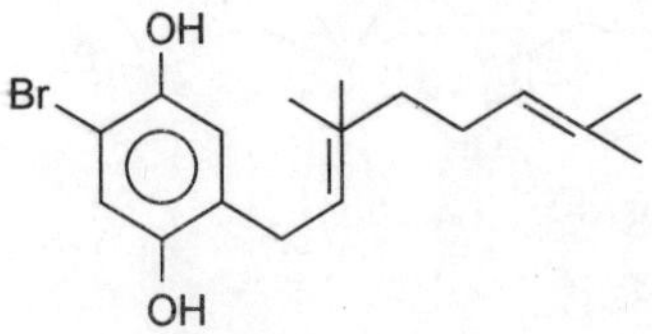

Fig. 42 *Cymopol.*

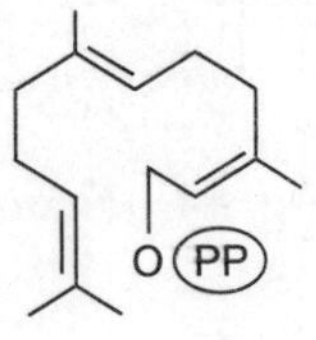

Fig. 43 *Trans-cis-farnesyl pyrophosphate.*

Acyclic sesquiterpenes are uncommon in seaweeds. Exceptions include flexilin (*Caulerpa flexilis*) and the related acetylene caulerpenyne (*C. prolifera*) presumably derived from farnesol or nerolidol.

Monocyclization of *trans-cis*-farnesyl pyrophosphate and subsequent addition of bromide and chloride to an enzyme-bound bisabolonium ion leads to caespitol, its *trans*-diaxial isomer isocaespitol (*Laurencia caespitosa, L. obtusa*), or related (iso-)furocaespitanes (*L. caespitosa*). Carbocyclization of a bromochloro intermediate may lead to chamigranes such as glanduliferol, spirolaurenone (*L. glandulifera*), and 10-bromo-α-chamigrane (*L. pacifica*). In the case of the *trans*-diequatorial bromochloro isomer, further rearrangements produce pacifenol (*L. pacifica, L. nidifica*) and related metabolites of *L. pacifica, L. johnstonii, L. okamurai, L. filiformis, L. tasmanica* and *L. intricata*, or alternatively nidificence and the related sesquiterpenes of *L. nidifica*. The *L. perforata* metabolites perforene and perforatone may also be biosynthetically related.

Alternatively, rearrangement of *trans*-diaxial bromochloro bisabolonium intermediates leads to (+)-chamigranes such as elatol (*L. elata, L. obtusa*), obtusol (*L. obtusa*), or possibly certain cuparanes (α-bromocuparene; *L. glandulifera, L. okamurai*). Other aromatic sesquiterpenes include rearranged chamigranes such as laurenisol (*L. nipponica*), laurinterol (*Laurencia* spp., *Marginisporum aberrans*), and aplysin (*L. okamurai, L. nidfica, M. aberrans*). Oxidative coupling dimers of laurinterol have been found in *L. decidua*.

Nonhalogenated sesquiterpenes have received less attention. Cadalanes (e.g. cadalene, (*Dictyopteris* spp.)) and an epoxidized muurolane from *Dilophus fasciola* have been reported from brown algae, while selinanes (e.g. dictyopterone, from *Distyopteris divaricata*) have been reported from both brown and red seaweeds. The red seaweed *Chondria oppositiclada* produces the cyclopropane sesquiterpene cycloeudesmol. Five known sesquiterpene-substituted hydroquinones (e.g. yahazunol), and the sesquiterpene-substituted 4-hydroxybenzoic acid, zonaroic acid from *Dictyopteris undulata* (= *D. zonariodes*) are reminiscent of ubiquinones.

Diterpenes

Diterpenes are C_{20} isoprenoids formally derivable from geranylgeranyl pyrophosphate. The most abundant diterpene in seaweeds in 7R,11-*trans*-phytol, a component part of chlorophylls *a* and *b*. *Cis*-phytol is accumulated by *Gracilaria andersoniana*. Other acyclic diterpenes include elaganolone (*Cystoseira elegans*), geranylgeranylglycerol (*Dilophus fasciola*), and the antibiotic trifarin (*Caulerpa trifaria*).

More complex diterpenoids may be considered to arise from selective cyclization of geranylgeraniol. A carotenoid-like C-15 → C-10 closure, for example, would lead to the retinol-like diterpene caulerpol found in *Caulerpa brownii*. A C-10 → C-2 cyclization would lead to the antibiotic dictyodial (*Dictyota flabellata*) and thence to acetoxycrenulatin (*D. crenulata*) and other *Dictyota* metabolites. Alternatively, C-10 → C-1 closure would lead to dilophol (*Dilophus ligulatus*) and to the related metabolites of *D. prolificans*. Similarly, pachydictyol A and related hydroazulenes of the Dictyotaceae (*Pachydictyon coriaceum*, *Dictyota* spp., *Dilophus ligulatus*, *Dictypteris membrancea*) may be considered to arise from further C-2 → C-6 cyclization of a dilophol-like intermediate.

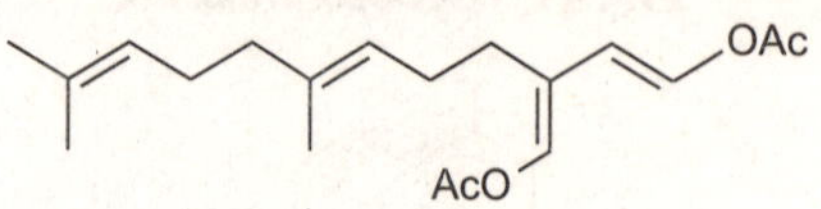

Fig. 44 *Flexilin.*

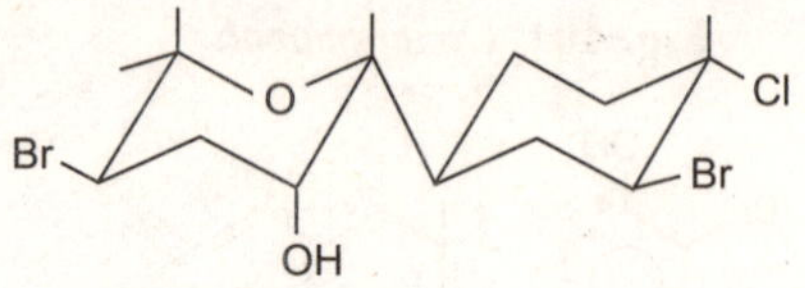

Fig. 45 *Caespitol.*

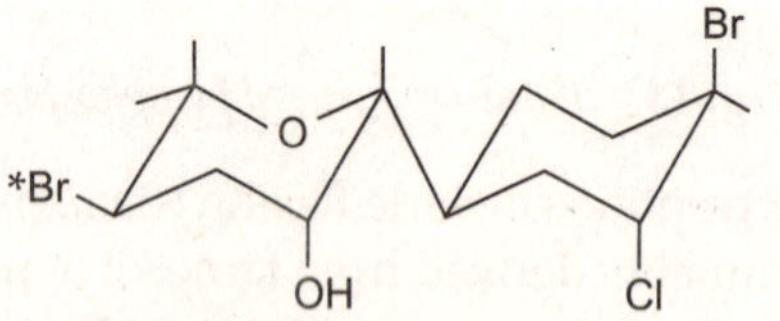

Fig. 46 *Isocaespitol.*

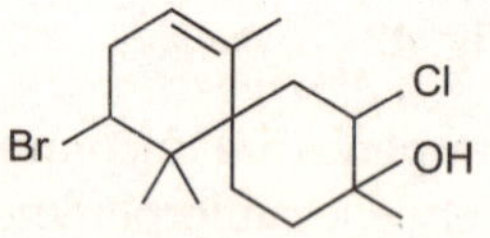

Fig. 47 *Glanduliferol.*

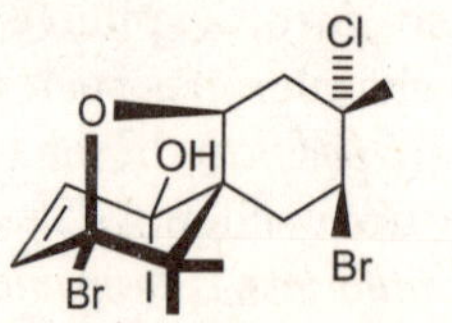

Fig. 48 *Pacifenol.*

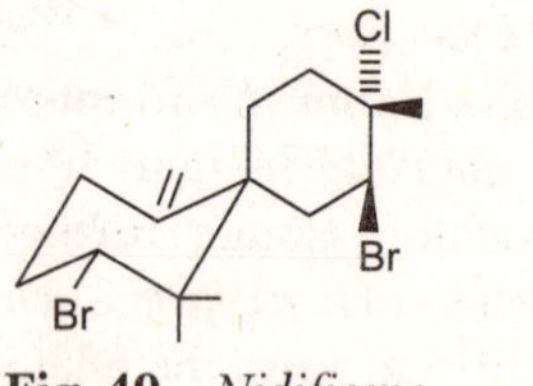

Fig. 49 *Nidificene.*

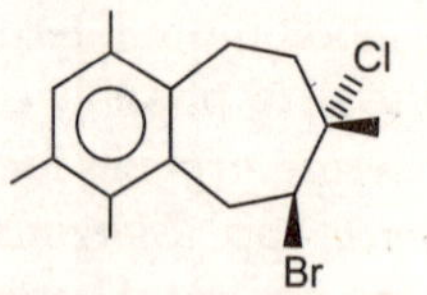

Fig. 50 *Perforene.*

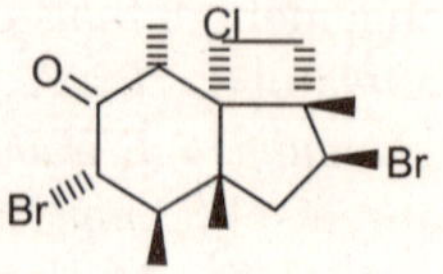

Fig. 51 *Perforatone.*

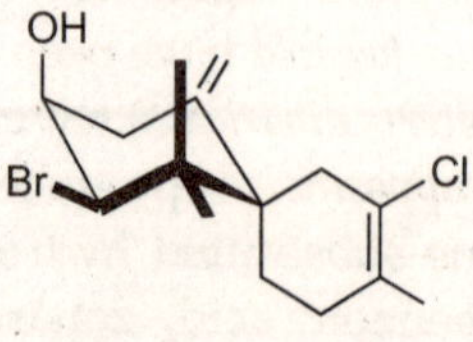

Fig. 52 *Elatol.*

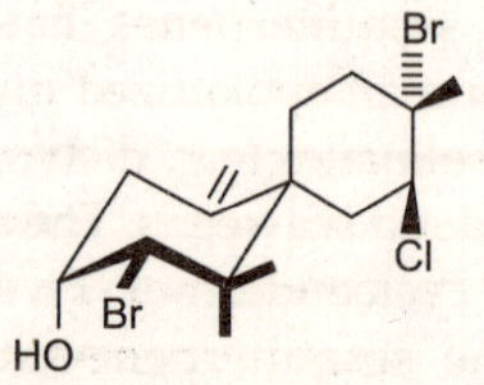

Fig. 53 *Obtusol.*

Fig. 54 α-Bromocuparene.

Fig. 55 Laurenisol.

Brominated diterpenes isolated from red seaweeds include concinndiol (*Laurencia concinna, L. snyderae*); metabolites such as irieol A from *L. irieii*; and bromosphaerol and related diterpenes from *Sphaerococcus coronopifolius*.

Fig. 56 Laurinterol.

Fig. 57 Aplysin.

Fig. 58 Cadalene.

Fig. 59 4, 10-Epoxymuurolane.

Fig. 60 Dictyopterone.

Fig. 61 Cycloeudesmol.

Fig. 62 Poitediol.

Fig. 63 Yahazunol.

Fig. 64 Zonaroic acid.

Fig. 65 Geranylgeranyl pyrophosphate.

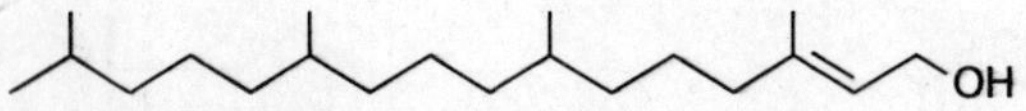

58 *A Textbook of Algae*

Fig. 66 *7R,11R-trans-phytol.*

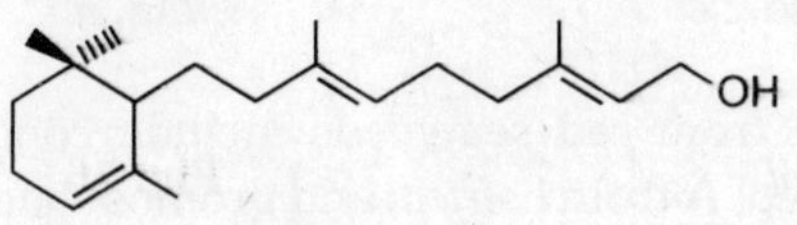

Fig. 67 *Caulerpol.*

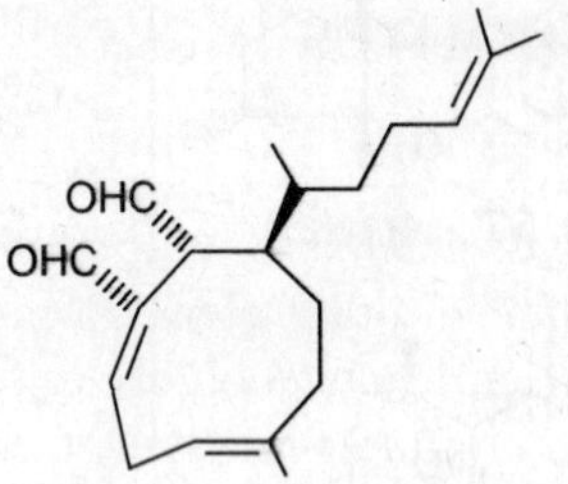

Fig. 68 *Dictyodial.*

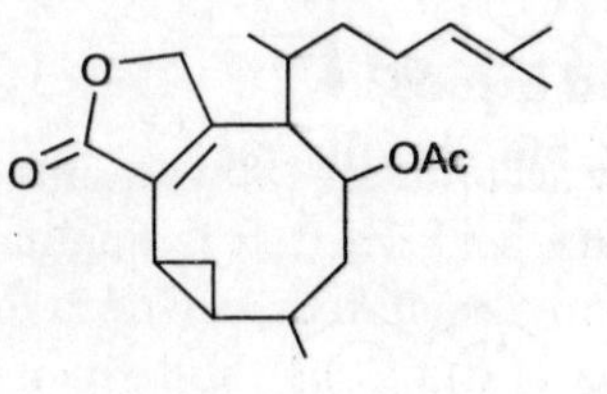

Fig. 69 *Acetoxycrenulatin.*

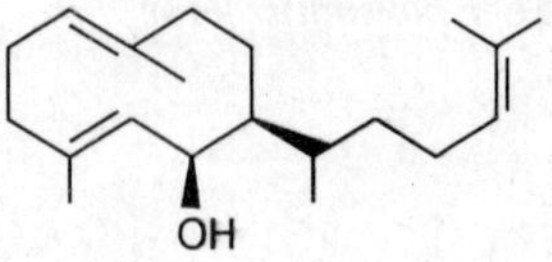

Fig. 70 *Dilophol.*

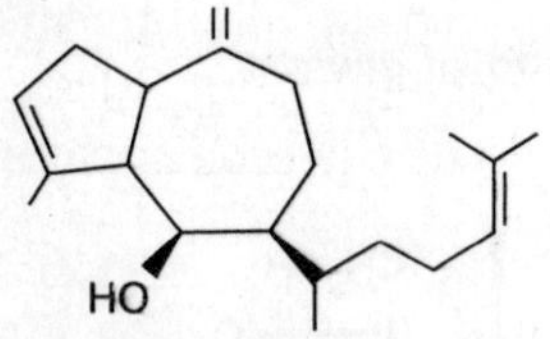

Fig. 71 *Pachydictyol A.*

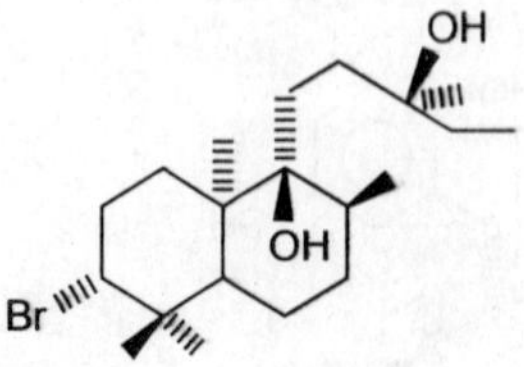

Fig. 72 *Concinndiol.*

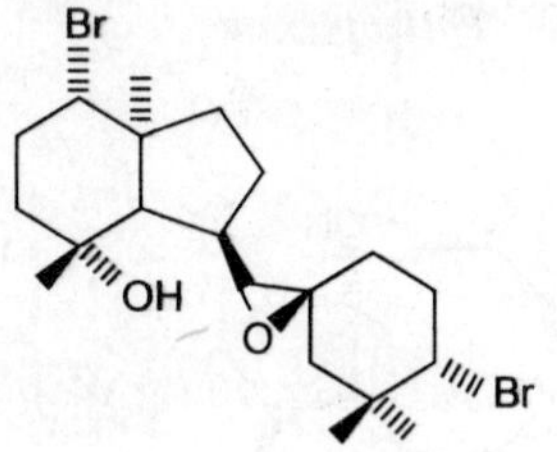

Fig. 73 *Irieol A.*

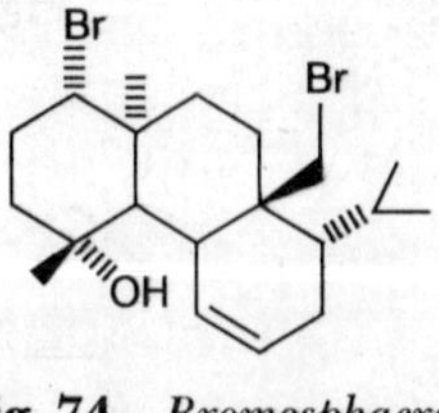

Fig. 74 *Bromosphaerol.*

Diterpene-substituted hydroquinone derivatives include: taondiol and a related carboxylic acid, atomaric acid, from *Taonia atomaria*: the coelenterate larval settling promoter δ-tocotrienol, and sargatriol from *Sargassum tortile* and stypotriol and related icthyotoxins from *Stypopodium zonale*.

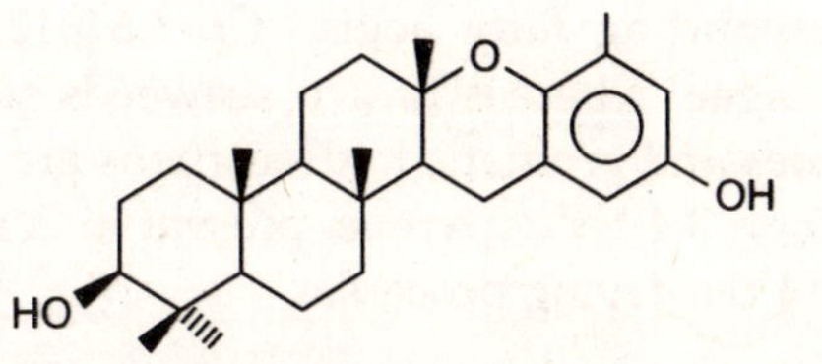

Fig. 75 *Taondiol*

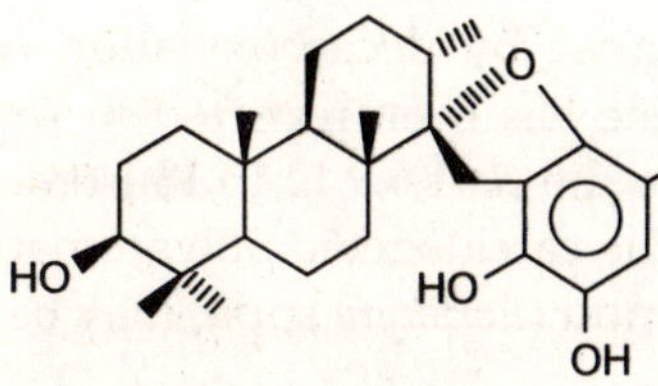

Fig. 76 *Stypotriol*

FATTY ACIDS, LIPIDS, HYDROCARBONS, ACETYLENES, AND OTHER LOW MOLECULAR WEIGHT LIPOPHILIC METABOLITES

Fatty Acids and Lipids

Seaweeds usually accumulate rather quantities of fatty and lipids. With few exceptions, the composition of their fatty acid fraction is qualitatively similar to that of other algae. Some interesting quantitative features exist: red seaweeds tend to be relatively enriched in highly unsaturated C_{20} fatty acids (20:4 ω6 and 20:5 ω3), and green seaweeds tend to accumulate relatively large quantities of *cis*-vaccenic (18:1 ω7), 16:4 ω3, and α-linolenic (18:3 ω3) acids.

The acyl lipids of seaweeds are for the most part unremarkable. Exceptions include the lysogalactolipid and lysosulpholipid hemolysins of *Ulva pertusa*, and the incompletely characterized glycosyl-diglyceride sulphuric, sulphonic and phosphoric esters of *Fucus serratus, F. vesiculosus* and *Pelvetia canaliculata*. Seaweeds apparently do not accumulate chlorosulpholipids.

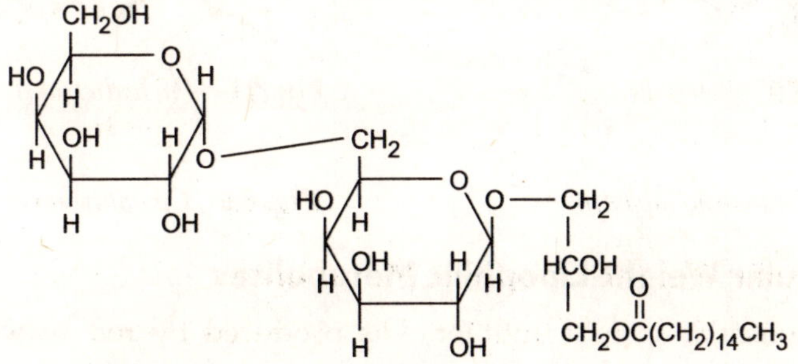

Fig. 77 Lysogalactolipid from *Ulva pertusa*.

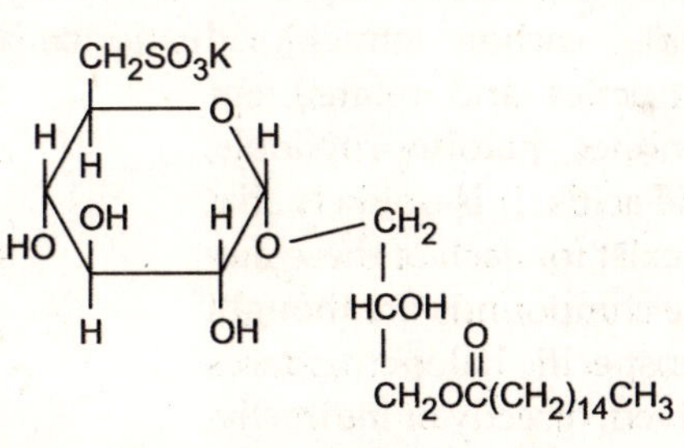

Fig. 78 Lysosulpholipid from *Ulva pertusa*.

Hydrocarbons

A variety of saturated and unsaturated hydrocarbons accumulate at low levels (usually <500 µgg^{-1} of dry matter) in seaweeds. Odd carbon number hydrocarbons (C_{15}, C_{17}, C_{21}) usually predominate, and probably arise by decarboxylation of the corresponding fatty acids. Cis-3,6,9,12,15,18-heneicosahexaene has been reported in several green algae, whereas brown seaweeds seem to accumulate only the cis-1,6,9,12,15,18 isomer. Phytadienes and aromatic hydrocarbons are minor consitituents. The carcinogenic polyaromatic hydrocarbon 3,4-benzopyrene, present in commercially available nori (*Porphyra* spp.), may be an artifact of the drying process.

Nonterpenoid Acetylenes

Acetylene cyclic ethers are known from eight species of *Laurencia*, and from '*Chondria oppositiclada*' (probably also a *Laurencia* sp.). Examples include laurefucin from *L. nipponica*, obtusenyne from *L. obtusa*, trans- and cis-rhodophytin from a *Laurencia* sp. and cis-maneonene A from *L. nidifica*.

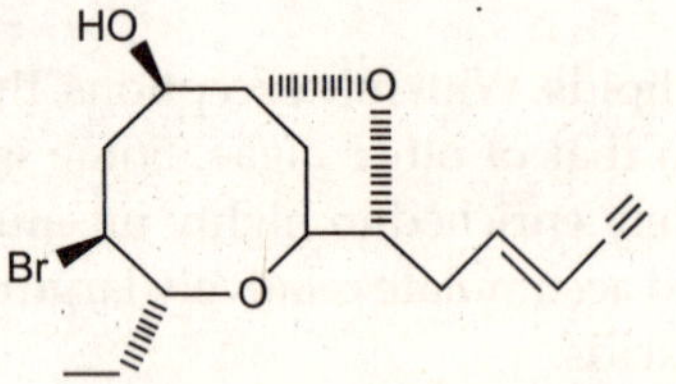

Fig. 79 *Laurefucin.*

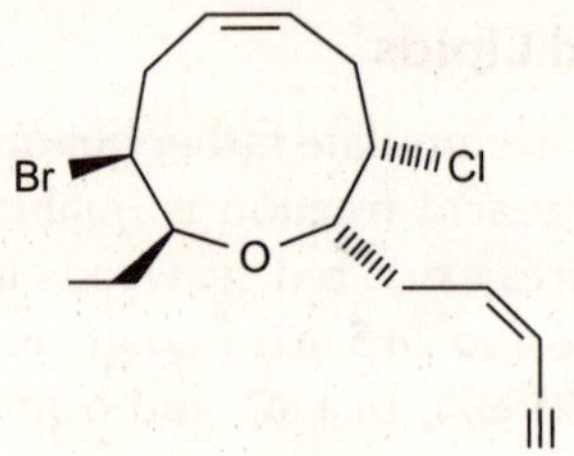

Fig. 80 *Obtusenyne.*

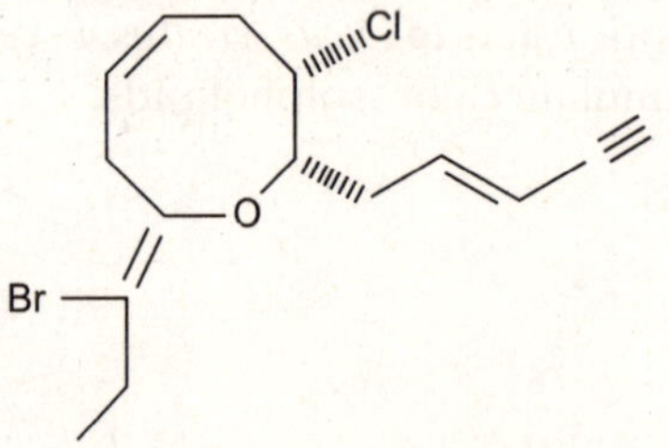

Fig. 81 *Trans-rhodophytin.*

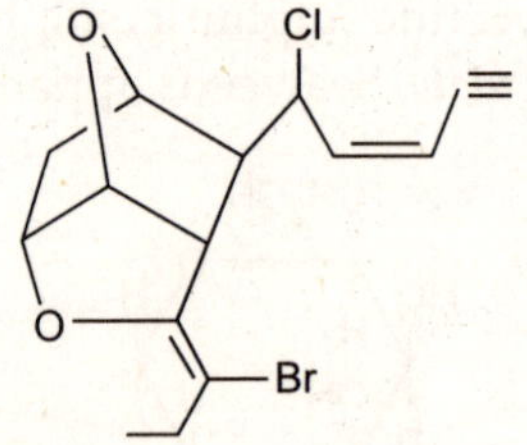

Fig. 82 *Cis-maneonene A.*

Other Low Molecular Weight Lipophilic Metabolites

A wealth of low molecular weight antibiotics is produced by red seaweeds of the genera *Asparagopsis* (and its tetrasporophyte *Falkebergia*). *Bonnemaisonia* (tetraspurophyle *Traillliella*). *Delisea, Plocamium* and *Ptilonia*. Among these compounds are methyl iodide, chloroform and other haloforms, carbon tetrabromide, carbon tetrachloride, iodoethanol, acetone, haloacetones, dibromoacetaldehyde, halopropenes and related epoxides and acetates, halobutenones, haloisopropanols, haloacetic acids, and haloacrylic acids. It is unlikely that specific halogenating enzymes exist for each of these metabolites; instead, many of these compounds are thought to be produced by relatively nonspecific haloperoxidases acting on carbon fragments derived, directly or indirectly, from acetate.

A group of halogenated lactones (beckerelides, fimbrolides) related to the structure shown in Fig. 85 contribute to the antibiotic activity of *Beckerella subcostatium*, *Delisea* spp., *Marginosporum aberrans*, and *Ptilonia australasica*.

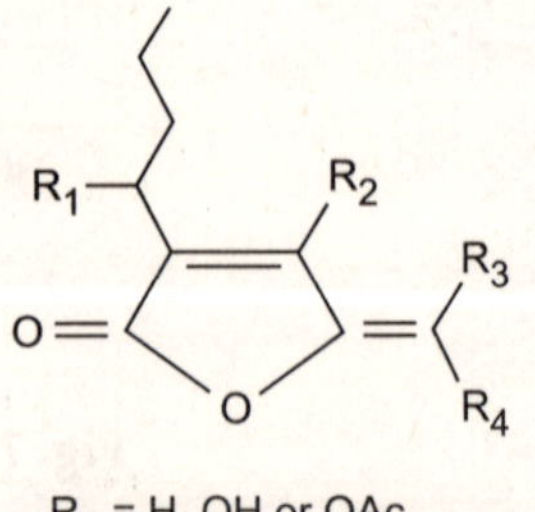

R_1 = H, OH or OAc
$R_2 - R_4$ = Br, Cl, I, H combinations

Fig. 83 *Halogenated lactones (showing alternative substitution patterns).*

NONTERPENOID PHENOLS

The nonterpenoid phenols of eukaryotic algae may be divided into three structural, and probably biosynthetic, groups. Two of these (below) are represented in seaweeds; the third, green algal (gall-?) tannins, are known only from the freshwater *Spirogyra arcta* and *Nitella* sp. The biosynthetic origin of the unusual phenolic antibiotic 3,5-dinitroguaiacol of *Marginosporum aberrans* is unknown.

Fig. 84 *3,5-Dinitroguaiacol.*

Phenols Related to Tyrosine

Phenols formally derivable from tyrosine are found in all three classes of seaweeds. Phenols with an intact C_6–C_3 unit include 3-iodo- and 3,5-diiodotyrosine, triiodothyronine, and thyroxine, present at low levels in many red and brown seaweeds. In addition, 3,5-dibromo-4-hydroxyphenylacrylic acid has been reported from *Halopytis incurvus* and *H. pinastroides*, and 4-hydroxyphenylpyruvic acid from *Odonthalia floccosa* and *Undaria pinnatifida*.

Fig. 85 *Triiodothyronine.*

Fig. 86 *3,5-Dibromo-4-hydroxyphenylacrylic acid.*

Eight-carbon (C_6–C_2) phenols such as 4-hydroxymandelic acid and 4-hydroxyphenylacetic acid (*O. floccosa*) can arise by decarboxylation of a C_6–C_3 precursor. Other C_6–C_2 phenols include 3,4-dihydroxyphenylethylamine (dopamine) from *Monostroma fuscum* and 3,5-dibromo-4-hydroxyphenylacetic acid from *H. incurvus* and *H. pinastroïdes*. Related C_6–C_1 phenols include 4-hydroxybenzaldehyde in *Dasya pedicellata* var. *stanfordiana*, *Marginosporum aberrans* and *O. floccosa*, as well as 4-hydroxybenzyl alcohol in *D. pedicellata*.

A variety of C_6–C_1-based brominated phenols, bromochloro-phenols and bromodiphenylmenthanes have been found in certain red seaweeds. *Halopytis pinastroïdes*

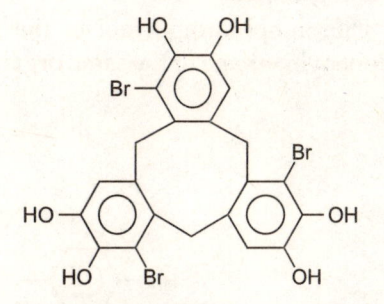

Fig. 87 *Cyclotribromoveratrylene.*

presumably produces the remarkable structure from a seven-carbon precursor. Many of the simple bromophenols probably exist *in vivo* largely as the sulphate esters; lanosol, for example, is found predominantly as the 1′,4-disulphate ester. However, until the recent development of liquid chromatographic methods for analysis of phenol sulphates, it has been very difficult to examine intact sulphate esters.

Claims of brown algal 'lignins' warrant further investigation.

Phenols Related to Phloroglucinol and Resorcinol

Brown seaweeds often accumulate large quantities of tannins structurally based on phloroglucinol. These phlorotannins are a mixture of polyphloroglucinols with molecular weights ranging to more

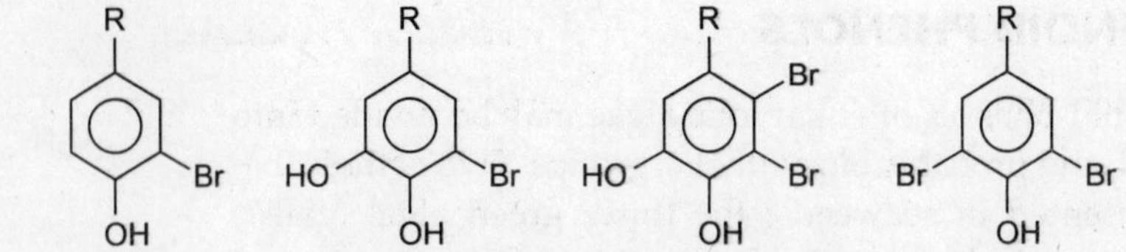

(90) (a) $R = CH_2OH$ (b) $R = CHO$ | (90) (a) $R = CH_2OH$ (b) $R = CHO$ | (90) (a) $R = CH_2OH$ (b) $R = CHO$ | (93) (a) $R = CH_2OH$ (b) $R = CHO$

Fig. 88 *Simple bromophenols produced by seaweeds.*

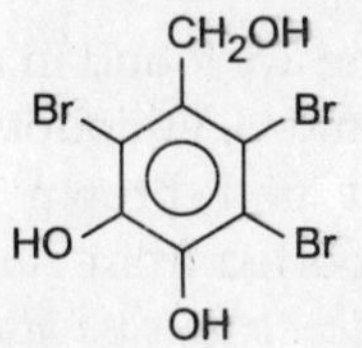

Fig. 89 *2,3,6-Tribromo-4,5-dihydroxybenzyle alcohol.*

than 10^5. Phloroglucinol-based 'oligomers' with biphenyl linkages, ether linkages, and/or combinations thereof have been reported in various brown seaweeds, and are thought to be indicative of the structures of the higher molecular weight tannins. Chlorophloroglucinols have been found in *Laminaria ochroleuca*. 1,2,3,5-Tetrahydroxybenzene 2,5-disulphate and/or related sulphate esters occur in many brown algae. The discovery of an alkenylphloroglucinol and related alkenylresorcinols in *Cystophora torulosa* suggests a polyketide biosynthetic origin for brown algal phloroglucinols.

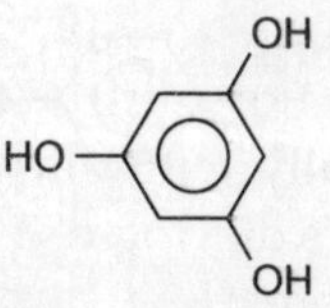

Fig. 90 *Phloroglucinol.*

Dibromophloroglucinol in the red seaweed *Rytiphlea tinctoria* is obviously related to the diphenylmethane of the same organism.

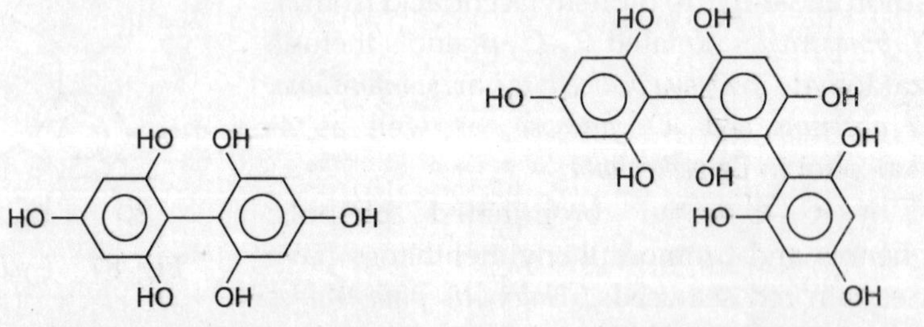

Fig. 91 *Examples of polyphloroglucinols with biphenyl linkages.*

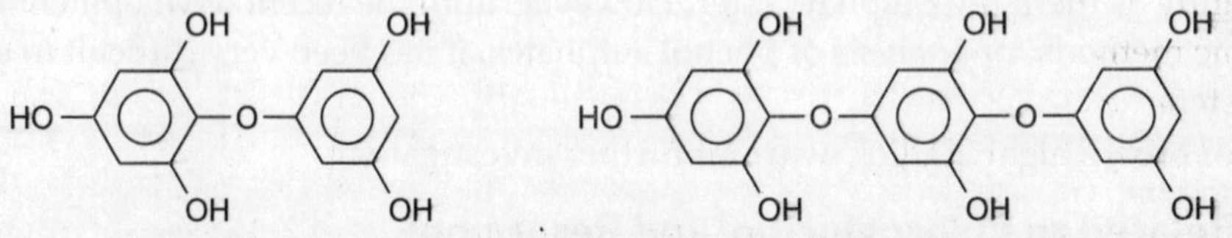

Fig. 92 *Examples of polyphloroglucinols with ether linkages.*

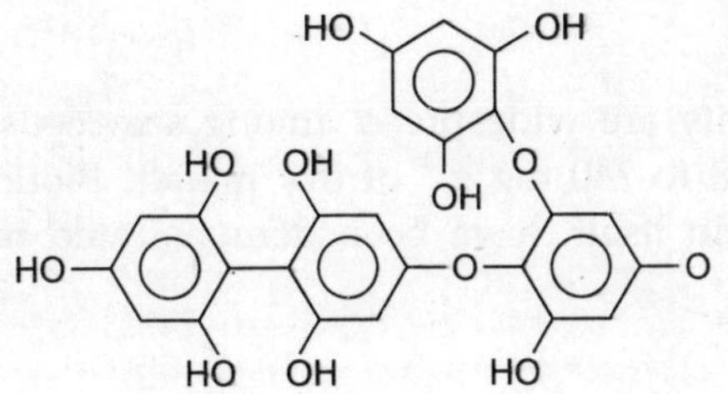

Fig. 93 *Example of a polyphloroglucinol with both biphenyl and ether linkages.*

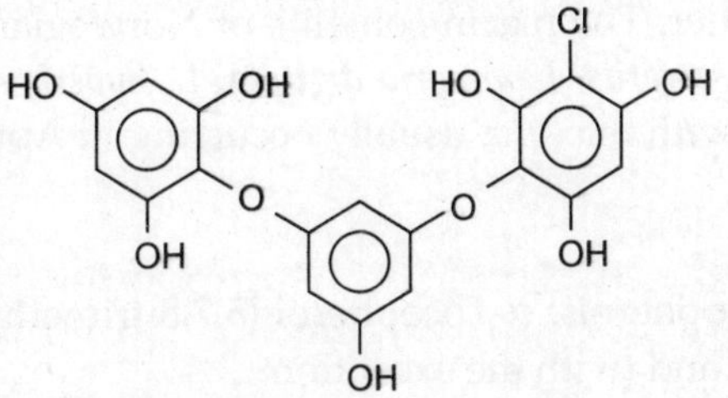

Fig. 94 *Chlorinated polyphloroglucinol from Laminaria ochroleuca.*

VITAMINS

Originally the term 'vitamin' referred to an essential factor required in small amounts in an animal's diet to prevent deficiency diseases. Many of these factors are known in seaweeds, including ascorbic acid, cobalamins (e.g. vitamin B_{12}), biotin, niacin, and tocopherols (including vitamin E).

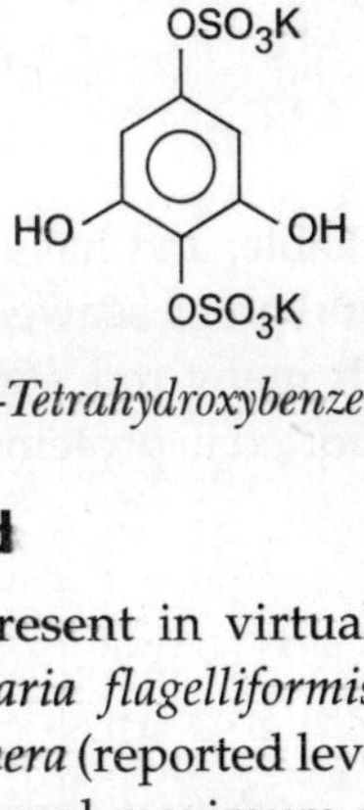

Fig. 95 *1,2,3,5-Tetrahydroxybenzene 2,5-disulphate.*

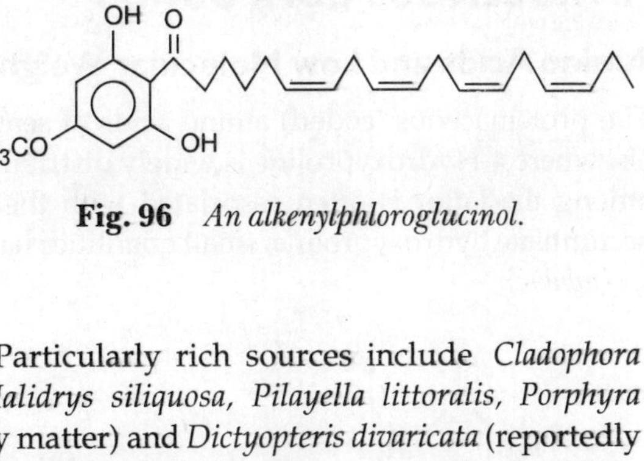

Fig. 96 *An alkenylphloroglucinol.*

Ascorbic Acid

Vitamin C is present in virtually all seaweeds. Particularly rich sources include *Cladophora rupestris, Chordaria flagelliformis, Chorda filum. Halidrys siliquosa, Pilayella littoralis, Porphyra umbilicalis, P. tenera* (reported levels > 5 mg g^{-1} of dry matter) and *Dictyopteris divaricata* (reportedly 10 mg g^{-1}). Seasonal maximum ascorbic acid concentrations in Norwegian *Ascophyllum nodosum, Fucus serratus* and *F. vesiculosis* occur in May and June, but earlier (January) in Adriatic *Pterocladia capillacea*.

Vitamin B$_{12}$

Vitamin B$_{12}$ activity has been reported to be widespread among seaweeds, generally occurring in higher concentrations in green and red algae than in brown algae. These activities often show seasonal variation, *Ulva rigida* : In *Rhodomela subfusca*, the vitamin B$_{12}$ activity could be resolved into at least three fractions presumably corresponding to different cobalamins. However, Ericson has suggested that epiphytic bacteria, and not the seaweeds themselves, biosynthesize the cobalamins.

Biotin

Substances with biotin activity are widespread among seaweeds; concentrations range from 18 to 740 mg g^{-1} of dry matter. Biotin sulphoxides, as well as biotin itself, have been demonstrated in brown seaweeds.

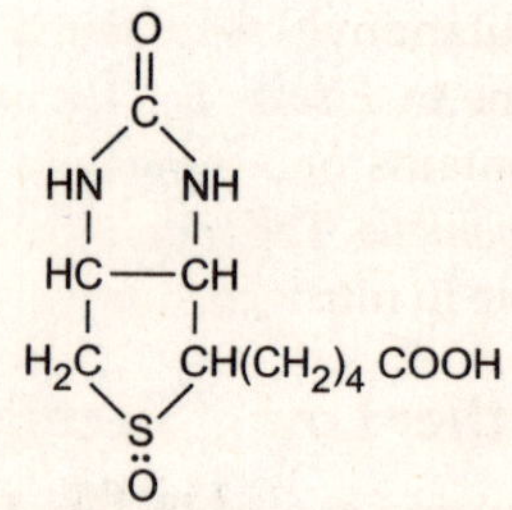

Niacin

Niacin (nicotinic acid) occurs widely among seaweeds in concentrations up to 68 µg g^{-1} dry matter. The niacin contents of Norwegian *Ascophyllum nodosum*, *Fucus serratus*, *Laminaria digitata*, *L. hyperborea* and *Palmaria palmata* show marked seasonal variations, with maxima usually occurring in April and May.

Fig. 97 *Biotin sulphoxides.*

Tocopherols

Vitamin E is a mixture of tocopherols. α-Tocopherol (5,7,8-trimethyltocol), is the only tocopherol known to occur in red, green, and (with the exception of the family Fucaceae) brown seaweeds. Fucaceae (*Fucus* spp., *Pelvetia canaliculata*, *Ascophyllum nodosum*) produce the γ-, δ-, and perhaps the β-isomers (7,8-dimethyl; 8-methyl; 5,8-dimethyl, respectively) in addition to α-tocopherol. Concentrations range from 7 to 80 µg g^{-1} of dry matter (except Fucaceae, 250 to 650 µg g^{-1}), and show seasonal variations.

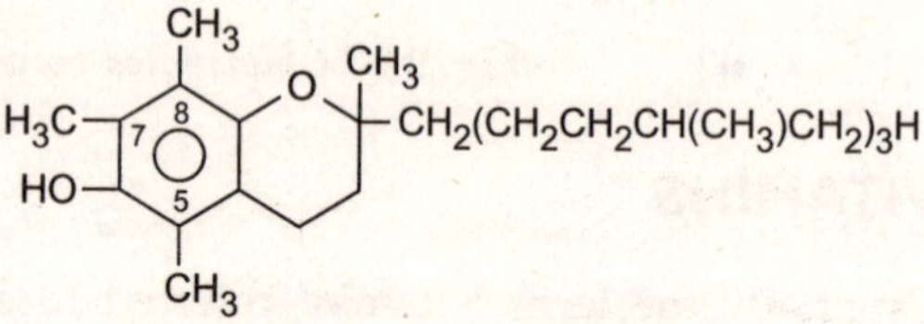

Fig. 98 α-*Tocopherol.*

NITROGENOUS COMPOUNDS

Amino Acids and Low Molecular Weight Peptides

The proteinaceous (coded) amino acids of seaweeds are unremarkable, and have been reviewed elsewhere 4-Hydroxyproline is widely distributed among brown and green seaweeds, and at least among the latter is often associated with the cell wall. Although many red algae seem not to accumulate hydroxyproline, small quantities have recently been reported in proteins from *Laurencia spectabilis*.

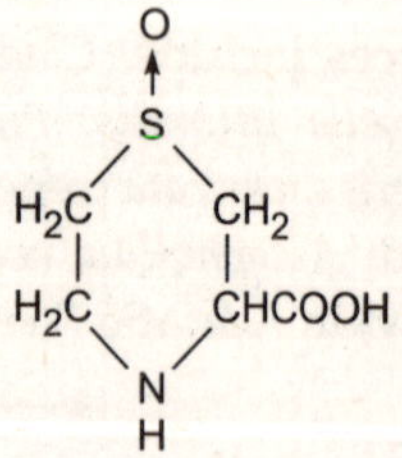

Fig. 99 *Chondrine (Yunaine).*

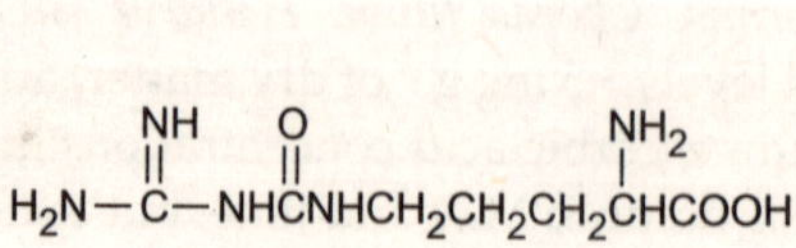

Fig. 100 *Gigartinine.*

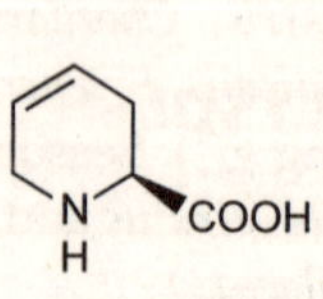

Fig. 101 *L-baikiain.*

Seaweeds produce a remarkable array of 'unusual' nonprotein amino acids, including chondrine (yunaine), gigartinine, gongrine, L-baikiain, methylated histidines, methionine sulphoxide and D-homocysteic acid. Low molecular weight peptides may also be accumulated: L-arginyl-L-glutamine in *Enteromorpha linza* and *Ulva pertusa*; β-L-aspartylglycine in *Ceramium rubrum*; L-citrullinyl-L-arginine in several red seaweeds; 'fastigiatine', L-(pyro-?)glutaminyl-α-L-glutaminyl-L-glutamine in *Pelvetia fastigiata*; 'eisenine', L-(pyro-?)glutaminyl-L-glutaminyl-L-alanine in *Eisenia bicyclis* and *E. cava*; and a partially characterized pentapeptide 'analipine', which contains one glutamata (or glutamine) and four aspartate (or asparagine) residues, in *Analipus japonicus*. The high nitrogen content of many of these amino acids and peptides suggests a possible role in nitrogen storage.

Other Low Molecular Weight Nitrogenous Compounds

Taurine is widely distributed among seaweeds; methylated taurines and related aminosulphonic acids have somewhat more restricted distributions. Trimethylamine and its *N*-oxide are also present in many red, brown and green seaweeds. Other nitrogenous compounds are known from only one or a few species: choline sulphate in *Gelidium cartilagineum*; the mycosporine palythine from *Chondrus yendoi*; polyhaloindoles from *Rhodophyllis membranacea* and *Laurencia brongniartii*; haloacetamides in *Asparagopsis taxiformis* and *Marginisporum aberrans*; a dihydrosphingosine diacetate from *Laurencia nidifica*, and *N*-acylsphingosines from *Amansia glomerata*; and caulerpicins and caulerpin in *Caulerpa* spp.

$H_2NCH_2CH_2SO_3H$

Fig. 102 *Taurine.*

$(CH_3)_3\overset{+}{N}CH_2CH_2OSO_3^-$

Fig. 103 *Choline sulphate.*

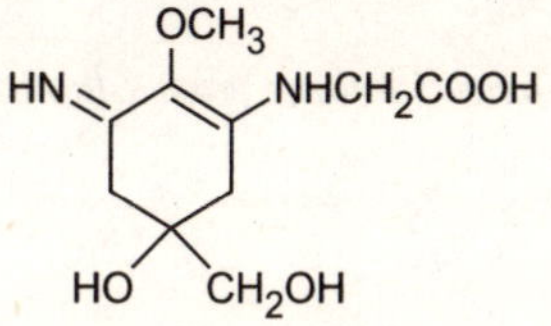

Fig. 104 *Palythine*

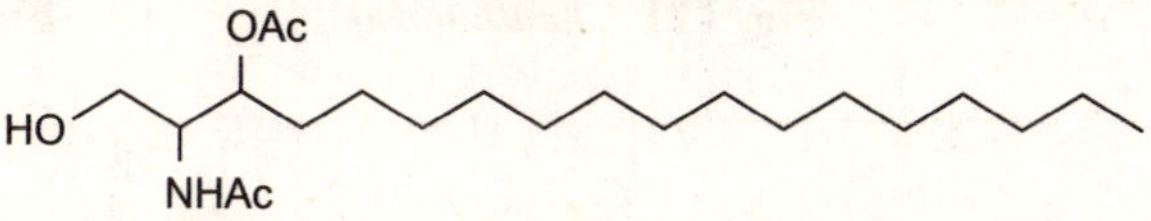

Fig. 105 *Dihydrosphingosine diacetate*

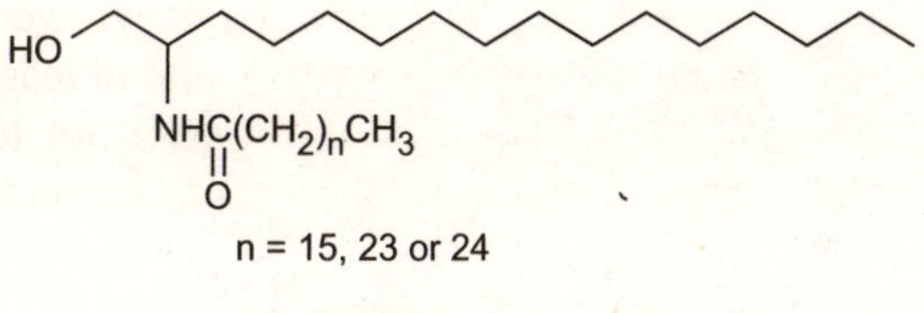

Fig. 106 *Caulerpicins.* n = 15, 23 or 24

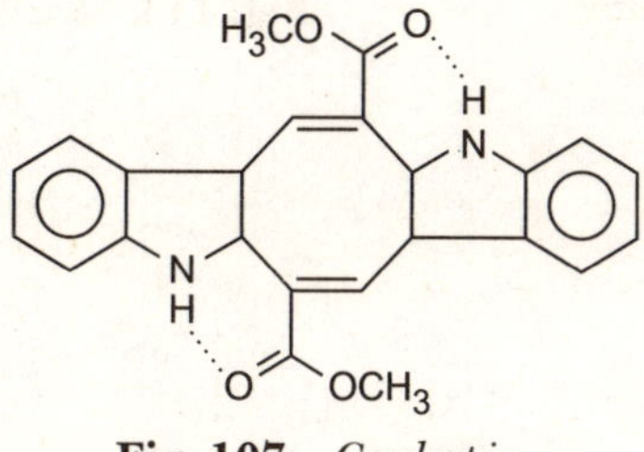

Fig. 107 *Caulerpin.*

POLYSULPHIDES AND POLYPHOSPHATES

A number of seaweeds (*Polysiphonia* spp., *Ceramium rubrum*, *Enteromorpha* spp., *Codium fragile*, *Monostroma* sp., *Spongomorpha* spp., *Cladophora* spp., *Ulva* spp., *Halidrys siliquosa*, *Pelvetia canaliculata*) liberate dimethyl sulphide when exposed to air. The immediate precursor to dimethyl sulphide is dimethyl-β-propiothetin, a metabolite of methionine.

$(CH_3)_2\overset{+}{S}CH_2CH_2COO^-$

Fig. 108 *Dimethyl-β-propiothetin.*

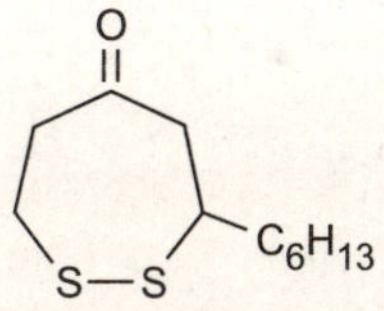

Fig. 109 *Cyclic disulphide from Dictyopteris spp.*

Dictyopteris australis and *D. plagiogramma* synthesize a cyclic disulphide and linear sulphides. The latter compounds, as well as the presumably related C_{11} sex hormones of certain brown algae, are presumably derived from fatty acid metabolism. The red seaweed *Chondria californica* accumulates a series of cyclic polysulphide antibiotics, primarily lenthionine, oxotrithiolanes, and a sulphone.

Cyclic metaphosphates containing three to seven phosphate residues, and linear condensed olige- and polyphosphates, serve as phosphorous reserves in a number of seaweeds, including *Ectocarpus siliculosus, Petalonia fascia, Pilayella littoralis, Ceramium deslongchampsii, C. rubrum, Rhodomela confervoides, Acetabularia crenulata, A. mediterranea* and *Cladophora* sp.

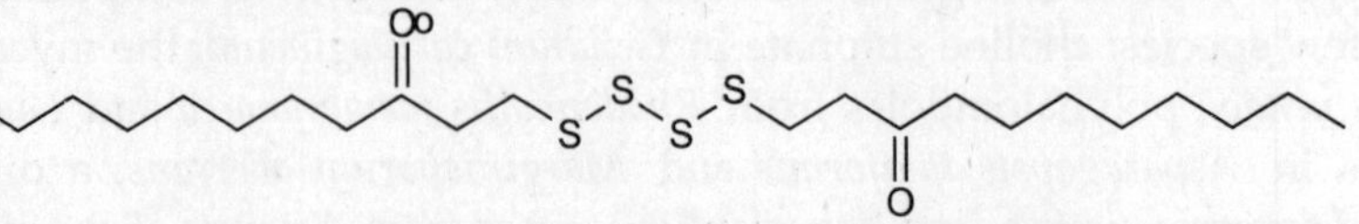

Fig. 110   *A linear sulphide from Dictyopteris.*

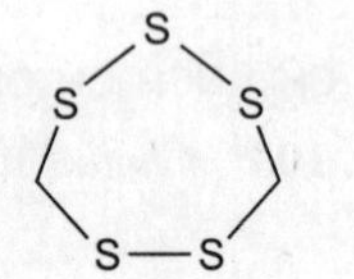

Fig. 111   *Lenthionine.*

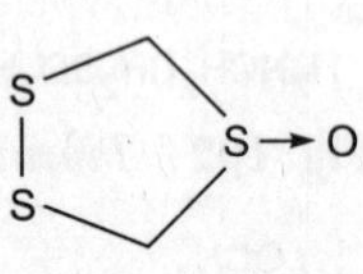

Fig. 112   *An oxotrithiolane from Chondria californica.*

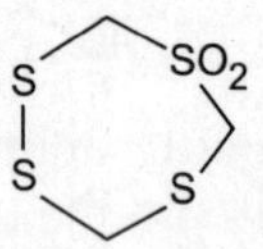

Fig. 113   *Cyclic sulphone from Chondria californica.*

5 Economic Importance of Algae

ECONOMIC IMPORTANCE

Algae have been intimately connected with human beings from times immemorial. They have also been connected with important industries like the production of iodine and agar-agar, to mention two of the most important products of algal industry. *Rhodymenia* and *Phyllophora*, members of Rhodophyceae are rich in iodine. The industrial utilisation of sea-weeds in various parts of the world dates back to many hundreds of years, particularly in China and Japan. *Porphyra vulgaris* and *Gelidium corneum,* are cultivated for food purposes in China and Japan.

Use of Marine Algae as Fodder

Sea weeds are also used as food for domestic animals in various parts of the world. In Norway, Scotland, France, America, and New Zealand the use of sea-weeds as animal food has been well recognised. Some countries have even factories to process sea-weeds into suitable cattle feed. The manufacture of cattle feed from sea-weeds is made principally from brown algae and the processed food is fed to cattle, poultry and even pigs. The sea-weed meal is used mainly as a protein concentrated and is usually fed to livestock in the proportion of 20 per cent of the total ration by weight by farmers in certain parts of the USA.

Algae as Food for Humans

Apart from China and Japan, in Malaya, Indonesia, Burma, Siam, Borneo, the Strait Settlements and Indo-China, the sea-weeds are used as food. *Ulva lactuca* was used in Scotland for the preparation of salad and soups. *Laminaria saccharina, Rhodymenia palmata* have been used as food in parts of Scotland and Ireland and the thin delicate sea-weed *Porphyra* is considered to be a tasteful culinary dish in many parts of England. Among the more important algal food industries may be mentioned carrageen. This is the product of several sea-weeds but principally *Chondrus crispus*. Carrageen is principally used for the production of blacmanges and moulds and is frequently seen in 'health stores'. It is used by soaking it in water and mixing it with milk. Carrageen possesses the properties of gelling which is one of the factors which enhances its usefulness. Apart from milk and water, fruit juice may also be mixed with it resulting in products like fruit jelly. Carrageen is also used in the preparation of ice creams and in the confectionery industry.

It is in the region of Far East and Australia that sea-weeds occupy an important position as human food. The inhabitants of the Hawaii islands consume large quantities of sea-weeds. The indigenous population of Chile use large quantities of *Durvillea antarctica* and some of the species of *Ulva*. The aborigines in Australia use *Sarcophycus potatorum* as food.

Although there are from 25 to 30 species of algae eaten by the Japanese 'amori' *(Porphyra)* and 'funori' *(Gloiopeltis)* are the important ones under cultivation. Bundles of bamboo or other material are sunk into muddy bottoms of water and these furnish lodging for the spores of the algae. The spores germinate, and plants grow rapidly, and within three or four months the harvest is made.

'Amori', 'kombu' (species of kelp), and tengusa *(Gelidium)* are considered the most important of the Japanese edible sea-weed. Amori may be eaten raw or preserved for future use by drying in the sun. It may be baked until crisp and then eaten with soup or made into a kind of sandwich known as 'succhi'. Kombu is one of the standard food of Japan and comes from various species *Laminaria, Arthrothamnium* and *Alaria* on the shores of the northernmost islands of the Archipelago. Kombu may be cooked with meat and soups, may serve as a vegetable, may be put on sauces and rice and may be used as tea.

The table below gives an idea of the nutritive value of different edible algae:

Sea-weed	Per cent water	Per cent raw protein	Per cent fat	Per cent starch, sugar	Per cent fibre	Per cent ash
Nostoc commune	10.6	20.9	1.2	55.7	5.1	7
Ulva lactuca and *U. fasciata*	1.7	14.9	0.04	50.6	0.2	15
Laminaria sp.	23.5	5.85	1.15	41.95	6.7	21
Porphyra tenera	17.1	28.0	0.8	40.1	-	10

Algae as Fish Food

Algae, both plankton and attached forms, in the sea as well as freshwater, provide the primary food for fish and other aquatic animals. In the sea, marine algae grow as attached forms on rocks near the sea-shore, or occur as floating plankton. They manufacture all the food material consumed by marine animals. The sea-shore constitutes the richest food producing region in the entire sea. The great fishing grounds of the oceans are found where algae grow on the sea-shore as attached forms, or where they are present in very large numbers in the form of plankton. *Zobrasoma flavescens*, the surgeon fish, feeds on the red alga, *Amansia glomerata*, and the brown alga, *Ectocarpus*. *Forcipiger longirostris*, the common fish of Hawaii, feeds on the blue-green alga, *Hydrocoleum cantharidosmum*.

In the Philippine Islands fish pond culture has been developed by the people on a scientific basis. The fry of the milkfish are taken from the sea and transferred to ponds in which the blue-green alga, *Lyngbya aestuari* grows. The gelatinous filaments of the alga serve as 'baby food' for the fry. After a period of about three months the fish are removed to a second pond in which *Cladophora* and *Chaetomorpha* grow. A few months later they are transferred to a third pond containing the green alga, *Enteromorpha*.

Diatoms are present in large numbers in the sea when the nutrients available are high, and produce phytoplankton during peak periods. They are followed by an equally abundant season for zoo-plankton. The plant and animal planktonic elements provide feed for pelagic fishes and variations in their numbers have repercussions on commercial fisheries. The close correlation between migration of sardines and diatoms of the genus *Fragillaria oceanica* on the Malabar coast has been well-established.

In freshwater lakes and ponds green filamentous algae, such as *Oedogonium, Spirogyra, Microspora. Ulothrix, Cladophora* and *Pithophora* serve directly as fish food, or are consumed by midge fly larvae and other insects which in their turn are eaten by the fish.

Apart from serving as food material, the physiological activity of algae plays an important role indirectly in influencing the water medium. By their photosynthesis they release oxygen in the aquatic medium and remove carbon dioxide. This is an asset to fish life since oxygen is vital for their existence and carbon dioxide in certain concentrations is lethal to fishes.

Nitrogen Fixation by Blue-green Algae

Soil is a living mass and apart from soil particles, there are in it a number of bacteria, fungi, algae and protozoa. Algae occupy a volume three times that of the bacteria and there may be about one million algae in a gram of manured soil. The blue-green algae show considerable vitality upto 70 years in the soil like *Nostoc muscorum* and *Nodularia harveyana*. Watanable (1951) collected more than 1,000 species of nitrogen fixing algae, from various regions of Asia. Out of these he found *Tolypothrix tenuis* from Borneo as a nitrogen fixer, and it was found that it could fix as much as 780 lb. of nitrogen per acre per year. Allen (1955) found that *Anabaena cylindrica* fixes 2,900 lb. of nitrogen per acre per year. Increased yields were obtained in rice using *Tolypothrix* cultures considerably in Japan.

Reclamation of Alkaline Usar Soils by Blue-green Algae

It has been found that some blue-green algae form a thick stratum on the surface of saline *usar* soils during the rainy season when other plants including crops fail to grow on such soils. Various species of *Nostoc, Scytonema, Anabaena*, etc. have been found to be plentifully growing on *usar* soils during the rains. These algae can be of use in the reclamation of usar lands. The actual process of reclamation is a series of successive growth of the algal crop on such soils in a water-logged condition. The water holding capacity also increases in the soils.

Mass Culture of Algae for Food

The culturing of algae with the aim of producing useful organic substances, food, feed and special organic material was done by Harder and von Witsch (1942), Spoehr and Milner (1947) and by Prof. Tamiya. Much work has been done on *Chlorella pyrenoidosa* on mass culturing. *Chlorella* possesses 50 per cent protein, which is four to five times the protein content of wheat grain.

Pollution of Water by Blue-green Algae

A number of species of blue-green algae occur as phytoplankton in the sea as well as in freshwater. *Oscillatoria erythraea* occurs in such large quantities that it gives the red colour to the so called Red sea, Species of *Microcystis, Chroococcus, Coelopshaerium, Spirulina, Oscillatoria, Anabaena, Nodularia, Nostoc, Aphanizomenon, Gloeotrichia* and *Rivularia* form blooms on the surface of water. Apart from unpleasant smell, these algae produce a large amount of suspended organic matter making the water unfit for consumption. Blue-green algae thrive on organic matter found in polluted water. Organic matter in ponds is contributed by human and animal excreta, dead and decaying leaves of trees, bushes and weeds, aquatic plants and algae. The presence of luxuriant growth of blue-green algae is an indicator of pollution. They lend objectionable taste and odour to the water, and also cause mortality among live-stock. When the surface of lakes or ponds is covered with decaying blue-green algae, oxygen content in water is decreased, and this causes the death of fish. They also choke the gills of fish.

Many types of bacteria are harboured in the gelatinous sheaths of blue-green algae. The best method to keep the water clean in water filter works is by scraping of the masonary wall of filter beds.

Lyngbya ochracea and the diatom *Synedra affinis* are found in abundance in the storage tanks and filter beds. Iron bacteria, such as *Leptothrix ochracea*, also grow in iron pipes, often choking them.

Damage to Salt Blue-green Algae

Blue-green algae grow in abundance in salt beds. They are also present in the salt lake at Sambhar in Rajasthan which affects the quality of salt and hence loss of revenue. The algae interfere with manufacturing operations by imparting an offensive smell and a pink rust red colour to the salt.

Agar Industry

Algae such as *Camplaephora, Pterocladia, Gracilaria, Ahnfeldita, Gelidium, Eucheuma, Chondrus, Gigartina, Phyllophora, Furcellaria* are used for the extraction of agar. Agar from *Gelidium cartilagineum* is known to consist of a chain of alternating D-galactose and 3:6, anhydro-L-galactose residue with half ester sulphate on about every tenth unit of the galactose. Pyuruvic acid, uronic acid, polysaccharides like agaropectin and agarose are also known to occur in agar-agar.

Agar is made from the algae by a series of processes consisting of bleaching, freezing at temperatures about 10°C, and then removing the water by melting. It is about 75 per cent gelose, 15 per cent water, 4 per cent ash, and smaller percentages of protein, fats, fibre, and silica.

Agar-agar is a favourite sea-weed in China and Japan for jelly and as a thickener of soups and sauces. Its chief commercial value lies in its ability to form a gel upon cooling, and so is used in the making of ice cream, desserts and various pastries. It is superior to gelatin in giving rigidity to soft canned fish in transport. It may be used in clarifying liquors, as sizing material, and in the preparation of bacteriological media. As a laxative it has the properties of absorbing much water, of becoming a lubricant, and of serving as a mild stimulant. It is also used in cosmetics, pharmaceuticals, leather and textile industries, emulsifiers and meat packing.

Alginic Acid

Alginic acid ($C_{21}H_{27}O_{20}$) is obtained from kelps (*Laminaria* etc.) by soaking in freshwater. Alginic acid is insoluble in cold water, absorbs many times its weight in water, and becomes hard and horny and resistant to solvents when dry. Many metallic alginates may be formed, but the most useful is sodium alginate. Alginates are used as sizing materials in water-proof varnishes, as dyes, and as rubber-algin substances such as knife-handles, buttons, and combs. Sodium alginate has found widespread use as a stabilizer in dairy products.

Iodine and Other Compounds

Some species of brown algae contain in the blades as much as 0.29 per cent iodine in *Macrocystis*, 0.38 per cent in *Pelagophycus*, 0.14 per cent in *Nereocystis*, and in whole plants of *Laminaria* about 9.49 per cent. The amount of vitamin C in *Sargassum myriocystium* is slightly more than that in the juice of a lime fruit.

Funoria and Funorin

Japan prepares a sizing agent and glue for textile called '*funori*' from *Gloeopeltis furcata* and for other inferior products, algae like *Iridaea, Grateloupia, Chondrus, Ahnfeldita* are used. The composition of funori is similar to that of agar but it lacks sulphate ester groups.

Diatomite (Khiesulghur)

Diatomaceous earth has been used variously from ancient times. It is alleged that Emperor Justinian in 532 A.D. gave directions that in the repair of the Church of Saint Sophia in Constantinople, bricks be made of diatomaceous earth, because of their lightness. It is extensively employed as an insulating material for coating steam pipes or pipes of refrigerating plants and for lining walls of blast furnaces. It is also used in the filtration of oils and syrups. As a polishing substance it has been largely replaced by such materials as carborundum, but is still used for polishing silverware. In making dynamite, diatomaceous earth was formerly used as an absorbent for liquid nitroglycerine. It is used in the manufacture of rubber, blottings, in the extraction of petroleum and in the insulation of boilers and blast furnaces. Diatomite is used as a cleansing agent in soap, toothpaste and metal polish industries. The sprinkling of diatomaceous earth on the floor and walls reduces the chances of explosion in coal mines.

Medicines

Brown algae like *Sargassum* and *Laminaria* species find medicinal value in checking goitre and other glandular troubles because of their high iodine content. *Gelidium* is used for other stomach disorders and for heat induced problems. Sodium laminarin sulphate and fucoidin are used as anticoagulants but carrageenan acts as a blood coagulant.

Chlorellin is extracted from *Chlorella* which inhibits growth of certain bacteria and a few algae. *Nitchia palea* checks the growth of *Escherichia coli*. *Microcystis* is known for its inhibitory action to *Staphylococcus*. Certain antibiotic properties are attributed to *Pelvetia, Rhodomela, Ascophyllum nodosum* etc.

Sewage Disposal

Algae play an important role in the oxygenation of sewage to a great extent. They form a surface film on sewage disposals which supplies oxygen and utilises nutrients to break down sewage. Examples of algae growing in sewage include: *Euglena, Chlorella, Scenedesmus, Chlamydomonas, Microactinum,* and members of Chlorococcales, Volvocales and Euglenophyceae etc.

Radioactive Uptake by Algae

Certain algae are known to play role in the uptake of radioactive wastes notably, *Porphyra laucinata* (Ru[106]), *Chlorella* and *Euglena* (Cesium[137]). *Scenedesmus* (Isotopes of Calcium), *Cladophora glomerata* (Cobalt, Iron, Rb, Sulphur), *Spirogyra* sp. (Zn, Sr).

Lens Paper

Japanese use *Spirogyra* in the manufacture of lens paper used for cleaning optical articles.

Harmful Effects

Toxins produced by *Apanizomenon* and *Microcystis aeruginosa* are poisonous to fish like *Crappis perch* and *Gambusia*. Certain blue-green algae such as *Anabaena, Nodularia, Gloeotrichia, Aphanizomenon, Lyngbya,* etc. are harmful to horses, cattle, sheep etc. even causing death of animals.

CHEMICAL CONSTITUENTS OF ALGAL SEAWEEDS

The chemistry of seaweeds is an exciting an rapidly expanding field. Interest in marine plants as potential sources of new antibiotics, pharmaceuticals and antitumor agents has been rewarded by the discovery of an unexpected diversity of new natural products.

Photosynthetic Pigments

Chlorophylls. Chlorophyll *a* is the most abundant chlorophyll, and the primary photosynthetic pigment, of sea-weeds. As a specialised chlorophyll-protein complex (P700), it is directly involved in the Photosystem I light reaction. Seaweeds probably synthesize chlorophyll *a* from 5-aminolevulinic acid. In certain unicellular protists *(Porphyridium purpureum, Chlorella vulgaris* etc.), cyanobacteria and higher plants, 5-ALA is produced from α-ketoglutaric acid via 4, 5-dioxovaleric acid.

Green and brown alga produce accessory chlorophylls in addition to chlorophyll *a*, chlorophyll *b*, probably synthesized directly from chlorophyll *a*, is found in green algae. Chlorophyll(ide)s c_1 and c_2, present in brown sea-weeds, may be derived from magnesium 2, 4-divinyl phaeoporphyrin a_5, an intermediate in the biosynthesis of chlorophyll *a*. Red algae do not produce chlorophylls *b* or *c*; a 'chlorophyll *d*', found in extracts of some red seaweeds, has not been detected *in vivo*.

Phycobiliproteins. Phycobiliproteins are open-chain ('linear') tetrapyrrole pigments found in Rhodophyceae, Cyanobacteria and Cryptophyceae. The bilin unit is derived from 5-aminolevulinic acid via protoporphyrin IX. Two types of bilins have been identified in seaweeds: phycocyanobilin and phycoerythrobilin. A third phycobilin, phycourobilin, may or may not be a structurally distinct entity. These bilins are covalently attached to polypeptides to form the major classes (α, β, γ) of subunit component proteins. Some of the commanly found biliproteins are: Allophycocyanin, Allophycocyanin β, C-phycocyanin, R-phycocyanin, phycoerythrocyanin, C-phycoerythrin, b-phycoerythrin, B-phycoerythrin and R-phycoerythrin.

Carotenoids. Carotenoids are isoprenoid polyene pigment present in all photosynthetic and many non-photosynthetic organisms. Bio-synthetically, carotenoids are derived from mevalonic acid via isopentenylpyrophosphate and lycopene and/or its isomer β zeacarotene. β, ε-Carotene and β-carotene are formed by enzymatic cyclization, further biosynthetic alterations (hydroxylation, epoxidation, inchain oxidation etc.) subsequently give rise to the variety of carotenoids and xanthophylls found in seaweeds. Most red seaweed exhibit a rather simple carotenoid pattern, accumulating mainly β-carotene, α-carotene and their dihydroxy derivatives, zeaxanthin and lutein. Most other carotenoids reported to be minor constituents of red seaweeds are closely related to these four major pigments: the corresponding monohydroxides β-and α-cryptoxanthin, lutein epoxide ('taraxanthin'), zeaxanthin mono-(antheraxanthin), and di-(violaxanthin), epoxides, and auroxanthin (violxanthin furanoxide). Neoxanthin, reported from *Antithamnion* and *Nemalion* is probably derived from a violaxanthin-like precursor. Fucoxanthin and peridinin, are also reported from certain red algae.

Brown sea-weeds accumulate β-carotene, fucoxanthin and violaxanthin (the closely related zeaxanthin), neoxanthin and fucoxanthinol (desacetoxy-fucoxanthin) are occasionally minor components, as is ε, ε-carotene. The acetylenic carotenoids diadinoxanthin and diatoxanthin are thought to be endogenous. Lutein and α-carotene have occasionally been reported.

Green seaweeds generally resemble higher plants in carotenoid composition. The major carotenoids are β-carotene, lutein, violaxanthin, antheraxanthin, zeaxanthin and neoxanthin. α-carotene, β, Ψ-carotene, β-cryptoxanthin; cryptoxanthin-5', 6'-epoxide, and lutein epoxide are minor or occasional constituents. The α-carotene derivative loroxanthin has been reported from *Cladophora* spp. and *Ulva rigida*.

Siphonaceous green seaweeds like *Avrainvillea, Bryopsis, Caulerpa, Codium, Halimeda, Pseudobryopsis, Udotea* etc. and members of Derbesiaceae *(Derbesia, Halicystis),* also accumulate the α-carotene derivative siphonaxanthin and its fatty acid ester(s), siphonein.

Steroids and Other Triterpenoids

Sterols and related compounds (4 β-methyl-and 4, 4-dimethyl sterols, keto steroids, steryl esters and steryl glycosides) are bio-synthesized from movalonic acid via isopentenylpyrophosphate and squalene. Cyclization of squalene 2, 3-epoxide produces cycloartenol, thought to be the precursor of all algal steroids. Sterols have been demonstrated in all red seaweeds consisting of about 120 species. C_{27} sterols such as cholesterol, desmosterol and 22-dehydrocholesterol are almost always predominant. Other phytosterols which may be present as minor constituents include: campesterol sitosterol stigmasterol; brassicasterol; fucosterol and 28-isofucasterol; 24-methyl-enecholesterol and liagosterol.

The predominant sterol of all investigated brown algae is fucosterol, and also cholesterol, 24-methylenecholesterol and 22-dehdrocholesterol are widespread. *Codium fragile* and *C. vermilara* differ considerably from other green seaweeds, accumulating only the unusual dienols clerosterol and codisterol.

Other Terpenoids

1. Monoterpenes. Monoterpenes are C_{10} isoprenoids structurally related to neryl pyrophosphate or its isomers. Five structural classes of monoterpenes have been described among seaweeds; *Chondrococcus (Desmia), Mcrocladia, Ochtodes* and *Plocamium*.

2. Sesquiterpenes. Sesquiterpenes are C_{15} isoprenoids formally derived from *trans, cis*-farnesyl pyrophosphate (or one of its isomers e.g. *Caulerpa flexilis* and *C. prolifera*).

3. Diterpenes. Diterpenes are C_{20} isoprenoids formally derivable from geranylgeranyl pyrophosphate. The most abundant diterpene in seaweeds is 7R, 11R-*trans*-phytol. *Cis*-phytol is accumulated by *Gracilaria andersoniana*. More complex diterpenoids may be considered to arise from selective cyclization of geranylgeraniol, like caulerpol found in *Caulerpa brownii;* and antibiotic dictyodial found in *Dictyota flabellata* etc.

6 Cytology and Ultrastructure

CHLOROPHYCEAE

The green algae have received extensive study from microscopists, but much of the existing information concerns freshwater members of the division. Several freshwater Chlorophyceae (e.g., *Chlamydomonas*) have been used extensively in cellular and developmental studies.

Characters such as whether the mitotic spindle is open or closed and whether cell division is associated with a 'phycoplast' or 'phragmoplast' have been used in recent phylogenetic schemes for the green algae (Lobban & Wynne, 1981).

Chloroplasts, Pyrenoids and Mitochondria. When viewed with the light microscope, green algal plastids are observed in a variety of shapes, and these are useful in algal identification. The plastids may be disc-shaped, form a reticulum or a parietal, cup-shaped plastids may be observed. The thylakoids of green algal plastids are arranged as two to many per lamella. In some, an irregular stacking results in a thylakoid arrangement reminiscent of the grana and intergrana regions of the plastids in the Charophyceae and higher plants.

Starch is synthesized and stored within the chloroplasts of most green algae, but plastids differentiate from a common stem organelle into chloroplasts and amyloplasts in some of the Caulerpales (Udoteaceae and Caulerpaceae), a condition known as *heteroplasty*. In contrast to the green algae, polysaccharide storage occurs in cytoplasmic vesicles in the red and brown algae.

Many chloroplasts of the coenocytic order (Caulerpales) of green algae contain structures termed the 'concentric lamellar system' (CLS) or 'polar thylakoid organizing body'. The CLS does not contain chlorophyll and appears to be continuous with the inner membrane of the chloroplast envelope. It is composed of concentric lamellae and may range from 3-25 pairs of thylakoids (Lobban & Wynne, 1981).

Chloroplast DNA was observed *in situ* in *Acetabularia*, *Batophora* (Dasycladales) and *Codium* (Caulerpales) using the DNA-specific, fluorescent dye, 4, 6-diamidino-2-phenylildole (DAPI).

Mitochondria in green algal cells resemble those in other eukaryotic cells. The inner membrane of the mitochondrial envelope enfolds to form tubular cristae. Most cells contain many mitochondria, but male gametes of *Bryopsis* have a single large mitochondrion. Several unusual mitochondrial conformations have been observed. Scott & Bullock (1976) observed an unusual mitochondrial morphogenesis in *Cladophora flexuosa* during the late stages in gamete formation. The mitochondria were often highly contorted, and regions of the mitochondria containing DNA were separated from the rest of the mitochondrial matrix by a double membrane. Mitochondria outside the differentiated gametangial area in *Derbesia* often have tubular paracrystalline inclusion in the matrix and concentrically arranged cristae.

Endomembranes (the Golgi Complex, Vesicles) Endoplasmic Reticulum, Microbodies).

Golgi bodies (i.e. dictyosomes) of green algae appear similar to those described in other plant cells. Very large vesicles form at the ends of Golgi cisternae, but there are no other conformational changes associated with increased Golgi activity. The typical Golgi body has about 7 to 10 cisternae. A process of cleavage in *Acetabularia* involving coated vesicles are produced by the golgi body. Cysts form in the cap through the alignment and fusion of large coated vesicles. Microtubules are associated with these vesicles and may be responsible for vesicle orientation.

Many cells of the green algae are highly vacuolated. This is especially true in coenocytic algae where the response to wounding involves gel formation by the vacuolar matrix as in *Caulerpa*. Protein bodies formed by the rough ER appear to be important in formation of a wound plug in *Bryopsis*. They break down into rod-like and granulo-fibrillar material when secreted into the vacuole. The association of microtubules with the protein bodies as they disperse is particularly interesting.

Microbodies are spherical, electron-opaque bodies with a finely granular matrix. They are surrounded by a single membrane. Microbodies may be closely associated with mitochondria, ER, lipiobodies and chloroplasts, and they are important in specific metabolic activities, e.g. glycolate metabolism. This organelle has been identified morphologically in a number of green algae including *Enteromorpha, Ulva, Ulothrix* and *Cladophora*. Microbodies are identified in *Acetabularia* also.

Cell Wall.

The composition of cell wall material and the microfibrillar architecture of its various layers are known in many marine Chlorophyceae. Some algae, such as *Valonia*, have highly cellulosic walls and microfibrils form well ordered arrays. Mannan and xylan are important wall constituents in many species belonging to the Caulerpales. The outer cuticle of the cell wall in several marine Chlorophyceae is composed of proteins. The differences in wall composition between the gametophyte and sporophyte of several green algae are of particular interest. The sporophytes of *Derbesia* and *Bryopsis* contain mannan as the main cell wall polysaccharide, whereas the gametophyte plants have xylan and cellulose as the main wall constituents. Similarly, in *Acetabularia*, the cyst walls are composed of cellulose, but the vegetative plant wall is composed of β-1, 4-mannan. Cytoplasmic continuity is maintained in some green and brown algae by plasmodesmata which traverse the crosswall between cells.

Flagella, Rhizoplast, and Phototaxis.

The gametes of green sea-weeds are biflagellate, whereas zoospores are bi-, quadri-, or multifiagellate. The flagellar surface is smooth, and insertion is typically apical. The transition zone between the 9-2 axoneme and the basal body in the flagella of some green algae has a stellate character; but in zoospores of *Urospora* this region has 'wings' attached to one tubule in each of the doublets. *Urospora* zoospores also have prominent striated fibrillar rootlets called rhizoplasts and microtubular rootlets with nine parallel microtubules per rootlet. Rhizoplasts are clearly plant 'muscles', contractile organelles intimately concerned with flagellar movement.

The chloroplast in many zoospores and gametes of the marine Chlorophyceae has a region differentiated as an eyespot. It was suggested that the photoreceptor is located in the outer membrane of the chloroplast envelope because of the increased number of intramembranous particles found in this membrane in the eyespot region.

PHAEOPHYCEAE

The plastids of brown algal cells occur in many patterns—as discs, parietal plates, or in stellate configurations and depending on the species, one to many plastids per cell are present. The plastid is enveloped by at least two unit membranes (i.e. the inner and outer chloroplast membranes), and two additional membranes often surround the plastid. These form the chloroplast endoplasmic reticulum, which is continuous with the outer membrane of the nuclear envelope.

Phaeophycean plastids typically have three thylakoids per lamella, and a girdle lamella is found near the chloroplast envelope. The lamellar arrangement is more irregular in dividing plastids. Both longitudinal and transverse divisions of the plastids occur within the same cell. The chloroplast matrix in many species of brown algae contains small electron-opaque globuli, most of which are presumed to consist of lipid or phenolic material.

The genophore of the brown algal plastid is arranged in a planar ring near the chloroplast envelope. The arrangement of chDNA was first demonstrated in *Sphacelaria* with electronmicroscopy and has been verified by fluorescence microscopy. The pyrenoid has a homogeneous matrix and usually lies on a short stalk at the periphery of the chloroplast. Pyrenoids are common in plastids of some orders e.g. Ectocarpales but are poorly developed in others, e.g. the eggs of some Fucales and in vegetative cells of Sphacelariales. Pyrenoids are observed in some members of the Dictyotales and Laminariales.

Algal mitochondria typically have tubular cristae, and in brown algal cells the cristae are usually perpendicular or irregularly arranged with respect to the longitudinal axis of the oraganelle. Sperm mitochondria in *Fucus*, however, have longitudinally-oriented cristae. Mitochondria are especially numerous in sieve elements of the Laminariales, reflecting their higher metabolic rates.

Endomembranes (the Golgi Complex, Endoplasmic Reticulum, Vesicles). The Golgi complex is generally perinuclear in brown algal cells, including specialized cells of the Laminariales (e.g. sieve elements of *Laminaria*), though Golgi bodies (dictyosomes) are present throughout the cytoplasm of many Fucalean cells. The Golgi body typically has six to eight cisternae which become horse-shoe shaped during vesicle production. The Golgi complex in gametangial cells of *Cutleria* is composed of 20 or more cisternae. In the brown algae, the Golgi complex has been identified as the site of polysaccharide sulphation and may be involved in mastigoneme production.

The endoplasmic reticulum (ER) is common throughout the cytoplasm of most brown algal cells. The ER is involved in at least some aspect of mastigoneme production. In some the ER is associated with the plastids, peculiarly.

Cell Wall. In the brown algae, cell wall formation has been studied in the Fucales. Wall formation in the Fucales follows fertilization, and the mature wall has three distinctive layers: an inner fibrillar layer (sulphated fucan and alginic acid), an outer fibrillar layer (alginic acid and cellulose) and an exterior amorphous layer (sulphated fucan and alginic acid). The exterior amorphous layer is thickened at the rhizoid tip. The sulphated fucan (fucoidan) in this outer layer is the 'glue' which holds the young embryo in place in the wave-swept intertidal zone.

Flagella. Mastigonemes decorate the anterior flagellum in the normally heterokont flagella or reproductive cells and are initially localized on one side at the flagellar base. The anterior flagellum is important in gamete recognition in the brown algae and may have a tightly coiled apical region which is undecorated. The posterior flagellum is smooth, and one of its most important characters is a more electron-opaque region of plasma membrane near its base, which faces a similar region on the adjacent plasma membrane of the cell. Centrioles move to the cell surface prior to formation of the flagella and serve as basal bodies. The plasma membrane extends to become the flagella sheath as microtubule nucleation occurs from the centriolar complex. A modified chloroplast, the eyespot, lies beneath the plasma membrane near the posterior flagellum, and flagellar rootlet microtubules are frequently associated with this organelle. Many brown algal gametes are negatively phototactic.

Rhodophyceae

The red algae are an ancient assemblage, and although members of this division may exhibit extreme complexity in their reproductive development and life histories, they also possess several primitive features. These include the presence of phycobilin pigments and lack of flagellated cells, both features of the prokaryotic blue-green algae. The red algae are therefore believed to have diverged from very primitive stock and are often postulated to be the oldest of eukaryotic cells.

Fine structural and cytological studies on the red algae have confirmed many of their unique and interesting features. *Chloroplasts* are surrounded by an envelope composed of two membranes and contain single, unstacked thylakoids. One or more peripheral thylakoids (inner limiting *discs*) may encircle the plastid. The water soluble accessory pigments, allophycocyanin, R-phyco-genophores, or electron-transparent regions presumed to contain DNA, are typically scattered in the stroma. The DNA fibrils are attached to the thylakoid membranes, which may be involved in their replication and segregation during chloroplast division. Small electron-dense 'globules' (lipid granules or plastoglubuli) are also observed in the stroma. Floridean starch is not found within red algal chloroplasts, but accumulates within the cytoplasm apparently in close association with the ER. Pyrenoids with a granular, crystalline, matrix have been observed in some red algae, particularly in the more primitive forms (Bangiophycidae and the Nemalionales).

The structure of the red algal mitochondrion is similar to that of most algae, consisting of a double membrane envelope tubular cristae, and electron-transparent regions with presumed DNA fibrils. Electron dense, spherical inclusions of unknown composition have been observed in some species (e.g. *Smithora*).

Golgi bodies typically possess six to ten cisternae and vary greatly in size according to the developmental state of the cell. Cisternae are normally separated from one another by a space of approximately 10 nm. Vesicles may arise by vesiculation at the periphery of cisternae, or alternatively, entire cisternae may detach and condense into single vesicles. There appears to be a uniform sequence of events whereby the Golgi apparatus secretes extracellular products during the differentiation of spores in several red algae. Golgi-derived vesicles are first active in mucilage formation followed later by the production of cored vesicles which are composed of glycoprotein, and appear to function in spore protection and/or adhesion following release. Lomasomes ('multivesicular bodies' or 'paramural bodies') often appear associated with the ER and plasma membrane, and have been implicated in cell wall formation.

Cell Wall. The wall matrix and outer mucilaginous layer are generally composed of a sulphated polysaccharide. Specific carrageenans characterize different phases in the life history and are observed in the first cells (carpospores and tetraspores) (e.g. *Chondrus*). A multilayered cuticle is observed in some species and has been demonstrated to be proteinaceous in *Porphyra*. $CaCO_3$ crystals (calcite or aragonite) are observed in the cell walls of several diverse taxa of the red algae, though little is known of the mechanism of calcification. Crystals are always arranged within an organic matrix, and are absent from the thin layer or wall material adjacent to the plasma membrane.

ALGAL CYTOLOGY

Chromosome studies in algae are sometimes difficult due to the large number of small chromosomes (e.g. desmids), or due to the entanglement of numerous thread like chromosomes (e.g. Dinophyta), whilst in many Phaeophyta large amounts of mucilage and other substances interfere with fixing and staining techniques. There has been a rapid publication in chromosome numbers of algae in spite of these difficulties. Much work has been done on the algal cytology by Prof. M.B.E. Godward and her colleagues in London, reporting the mitotic and meiotic studies of various groups in algae, effects of radiation on chromosomes, electron microscope studies etc.

Polyploidy is recorded in some species of *Cladophora* and *Chlamydomonas reinhardtii*, Colchicine treatment induces increase in cell diameter and nuclear mass in *Chlamydomonas*, and nucleic appear polyploid, but when transferred back to colchicine free medium, there is a reduction to the normal haploid state.

Genetic Analysis

Tetrad analysis has been first reported from *Chlamydomonas* by Pascher in the early twenties. Mutants may be produced spontaneously or by UV-light or X-radiation and these may be non-photosynthetic, have paralysed flagella, eye spots absent, form unusual colony growths, be resistant to changes in pH or streptomycin, require organic nutrients such as thiamine or p-aminobenzoic acid. Many of these characters are determined by single pair of alleles, are segregated in 2:2 ratio and result from single gene mutations. Chromosome mapping has been done using tetrad analysis and recombination frequencies.

Cytoplasmic inheritance has been investigated in a *Chlamydomonas reinhardtii* mutant, which is resistant to streptomycin.

Interactions between cytoplasm and nucleus have been studied by Hammerling (1953) in a siphonaceous alga *Acetabularia mediterranea* and *A. crenulata*. Grafts between nucleated and anucleated parts of different species always form caps whose structure is that of the nucleated portion, that is the nucleus of *A. mediterranea* can influence cytoplasm of a different species *A. crenulata* to form *A. mediterranea* and *vice versa*. Binucleate grafts with nuclei derived from different species form intermediate type caps, whilst trinucleate grafts form caps which are closest in form to those of the type formed by the species contributing two of the nuclei.

Chromosome numbers of few representative species of algae are given below and for details the readers are referred to *Chromosomes of the Algae* by M.B.E. Godward (1966).

Species	Chromosome Number (n)
Chlamydomonas eugametos	10
C. reinhardii	16, 8, 11
Gonium pectorale	17
Eudorina elegans	12, 10
Pandorina morum	10
Volvox aureus	14
Ulothrix zonata	10
Cylindrocapsa involuta	16
Enteromorpha compressa	10
Ulva lactuca	10
Stigeoclonium amoenum	12

Contd.

Draparnaldia plumosa	14
Trentepohlia aurea	18
Coleochaete scutata	36, 42
Oedogonium geniculatum	32
O. vaucheri	16
Cladophora glomerata	48
Chaetomorpha linum	36
Rhizoclonium riparium	36
Pithophora Kewensis	24
Spirogyra crassa	12
Closterium acerosum	60-220
Cosmarium botrytis	18-52
Zygnema cylindrospermum	15-18
Mougeotia viridis	51
Euglena spirogyra	86
Cryptomonas ovata	86-209
Laminaria angustata	22
Nereocystis luelkeana	31
Dictyota dichotoma	32
Padina japonica	32
Fucus evanescens	32
Sargassum patens	32
Gelidium corneum	4-5
Batrachospermum moniliforme	10
Nemalion helminthoides	8
Corallina officinalis	24
Polysiphonia elongata	37
Porphyra linearis	2

7 Thallus Organization

Thallus shows several variations in the unicellular as well as multicellular forms. In the unicellular algae, cell division constitutes the reproductive phase. But in the multicellular forms cell division occurs both in somatic and reproductive phases. The following are some of the main types of thallus organization.

UNICELLULAR HABIT

Unicellular forms may be motile or non-motile.

(a) **Motile unicellular forms** Motile unicellular forms occur in several classes of algae. They are mostly flagellated forms having a definite cell wall. But some are periplastic as in *Euglena*. The flagella may be two in number, and they are equal *(Chlamydomonas)* or unequal *(Cryptomonas)*. In some like *Chromulina*, only one flagellum is present. A few forms are encapsulated having calcareous envelope. Some forms have fine protoplasmic projections called rhizopodia and they show amoeboid movement e.g., *Chrysamoeba*.

(b) **Non-motile or coccoid forms** These forms are non-flagellate having a rigid cell wall. They are of several shapes and sizes e.g., *Chlorella* (Chlorophyceae), *Chroococcus* (Cyanophyta).

COLONIAL HABIT

A colony is an aggregate of several cells having a mucilage envelope. The colonial forms can be distinguished into coenobial, palmelloid, dendroid and rhizopodial types.

(a) Coenobium A coenobium is a colony of a definite integration having a fixed number of cells. During the growth of the colony, the cells increase in size but they do not multiply by cell division. The coenobium may be motile having flagella *(Volvox)* or they may be non-motile *(Hydrodictyon)*. In a colonial form, all the cells may be similar or the reproductive cells may be distinguished into asexual gonidial cells and antheridia (male) and oogonia (female) as in *Volvox*.

(b) Palmelloid, dendroid and rhizopodial colonies These colonies are not constant in shape and size as cell division takes place during the somatic phase. In palmelloid colonies, the cells are embedded in a mucilage matrix of irregular shape and size e.g., *Tetraspora* (Chlorophyceae). In the dendroid forms, the cells are joined by mucilage stalks giving a branching appearance to the whole colony e.g., *Ecballocystis*. In the rhizopodial colonies cells are united through rhizopodia.

FILAMENTOUS HABIT

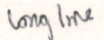

Filamentous habit is very common in algae. It must have originated from unicellular forms by transverse septation in the vegetative phase without gelatinisation of cell walls. In a filament, the cells are in a row joined end to end through the middle lamella. In Myxophyceae, a filament consists of a trichome of uniseriate cells and its mucilage sheath. Many filamentous forms produce motile swarmers during reproduction. The mode of origin of filamentous habit is recapitulated whenever a swarmer settles down and divides transversely forming a filament.

The filaments are unbranched *(Ulothrix, Nostoc)* or branched. Branched filaments are found in all classes of algae where filamentous habit has appeared. When the branched filaments are distinguished, into prostrate and erect systems, it is known as heterotrichous branching.

(a) Simple unbranched filaments The simpler filamentous forms are unbranched and all the cells are capable of cell division, growth and reproduction e.g., *Spirogyra, Ulothrix, Nostoc*. In some, the basal cell may be differentiated into a hapteron for attachment. Some filamentous forms show distinct polarity with the trichomes tapering towards the tip e.g., Rivulariaceae.

(b) Branched filamentous forms Branched filaments are common in all classes where filamentous habit has appeared. In the branched forms, cell division and growth may be restricted to the end cells *(Cladophora)*. But in some forms, filaments terminate in colourless hairs and an intercalary meristem occurs at the base of the hair. Growth due to cell division of such a meristem is known as trichothallic growth e.g., *Ectocarpus*. In most of the forms, true branching is due to lateral outgrowths developing into branches. But false branching occurs in Scytonemataceae (Myxophyceae) where the broken ends of trichome grow out of the mucilage sheath and appear like branches.

(c) Heterotrichous habit It is the most highly developed type of filamentous habit. It consists of a prostrate system of creeping filaments and an erect system of several branched filaments e.g., *Coleochaete* (Chlorophyceae), *Ectocarpus* (Phaeophyceae), *Batrachospermum* (Rhodophyceae) and *Stigonema* (Cyanophyceae). In some forms, the erect system is eliminated and the prostrate system forms a discoid thallus as in *Coleochaete scutata*. In forms like *Draparnaldiopsis*, the prostrate system is suppressed and the erect system is elaborated. Fritsch considers that the first land plants might have arisen from algae exhibiting heterotrichous habit.

SIPHONEOUS HABIT

Siphoneous forms are coenocytic and they lack septa. But septa may be formed during the formation of reproductive organs. The siphoneous habit is found in Siphonales (Chlorophyceae) and *Botrydium* (Xanthophyceae). In *Vaucheria*, the branched filaments are siphoneous. Siphonales are paralleled by Phaeophyceae and Rhodophyceae. According to Fritsch, the tendency of cells of Chlorococcales to become coenocytic prior to reproduction could have led to the evolution of siphoneous habit. Oltmanns has suggested that siphoneous habit is derived from filamentous habit by complete loss of power of septation. Thus, he considered that Siphonales are derived from Cladophorales but Fritsch regarded Cladophorales as an offshoot of Ulotrichales and that it is not related to the evolution of Siphonales.

PSEUDOPARENCHYMATOUS AND PARENCHYMATOUS HABIT (Figs. 7.1 to 7.8)

The filamentous forms evolved in two directions resulting in pseudoparenchymatous and parencymatous forms. The pseudoparenchymatous thalli may be formed by juxtaposition of the branch system of a single main axial thread or many axial filaments *(Nemalion)*.

The parenchymatous habit is derived from the filamentous one as a result of cell divisions taking place in more than one plane. The development of germlings of *Ulva* into a foliaceous thallus

82 *A Textbook of Algae*

recepticulates the mode of origin of parenchymatous habit. The foliaceous parenchymatous thalli are of a very large size in the brown seaweeds (Phaeophyceae). These thalli show considerable anatomical differentiation. Foliaceous plant body is also found in *Ulva* (Chlorophyceae) and *Porphyra* (Rhodophyceae).

Fig. 7.1 *Sargassum filipendula*

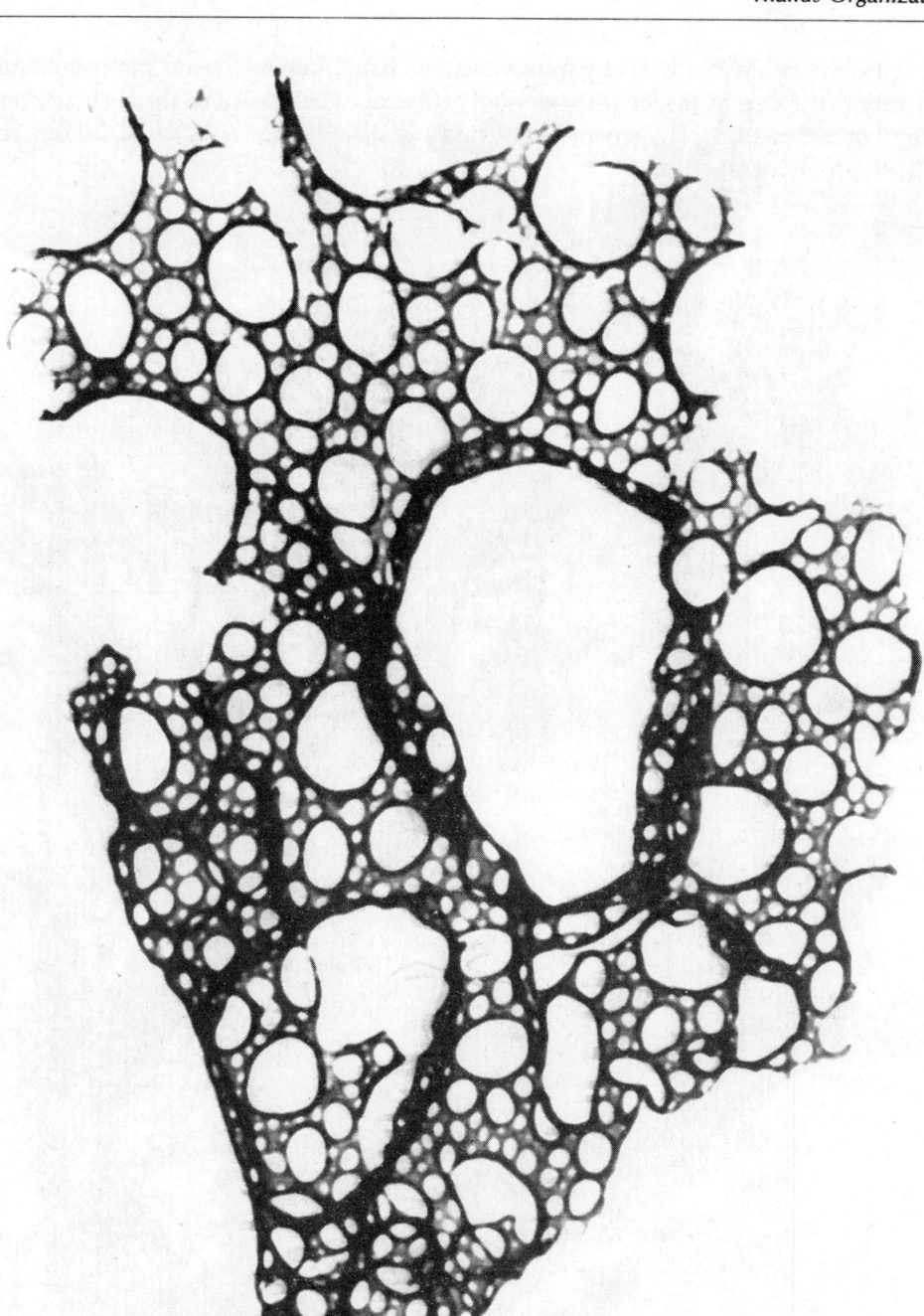

Fig. 7.2 *Hydroclathrus clathratus*

Fig. 7.3 *Coilodesme plana epiphytic on Cystoseira osmundacea*

Fig. 7.4 *Padina tenuis*

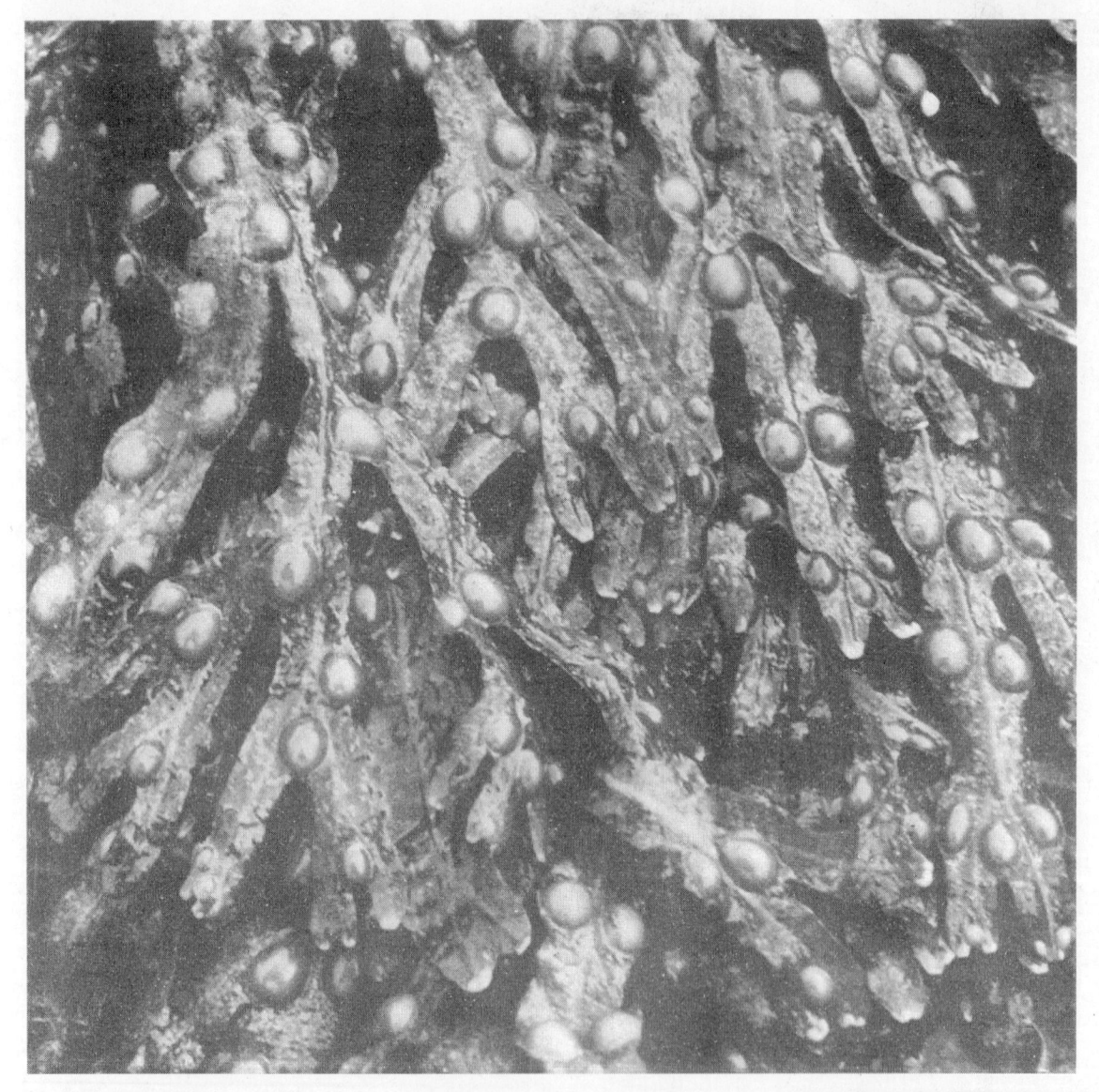

Fig. 7.5 *Fucus vesiculosus*

Fig. 7.6 *Costaria costata*

Fig. 7.7 *Nereocystis lutekeana*

Fig. 7.8 *Macrocystis integrifolia*

8 Life Cycles: Phylogeny–Sexuality

LIFE CYCLES

Algae exhibit considerable diversity in their life cycles. The process of fertilization leads to the development of diploid phase and subsequent meiosis leads to the haploid phase. The relative importance of the two phases varies in different algae and depending upon this, different life cycles are recognised. But in Myxophyceae there is no sexual reproduction and hence the life cycle is simple involving various methods of vegetative and asexual reproduction.

The following are the various types of life cycles:

Haplontic or Haploid Type

Most of the members of Chlorophyceae are haploid. Asexual reproduction takes place through zoospores or aplanospores. Sexual reproduction is through gametes formed in any of the cells (*Ulothrix*) or in special reproductive structures called antheridia and oogonia (*Oedogonium* etc). The gametic union ranges from isogamy to oogamy through anisogamy. The zygote is diploid and it develops a thick wall. It crosses over the unfavourable period. During germination it divides meiotically forming four haploid zoospores or aplanospores which give rise to the adult plants. In Conjugales, after meiosis three of the nuclei degenerate and the zygote germinates directly producing a filament (*Spirogyra*). Thus, in this life cycle, the plant is haploid except the zygote which is diploid and all other stages are haploid. So it is called *haploid* life cycle.

Diplontic or Diploid Type

This type of life cycle is less common than the haploid type. In this case, the plant is diploid as in *Caulerpa* and *Cladophora glomerata*. Meiosis takes place during the formation of gametes. Gametic union is generally isogamous and it results in the diploid zygote.

The zygote divides mitotically and gives rise to a diploid plant. Asexual reproduction may take place through diploid zoospores as in *Cladophora glomerata*. Thus, in the life history of these forms, the gametes alone are haploid and all other stages are diploid. So, the life cycle is described as diploid or diplontic type.

In *Sargassum* which is diploid, meiosis takes place in the antheridia and oogonia during the formation of antherozoids and the egg respectively. Sexual union is oogamous and the diploid zygote directly develops into a *Sargassum* plant. Diatoms also show diploid type of life cycle.

Isomorphic Type

In some algae, the life cycle involves two generations namely the haploid gametophytes and diploid sporophytes which are morphologically similar e.g., *Ulva* and *Cladophora*. The diploid plants form zoospores after meiosis. So, the zoospores are haploid and they give rise to the gametophytes which produce the gametes. Gametic union is isogamous and it results in a diploid zygote. It undergoes mitotic divisions and gives rise to a diploid plant. The life cycle is described as isomorphic as the sporophytic and gametophytic generations are similar. They are complementary to each other for the completion of the life cycle.

In *Ectocarpus*, the life cycle is complex. The diploid plants bear plurilocular and unilocular sporangia. But the haploid plants bear only plurilocular sporangia. The diploid plants multiply asexually by forming zoospores within the plurilocular sporangia. The unilocular sporangia produce 64 haploid zoospores after meiosis. These zoospores give rise to the gametophytes. The plurilocular sporangia of the gametophytes function as gametangia and produce the gametes. The gametic union is isogamous and it results in a diploid zygote. The zygote divides mitotically and gives rise to the diploid sporophytes. Gametophytes may also multiply by parthenogenic development of gametes.

In *Dictyota*, the male and female gametophytes are similar to the diploid tetrasporic plants. Sex organs are antheridia and oogonia and they form sori. Sexual union is oogamous. The zygote gives rise to the diploid tetrasporic plant which bears tetrasporangia. Meiosis takes place during the formation of tetraspores from the tetrasporangia. The tetraspores develop into male and female plants.

Haplobiontic Type

This type of life cycle occurs in Nemalionales *(Batrachospermum)*. During sexual reproduction the male nucleus of the spermatium unites with that of the egg. The zygote nucleus immediately divides by meiosis so that the post-fertilization stages are haploid. The cystocarp (carposporophyte) has a sterile sheath enveloping the gonimoblast filaments and carposporangia. The carpospores are haploid and they give rise to **chantransia** stages. The cystocarp is considered to be a second haploid generation. Some special shoots of chantransia stages grow into *Batrachospermum* plants. The chantransia stages can multiply by monospores formed from monosporangia. The life cycle is haplobiontic as the *Batrachospermum* plant and carposporophyte are haploid and zygote nucleus alone is diploid.

Diplobiontic Type

In *Polysiphonia*, the male and female gametopytes and tetrasporophytes are morphologically similar. During sexual reproduction, the spermatial nucleus unites with that of the egg. The resulting zygote nucleus divides mitotically so that the carposporophyte is diploid. This is considered to be an additional diploid generation. The carposporangia produce carpospores which give rise to the diploid tetrasporophytes. The tetrasporophytes produce tetrasporangia in which meiosis takes place forming tetraspores. These tetraspores grow into male and female gametophytes. Thus, in the life cycle, there are two diploid generations, the tetrasporophyte and carposporophyte and only one haploid somatic generation. So, the life cycle is described as diplobiontic. It is also called as triphasic as the life cycle involves three generations.

Heteromorphic Type

This type of alternation of generations is found in Sporochanales, Desmarestiales and Laminariales. The gametophytes are filamentous and microscopic in contrast to the elaborate macroscopic sporophytes *(Laminaria)*. As the two alternating generations are dissimilar, the life cycle is described as heteromorphic.

PHYLOGENETIC RELATIONSHIPS IN ALGAE

A phylogenetic classification of any group of organisms aims to present the evolutionary relationships between members of the group. The classification of algae into major phyla are based on photosynthetic pigments which is supported by other criteria as storage products, cell wall constituents, nature of flagella and details of cell structure, suggests that the classification is natural.

From scanty fossil evidence it is suggested that the dominant kind of pre-Cambrian life was blue-green algal-like, thus supporting the idea that blue-green algae are more primitive than the others. Observations of fossil diatoms, Dinoflagellates and Dasycladales etc. have resulted that fossil evidence is important within relatively small taxa, for phylogenetic considerations, but for general discussion in larger phyla it is of less use due to lack of fossil evidence.

In general, three criteria are commonly used to establish whether a particular type of alga is evolutionarily more advanced than another: (1) Vegetative form—unicellular-colonial-filamentous-parenchymatous. Within multicellular forms the method of growth-diffuse-intercalary or apical, structural differentiation of the thallus, (2) Specialization of the sexual process—isogamous—anisogamous to oogamous, (3) Life history—vegetative plant haploid and the zygote diploid to vegetative plant diploid and gametes haploid—intermediate stages isomorphic and heteromorphic alternation of generations.

Origin of the Filamentous Forms in the Chlorophyceae

There are two views regarding the origin of filamentous habit in Chlorophyceae, mainly, (1) Origin from unicellular types, (2) Origin from palmelloid forms. The first was propounded by Fritsch and the second by Smith, Gupta and Nair. A third view suggests that the series of filamentous (and parenchymatous) forms are uninucleate cells arose from the motile unicells whereas the multinucleate series of forms arose from the Chlorococcales. It is probable that all the three possible methods happened during evolution.

Two main lines of affinities among the major phyla of algae are as follows: (1) On the basis of absence of flagella and presence of biloproteins, the Myxophyceae and Rhodophyceae may be closely related, (2) Because Chlorophyll *b* is present in the Chlorophyceae and Euglenophyceae, some kind of relationship between these can be suggested, (3) The presence of chlorophyll *c* (together with some common carotenoids) in Chrysophyceae, Bacillariophyceae, Cryptophyceae, Dinophyceae and Phaeophyceae, suggests a close affinity amongst them.

Christensen (1962) classified algae into Procaryota and Eucaryota. Eucaryota are further divided into *Aconta* (no flagellate stages Rhodophyceae) and *Contophora* (with flagellate stages). He divided Contophora into *Chlorophyta* (chlorophylls dominate) and *Chromophyta* (carotenoids predominate) but this classification has not received much attention as it is supported by a single character and ignoring many other characters (Morris, 1973).

ALGAE AND THE ORIGIN OF LAND FLORA

Algae are the primitive forms in the entire plant kingdom and are mostly aquatic either freshwater or marine in habitat. There have been several theories regarding the origin of first formed land plants with heterotrichous habit, i.e., prostrate and erect systems in the plant. One of those theories was first put forward by F.E. Fritsch (1937 and 1945) that two Chlorophyceae members having heterotrichy viz., *Coleochaete* and *Fritschiella* might have given rise to the terrestrial heterotrichous primitive plants. These two algae are terrestrial in habitat and have tendencies towards and plants in than structure and habitat. According to Fritsch Chaetophorales represent the surviving descendants of forms which migrated to the land in the remotest past. He visualized that the prostrate gametophyte and the erect aerial sporophytes of land plants evolved from the heterotrichous terrestrial green algae, the former by the suppression of erect system and the latter by the suppression of the prostrate system. The discovery of *Fritschiella tuberosa* by M.O.P. Iyengar in 1932, which is a terrestrial alga with heterotrichous habit gave strong support to this hypothesis for the origin of land flora.

SEXUALITY IN ALGAE

Sexuality may be considered as one of the fundamental characteristics of living organisms. In eukaryotes it involves the events of syngamy and meiosis, which are integrated into the life histories of the organisms in various ways. A good amount of variability exists in algae, some green algae exist in the haploid phase, with meiosis following immediately upon syngamy. Another extreme is shown by *Fucus* where the organism is diploid and meiosis takes place during the formation of gametes. The majority of macroscopic green, red and brown algae have haploid and diploid phases in their life history which are connected by syngamy and meiosis.

Sexuality creates genetic recombination in the offspring and hence variability in the population. Maximum efficiency is achieved when dioecism guarantees outbreeding. Monoecism, as found in many marine algae, may be considered as a generation step which might nevertheless offer benefits in other respects. Secondary mechanisms to block inbreeding, such as incompatibility reactions in fungi and higher plants, are not known in marine algae.

Culture studies revealed that 14 North American Pacific brown algae examined, 12 seem to completely lack sexuality. This condition obviously is a secondary phenomenon, and in several cases it can be demonstrated that the same species behaves differently on different coasts. As an example, *Scytosiphon lomentaria* is asexual in Pacific North America, whereas functional gametophytes are known from Japan. Culture studies in *Ectocarpus siliculosus* and *Ulva mutabilis* show that algae may have a great potential for reproduction in addition to the possibilities connected with sexuality. Even polyploidy level changes may occur from sexuality.

Algae have developed a variety of mechanisms in order to coordinate the activity of individuals engaging in sexual reproduction. In freshwater Volvocales cases are known where messenger substances are employed to coordinate gametogenesis. In *Volvox carteri* a glycoprotein is produced by a spontaneously sexualized male colony. This inducer is highly active and changes the potential of vegetative male and female colonies to form offspring with eggs and spermatozoid packets. Within a short time, the entire *Volvox* population culminates in the formation of zygotes. Various mechanisms linked to environmental factors have been developed by marine algae to coordinate gametogenesis and synchronize the discharge of gametes. A unique synchronization pattern apparently restricted to the marine environment is the coupling of sexual activity with specified lunar phases. Experimental work with *Dictyota dichotoma* indicated that periodic illumination by the full moon is a synchronizing factor.

A number of studies have indicated that environmental factors influence or control life histories in some species of green algae. Some of the factors that have been studied are temperature, light quality and period, exposure, salinity and levels of nutrients or growth factors. In species of *Ulothrix* and *Urospora* the alternation between morphological phases has been shown to be regulated in culture by temperature. Temperature has also been suspected as the regulating factor for a number of green algae, such as *Collinsiella* and *Prasiola,* that show a regular seasonal alternation in their life histories. Photoperiodic phenomena involving a phytochrome-like system have been demonstrated for both red and brown algae. In some green algae light quality influences reproduction. It has been observed that blue light related phenomena may explain the induction of reproduction by transfer from dark to light or vice versa observed in a number of marine algae. In *Prasiola stipitata* the degree of wetting appears to influence the percentage of plants that undergoes meiosis and sexual reproduction. In some filamentous green algae the degree of sexuality appears to be related to salinity or stability of the environment.

Changes in temperature, light intensity and photoperiod, culture medium or combination of these factors includes reproduction in many species of marine green algae and can act as an intraining stimulus for endogenous rhythms. In nature, reproduction is often periodic, generally in relation to tidal cycles.

There is impressive coordination chemically on the cellular level in many brown algae. Chemical substances secreted by one cell and causing specific effects in a target cell are termed hormones. The male attracting substances are termed hormones in the present context.

Using clonal laboratory cultures it was found that the male cells in *Ectocarpus siliculosus* respond to a highly volatile compound secreted by receptive female gametes. This substance was named as *ectocarpen.*

Gametophytes of *Cutleria multifeda,* like those of *Ectocarpus siliculosus* are found in every spring on the coasts of the Mediterranean sea. Male and female plants are isomorphic, forming rigidly branched thalli 20 cm long. Gametes are markedly different in size. From bulk suspensions of female *Cutleria* gametes, a compound called *multifiden* was extracted which is a male-attractant apart from *aucanten* and *ectocarpen* which are also included.

Similarly, in *Fucus serratus,* a male attractant called *fucoserraten* was extracted from eggs. In the case of *Laminaria digitata* also two compounds were identified viz., *ectocarpen* and *n-pentadecane* as male-attracting compounds. (Lobban & Wynne, 1981).

ORIGIN AND EVOLUTION OF SEX IN ALGAE

Origin of sex in algae goes hand in hand with the origin of gametes from the lower most forms to advanced forms through different ways. Algae reproduce asexually when the conditions are favourable and sexually when conditions are unfavourable. This is because the same vegetative cell which had hitherto been responsible for zoospores formation, gives rise to gametes, which are smaller in size. The gametes are exactly similar in their morphology to zoospores but different in their behaviour. Zoospores can give rise to daughter plants directly, but gametes do so after fusion of two different gametes. The gametes fuse to form a zygote which develops into a new individual.

Simplest forms are to be found in Cyanophyceae which reproduce only by asexual methods. *Protococcus* reproduces entirely by fission. The fact that gametes arose from zoospores is borne out by their almost similarity, in *Chlamydomonas debaryanum,* and the presence of intergrades of zoospores and gametes in species of *Chlamydomonas* and *Ulothrix* and other isogamous algae forms. In *Ulothrix zonata,* three kinds of zoospores are reported viz., (1) Quadriflagellated macrozoospores formed in small numbers per cell, (2) quadriflagellated microzoospores formed in large numbers per cell, and (3) biflagellated microzoospores formed in still large numbers per cell. The gametes in *Ulothrix* are also biflagellated and are produced in large numbers. The zoospores are thought to have given rise to sex by the accidental fusion favouring under unfavourable conditions to the plant.

Life Cycles: Phylogeny–Sexuality **95**

IMPORTANT QUESTIONS

Essay type

1. Give an account of the range of thallus organisation in algae studied by you.
2. Give an account of the origin and evolution of sex in algae.
3. Give an account of the asexual reproduction in algae studied by you.
4. Give an illustrated account of various life cycle patterns found in Algae.
5. Describe various modes of reproduction in Algae.
6. Write an essay on economic importance of Algae.
7. (a) Give an account of various types of chloroplasts in green algae.
   (b) Give an account of asexual reproduction in Chlorophyceae.
8. Give a brief account of various types of habitats occupied by algae.
9. Comment upon the evolution of thallus in algae.
10. Describe briefly with suitable examples and illustrations the alternation of generation in Algae.
11. What are Algae? With the help of suitable examples and illustrations trace the origin and evolution of sexuality in Algae.
12. Describe the various modes of perennation in Fresh water algae studied by you.
13. Write an account of origin and development of sex in algae.
14. Describe the methods of sexual reproduction in Algae giving suitable examples.
15. Give a detailed account of algal pigments and their significance in algal classification.
16. Give a detailed account of the benefits we obtain from algae.
17. Explain the evolution of thallus in algae giving suitable examples from the genera you have studied.
18. Discuss the economic importance of algae.
19. Write a short essay on the industrial uses of algae.
20. Trace the origin and evolution of sexuality in green algae. Illustrate your answer with suitable sketches and examples.
21. Write what you know about various types of life cycle in algae.

Short Answer Type

22. Write detailed/short notes on the following:
    (i) Algae as food.
    (ii) Pigments in Algae
    (iii) Algae in agriculture/use of algae in agriculture.
    (iv) Plastids in Algae
    (v) Alternation of generation
    (vi) Flagella in Algae.
    (vii) Water bloom
    (viii) Importance of reserve food materials in the classification of algae.
    (ix) Use of algae in industry.
    (x) Algae as nitrogenous fertilisers
23. Describe the role of algae in industry and as manure.

24. Differentiate between:
    (a) Zoospore and Oospore
    (b) Aplanospore and Akinete
    (c) Aplanospore and Oospores
    (d) Aplanospore and Coospore
    (e) Parasitic algae and Epiphytic algae
    (f) Isomorphic and Heteromorphic alternation of generations
    (g) Zoospore and synzoospore
    (h) Hypnospore and Akinetes
    (i) Coenobium and Colony
    (l) Cystocarp and Spermocarp
25. Describe the role of algae in agriculture with special reference to (i) Nitrogen fixers (ii) Soil reclamation and (iii) Fertilisers and manures.
26. Name four industrial products obtained from algae and mention the class to which they belong.
27. What are the harmful effects of algae.
28. Write an explanatory note about algae as biofertilisers.
29. Give the salient features of alternation of generation.
30. Name an alga which is used in reclamation of alkaline user soils.
31. Comment on the following:
    (i) Economic importance of Algae.
    (ii) Range of vegetative structures (thallus organisation) in Algae.
    (iii) Shapes of chloroplasts in Algae.
    (iv) Biofertilisers or Algae as fertilisers.
    (v) Evolution of sex in Algae.
    (vi) Haplobiontic life cycle.
    (vii) Vegetative reproduction in algae
    (viii) (a) Diplohaplontic life cycle with examples and graphic life cycle
        (b) Asexual reproduction in algae.
    (ix) Role of algae in medicines
    (x) Importance of algae as food and in industry
    (xi) Origin of sex in algae.
    (xii) Evolution of thallus in green algae
    (xiii) Fossil algae
    (xiv) Culture of algae
    (xv) Chloroplasts in algae

Objective Type

32. Select the correct answer:
    (i) Auxospore formation takes place in
        (a) Desmids   (b) Diatoms
        (c) Green algae   (d) Red algae
    (ii) A motile flagellated cell that reproduces asexually is called as
        (a) sperm   (b) zoospore

(c) oospore
(d) Zygospore

(iii) Which of the following part of an algal cell is alive?
- (a) Cell wall
- (b) Chloroplast
- (c) Vacuole
- (d) Starch.

(iv) The Red sea gets its name because of the presence in its water red coloured algae belonging to class
- (a) Rhodophyceae
- (b) Chlorophyceae
- (c) Phaeophyceae
- (d) Myxophyceae.

(v) Fusion of mature individuals which act as gametes is called
- (a) Isogamy
- (b) Conjugation
- (c) Hologamy
- (d) Autogamy.

(vi) Chloprophyll-e is present in
- (a) Chlorophyceae
- (b) Xanthophyceae
- (c) Phaephyceae
- (d) Rhodophyceae.

(vii) Reproduction through hormogonia is found in
- (a) *Spirogyra*
- (b) *Oscillatoria*
- (c) *Ectocarpus*
- (d) *Sargassum*

(viii) In Cyanophyceae the site of nitrogen fixation is
- (a) Akinetes
- (b) Incipient nucleus
- (c) Heterocysts
- (d) Plasma membrane.

9 Class: Chlorophyceae: General Characters and Type Study

CHLOROPHYCEAE

The *Chlorophyceae* consists of mostly freshwater forms and only a few are marine. They are widely distributed. A few forms are terrestrial. The Ulvaceae and a majority of Siphonales are marine. Members of other orders are predominantly freshwater forms. They are mostly found submerged in shallow waters but some can grow on bark, rocks, etc.

The characteristic features of Chlorophyceae are:

1. The pigmentation is like that of higher plants. Chlorophyll *a* and chlorophyll *b* are present along with *carotenes* and *xanthophylls*.
2. Presence of *pyrenoids* in the chloroplasts.
3. Formation of *starch* as reserve food and its aggregation around pyrenoids.
4. Cell wall is formed of cellulose.
5. Motile stages usually have equal flagella and they are acronematic.

Thallus

The thallus shows several variations. The plant body in the simplest forms is unicellular (*Chlamydomonas*). *Pandorina* and *Volvox* are colonial forms with individual cells embedded in mucilage. Another line of development shows simple (*Ulothrix*) or branched filaments (*Cladophora*). The heterotrichous forms (*Coleochaete*) have distinct erect and prostrate systems. Sometimes, the thallus is foliose (*Ulva*). In some, the filaments are siphonaceous or pseudoparenchymatous as in Siphonales.

Structure of a Cell

The cells have a definite cell wall. The innermost layer is mostly formed of cellulose. The mucilage envelope is best seen in conjugales. In Desmids, the mucilage is secreted through the complex pore organs. In Charales, it is encrusted with carbonate of lime.

Chloroplasts and Pyrenoids

The chloroplast is a plastid bearing chlorophylls and other pigments. The chlorophylls *a* and *b* predominate, but carotenes and xanthophylls also may be present. The proportion of the pigments is variable from species to species. The chloroplast is very characteristic in different forms. Usually, there is a single chloroplast in each cell except in some Conjugales, Siphonales and Charales. In *Chlamydomonas* typically, chloroplast is cup-shaped but it varies in different species *viz.* spiral *(Spirogyra)*, stellate *(Zygnema)* or reticulate e.g., *Oedogonium*. The chloroplast is usually parietal but may be central or axile as in *Zygnema*. The desmids show a great range of variation in the structure of chloroplasts. The chloroplast has no grana.

Pyrenoids occur singly when the chloroplasts are small. But when chloroplasts are elaborate, they may be several as in *Oedogonium* and *Spirogyra*. Pyrenoids may be fixed in number even in a young chloroplast or they may increase as the chloroplast grows *(Hydrodictyon)*. Pyrenoids may increase by division or by arising *de novo*. In Siphonales and Charales, the chloroplast may be with or without pyrenoids.

Nucleus

The nucleus has a distinct organization. The cells may be uninucleate or multinucleate or coenocytic *(Cladophora, Hydrodictyon)*.

Flagella and Eye-spot

The flagella and eye-spot or stigma are found in primitive unicellular and colonial forms. In most of the other forms which have no motile vegetative phase, the flagella and eye-spot are found in the reproductive cells only. The flagella which are of whiplash type help in locomotion and the eye-spot in its phototactic responses.

Vacuole

In algal cells, vacuoles may be few or they may coalesce to form a single large central vacuole, which may be traversed by cytoplasmic strands. Contractile vacuoles which are excretory in function, are found in members of Volvocales *(Chlamydomonas)*.

Reproduction

It takes place by vegetative, asexual and sexual means. Vegetative reproduction takes place in some filamentous forms by fragmentation or by special structures *(Chara)*.

Asexual Reproduction

Usually, it takes place by formation of zoospores. These may be formed in vegetative cells. In colonial forms like *Volvox*, zoospore-formation and organisation of daughter colony is restricted to gonidial cells. Zoospores are motile, biflagellate (Volvocales), quadriflagellate *(Ulothrix)* or multiflagellate *(Oedogonium)*. Usually, they are produced in numbers of 4, 8 and 16 per cell. Cells with coenocytic protoplasts may undergo cleavage and form usually a large number of zoospores. In some cases, only a single zoospore may be formed in each cell *(Oedogonium)*.

The zoospores are usually liberated through a pore in the wall directly into water or first into a vesicle and then into water. It may survive for a few minutes or for several days *(Ulothrix)*. zoospores may be similar or may be distinguished into larger macro- and smaller microzoospores *(Ulothrix)*. In Conjugales zoospore formation is completely absent.

Zoospores are naked but after a swimming for a period they secrete a wall, settle on a rock and develop into the adult plants. In colonial forms, zoospores may not be liberated, they remain within the parent cell and organise into a colony.

In some green algae, *aplanospores* are formed. These are non-motile supressed zoospores. The zoospore is naked but aplanospore has a distinct wall different from that of the parent cell wall. They may be formed singly or in large numbers. Aplanospores are described as *hypnospores* if the wall is greatly thickened.

Some green algae show formation of *akinetes*. Akinetes are thick-walled vegetative cells with abundant reserve food. In an akinete, the parent cell wall is not distinct. The aplanospores and akinetes can grow directly into new plants.

Sexual Reproduction

Sexual reproduction takes place by forming gametes. The gametes are usually free-swimming and flagellate. But in Conjugales, *aplanogametes* are produced which are without flagella and they show amoeboid movement; they are not liberated into the water and they meet together by the establishment of a conjugation tube. Usually, gametes from different plants only unite (heterothallic) but sometimes, gametes from the same plant fuse (homothallic). In the simplest forms, the fusing gametes are exactly similar and the union is termed as *isogamous*. But when the gametes are unequal the union is described as *anisogamous*. In the more advanced forms, male gamete is flagellate and the female gamete is non-flagellate. So, the non-motile egg is retained within the oogonium where fertilization takes place. This union is known as *oogamy* e.g., *Oedogonium*.

Thus, there is a progressive evolutionary change from isogamy to oogamy. This should have occurred several times in algal evolution. In fact, even a single genus like *Chlamydomonas* shows all the three types.

Species which show isogamy or anisogamy have no special reproductive cells for the formation of gametes. They are produced by the vegetative cells. But in oogamous forms *(Oedogonium, Coleochoete)* the egg is singly produced in the oogonium and the antherozoids are formed in the antheridia. Some species show *parthenogenetic* development of gametes i.e., gametes which fail to fuse, develop into new plants.

In the isogamous species, sexual union of gametes is outside the cell and so zygotes are formed in the water. But in oogamous forms, fertilization takes place in the oogonium and hence the zygote becomes free only after the disorganization of oogonial wall.

Zygote

Zygotes with thin walls germinate within a day and thick-walled ones after a longer time. In isogamous and anisogamous species, the zygote may remain motile and flagellate for some time, before coming to rest. But in Conjugales and in oogamous forms, zygote is non-motile from the beginning. As the *zygote* ripens, the green colour will be masked by the *hematochrome* and starch may get converted into oil.

In *isomorphic* forms, the zygote may develop into diploid asexual plant (*Ulva, Cladophora*). In all others, where plants are *haploid*, zygote germinates after *meiosis*. The resting period of the zygote may be of a few day's duration or of several months. Meiosis will be followed by formation of 4 zoospores (*Chlamydomonas*) or aplanospores (*Ulothrix*). Sometimes, zoospores may not be formed and after degeneration of some of the haploid nuclei, zygote may directly produce a new filament (*Zygnema*). In desmids, from the zygote, 2-4 individuals are formed. Sometimes, germination of zygote results in 4 zoospores and the zoospores form polyhedral tetrahedron stages (*Hydrodictyon*). The protoplast of the tetrahedron stage forms a new net within it. In *Coleochaete*, the zygote divides forming 16 or 32 wedge-shaped cells and the fruiting body is known as *spermocarp*. Each cell of spermorcarp produces a single zoospore which can give rise to the *Coleochaete* plant.

Life Cycle

Members of Volvocales, *Ulothrix, Coleochaete,* etc. are haploid and only the zygote represents the diploid phase. They show *haploid* type of life cycle. In Siphonales *(Caulerpa),* the vegetative plant is diploid and meiosis takes place during the formation of gametes. These haploid gametes unite to form the zygote that give rise to the diploid plant. This is known as *diploid* life cycle. In *Ulva* and *Cladophora,* the diploid sporophyte alternates with a haploid gametophyte which is morphologically similar. This is known as *isomorphic* life cycle. In *Urospora*, the sporophytic and gametophytic generations are free-living and they are morphologically dissimilar. This is known as *heteromorphic* life cycle.

EVOLUTION OF THALLUS ORGANIZATION IN CHLOROPHYCEAE

The thallus of Chlorophyceae shows various levels of organization. It shows evolutionary advance from simple to more complex types of construction.

Unicellular

The simplest type of construction is exhibited by *Chlamydomonas* which is unicellular, and free swimming. The two flagella are of equal size and they are responsible for the locomotion. The unicellular Desmids can be considered as reduced forms but not primitive.

Colonial

A unicellular individual, further evolved in the direction of a free-swimming colony. In a colony, a number of individuals are aggregated together in different ways, within a mucilagenous sheath. In the simplest colonial forms *(Pandorina),* all the cells are almost functionally alike, but in the highest developed colonial forms *(Volvox)* where there are thousands of cells, only a few are reproductive and others are somatic. All the cells of a colony function as a homogeneous entity, and the movement of the colony is affected by the combined action of all the flagella. The colonial line of development is an experiment in evolution, which did not progress further than the type of structure exhibited by *Volvox.*

Palmelloid and Dendroid Types

In some species of *Chlamydomonas,* asexual reproduction is slightly modified so that palmella-stage is formed as a temporary stage in the life cycle. But in some genera, such a tendency has become permanent *(Tetraspora).* In some cases, by a slight variation from the former, dendroid colonies are formed *(Ecballocystis).* In dendroid forms, individual cells readily detach and assume a motile habit. Fritsch classifies the palmelloid and dendroid forms under different sub-orders of Volvocales.

Coccoid Habit

The motile unicellular individuals *(Chlamydomonas)* withdraw flagella and become non-motile before reproduction. This becomes a more permanent feature at the expense of the period of free movement. This secondary condition (coccoid phase) is seen in *Chlorococcum humicolum.*

Filamentous Habit

Ulothrix exhibits the simplest type of filamentous structure. The filamentous habit is the most common one in Chlorophyceae. It must have originated from unicellular forms by transverse septation in the vegetative phase not accompanied by gelatinization of cell walls. Thus, the cell division observed in unicellular forms, in connection with reproduction has been relegated to the vegetative phase resulting in multicellular filamentous habit.

The simple unbranched filament further evolved in two directions. On one hand, it developed into a branched filament (*Cladophora*, etc.,) and on the other into a flat foliose thallus. The branched filamentous habit is the most common one. Some of the more advanced forms show heterotrichous habit.

Foliaceous Plant Body

The second line of evolution of the filament led to the flat leaf-like expanses, as seen in *Ulva*. In fact, the early stages of *Ulva* resemble the germlings of *Ulothrix*, thereby indicating that they are derived from filamentous forms. The foliaceous thallus has evolved due to transverse as well as longitudinal septation of the filament.

Heterotrichous Habit

The filamentous body, in its highest development, shows the *heterotrichous* habit. This is characteristic of Chaetophorales and it also occurs in other classes of algae. The thallus consists of creeping prostrate portion and an erect system consisting of a few branched filaments. In *Coleochaete scutata*, the erect system is completely suppressed and the prostrate system of filaments form a compact, flat, disc-like structure.

In some other cases, the prostrate system may be suppressed and the erect system may be elaborated (*Draparnaldiopsis*). Many heterotrichous forms of Chaetophorales can adapt themselves to terrestrial habit (*Fritschiella*) and according to Fritsch the higher terrestrial plants might have arisen from algae exhibiting heterotrichous habit.

Siphoneous Habit

In *Cladophora*, the cells are multinucleate and this must have been derived from filamentous forms with uninucleate cells, due to free nuclear divisions and lack of corresponding septum formation. Such a tendency might have resulted in the complete loss of power of septation resulting in coenocytic siphoneous filaments as seen in *Vaucheria* (Oltmanns).

But Fritsch views that siphoneous filament must have originated from a unicellular form of Chlorococales.

Fritsch considers Cladophorales as an offshoot from Ulotrichales and that it is not related to the evolution of Siphonales. Siphonales which are marine, exhibit an elaborate construction, but the thallus is generally not as big as those of the members of Rhodophyceae and Phaeophyceae.

Charales

In Charales, the plant body is highly organized and sex organs show a very high degree of specialisation. Some people consider it as a separate class (H.C. Bold, 1969). But Fritsch considers the order Charales as merely a highly specialised line of Chlorophyceae since, like many other green algae, they are haploid, contain chloroplasts, and store starch as reserve food.

CHLAMYDOMONAS

Class : Chlorophyceae

Order : Volvocales
Family : Chlamydomonadaceae

Occurrence

Chlamydomonas is one of the most primitive eukaryotic organisms in the plant kingdom. It is a widely distributed freshwater, free swimming green alga.

Thallus Structure

The plant body is a thallus which consists of a single biflagellated cell (Fig. 9.1). It is a microscopic, unicellular organism and it exhibits a very primitive type of structure amongst the algae. They are

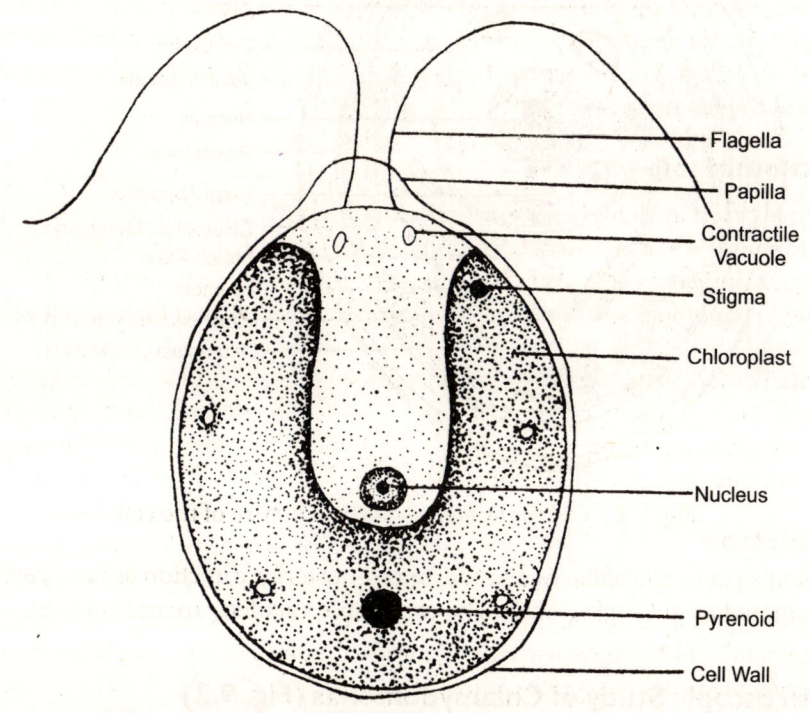

Fig. 9.1 *Chlamydomonas*—Plant body.

ovoid, spherical, ellipsoidal or pyriform in shape. The two flagella are of equal size and they are at the narrow anterior end which may be produced into a papilla. The two flagella or cilia propel the organism by their lashing movements in water. The cell wall is thin, transparent and cellulose in nature. Electron microscope studies reveal that the cellulose layer is finely striated with parallel cellulose fibrils (Fig. 9.2). Internally, the cell encloses a single nucleus, a single large cup-shaped chloroplast with one or more pyrenoids, two contractile vacuoles, a red eye spot and two flagella. The pyrenoid stores reserve starch in the form of layers around it. At the anterior end, towards a side, the stigma or eye spot is present. The stigma is oval or circular and it is sensitive to light. The eye spot consists of two parts, *viz.*, a curved pigmented plate and a biconvex, hyaline photosensitive substance which functions as a lens. The eye spot is sensitive to light and directs the movements of swimming cells.

Chlamydomonas cells are provided with two anterior *flagella*. Each flagellum arises from a basal granule, the *blepharoplast*. The flagella are of equal length and longer than the length of the cell.

The structure of the chloroplast varies in different species of *Chlamydomonas*. It is reticulate in *C. reticulata*, H-shaped in *C. biciliata* and laminate in *C. sphagnicola*. Similarly, the pyrenoids may be

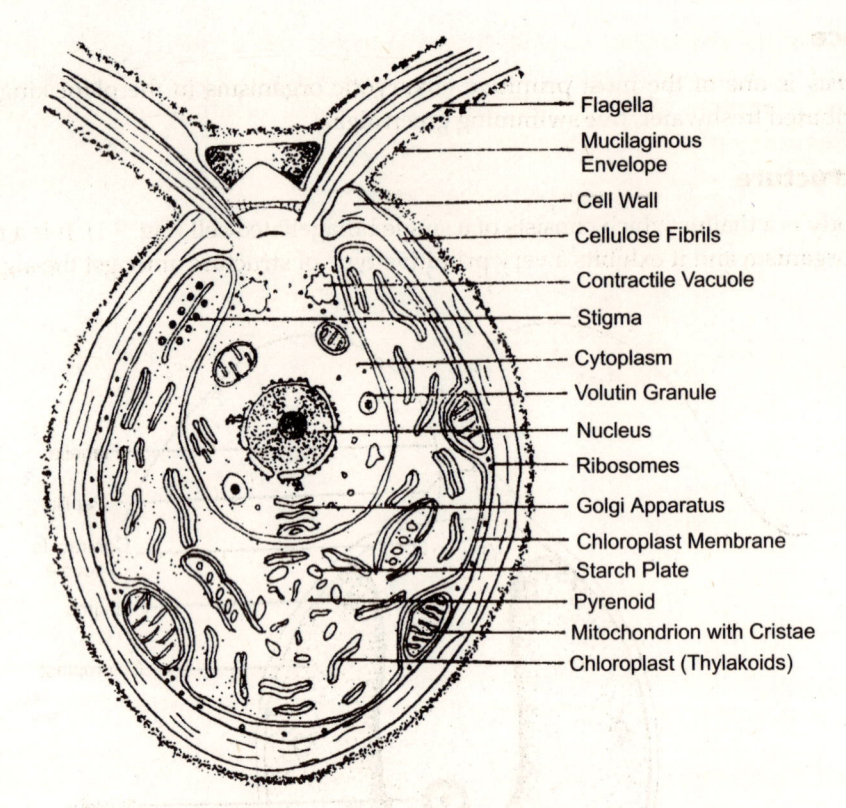

Fig. 9.2 *Chlamydomonas*—Ultrastructure of the cell.

altogether absent as in *C. reticulata*, two in the antero-posterior direction as in *C. pertusa*, or may be several and scattered as in *C. sphagnicola*. The contractile vacuoles are not constant in the different species.

Electron Microscopic Study of Chlamydomonas (Fig. 9.2)

Many details of the internal structure of the cell have been revealed by the electron microscopic studies. The cytoplasm is bounded by cytoplasmic membrane which consists of two electron opaque layers. At the periphery of the cytoplasm are seen few strands of granular endoplasmic reticulum, free ribosomes and mitochondria. Inside the chloroplast lie the mitochondria, volutin granules, a few strands of granular tubules, ribosomes, Golgi bodies and the nucleus. The mitochondria have plate-like *cristae*. The chloroplast is distinguished into granular matrix called *stroma*, which contains stacks of tube-like structures called discs or *thylakoids*. The eyespot is composed of two or three rows of globules at the anterior end.

Reproduction

Chlamydomonas reproduces both by asexual and sexual methods.

Asexual Reproduction. When the cell matures, under favourable conditions it multiplies by the formation of *zoospores*. First the parent cell comes to rest, and the flagella are cast off or reabsorbed. The contractile vacuoles disappear and the protoplast withdraws from the cell wall (Fig. 9.3). The cell undergoes a longitudinal division giving two daughter protoplasts. Another longitudinal division giving two daughter cells into four. All the divisions are mitotic. During the division of the protoplast, the chloroplast as well as the pyrenoids also divide. The daughter protoplasts form their own cell wall, and flagella and develop into zoospores. The zoospores are liberated into the water by the rupture of the parent cell wall or by the gelatinization of the cell wall. The zoospores later develop into *Chlamydomonas* cells.

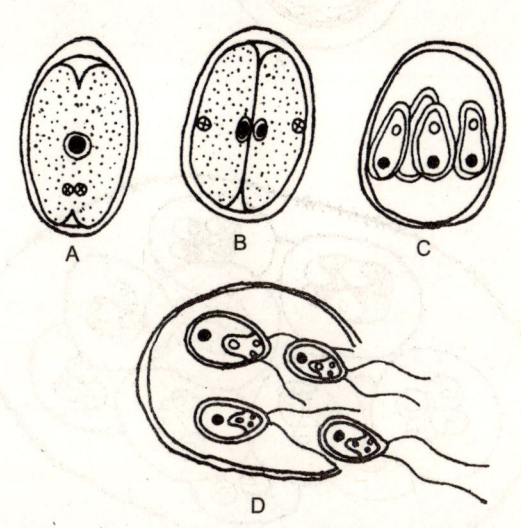

Fig. 9.3 *Chlamydomonas*—Asexual reproduction. Formation of zoospores.

Aplanospore Formation. *Chlamydomonas* also reproduces asexually by the formation of *aplanospores* or *hypnospores*. Thus in *C. caudata* the whole protoplasm rounds up to form a single aplanospore. In *C. nivalis*, thick-walled aplanospores (hypnospores) which are coloured red due to hematochrome are produced and they cause the formation of red snow in the arctic regions.

Palmella Stage. Under unfavourable conditions the motile cells come to rest and lose their flagella. The parent cell comes to rest and divides into 4 or 8 daughter protoplasts. The daughter cells do not develop flagella but remain within the gelatinized parent cell wall. Progressive division and redivision of daughter protoplasts followed by the gelatinization of the walls results in a colony of several hundreds of cells embedded in a gelatinous matrix. This aggregation of the cells is known as the *Palmella stage* (Fig. 9.4).

Originally when it was first discovered by algologists, they were mistaken for a species of *Palmella* and hence the name *palmella* stage. Exceptionally, in *C. kleinii* palmella stage is a rule in its life cycle, otherwise it is of rare occurrence in other species of *Chlamydomonas*. When the damp soil gets flooded with water, the individual cells readily develop their cilia and develop into free swimming zoospores, which later develop into *Chlamydomonas* cells.

Sexual Reproduction. Sexual reproduction in *Chlamydomonas* takes place by *isogamy*, *anisogamy* and *oogamy*. It takes place when conditions are unfavourable to growth. The cells may withdraw their flagella before reproduction. Certain species are *homothallic* and some *heterothallic*, depending upon whether the two gametes taking part in fusion are derived from the same parent or different parent respectively. Majority of species of *Chlamydomonas* show isogamous type of sexual reproduction. The protoplast of each cell divides longitudinally several times forming 8 to 32 biflagellated gametes. These gametes are biflagellated and without a cell wall look exactly like zoospores except for their smaller size. After liberation, gametes derived from different individual cells usually fuse

106 *A Textbook of Algae*

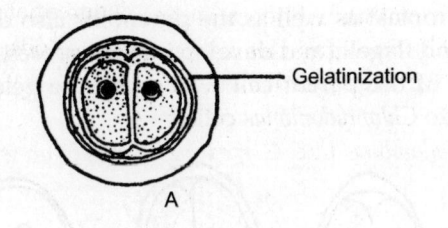

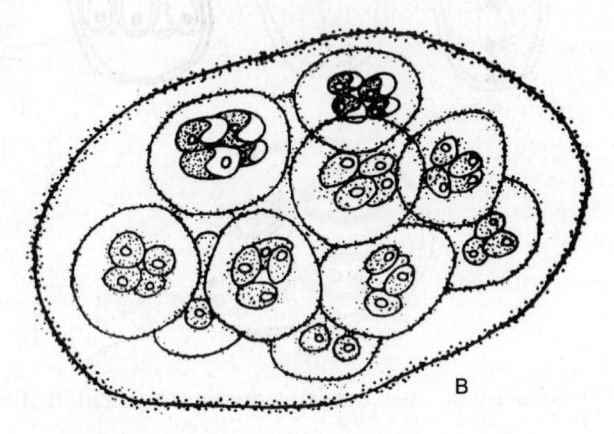

Fig. 9.4 *Chlamydomonas*—Palmella stage.

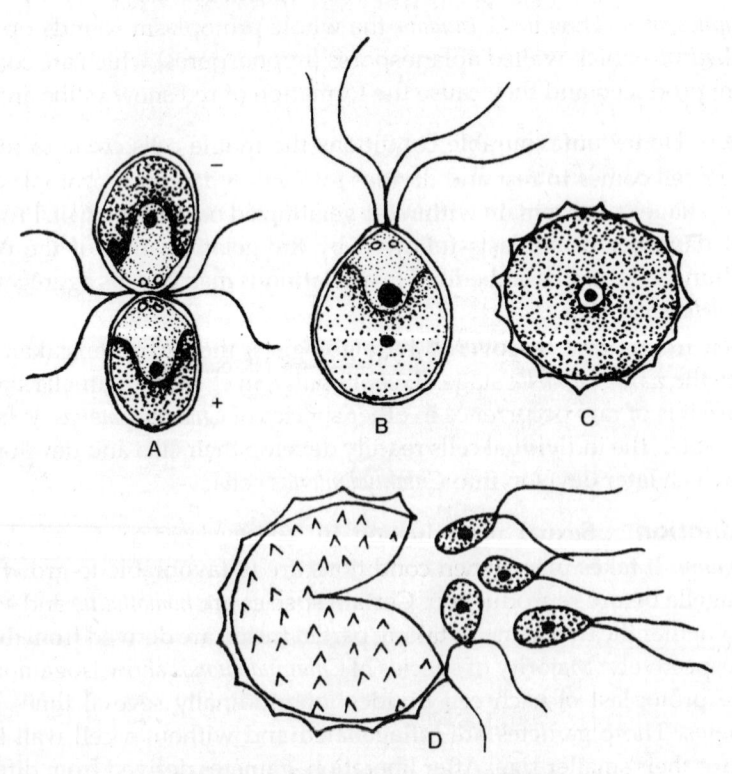

Fig. 9.5 *Chlamydomonas*—Isogamy.

by their anterior ends i.e., *heterothallic*. The gametes are exactly alike in their size, structure and physiological activity (Fig. 9.5). Such gametes are known as *isogametes* and their fusion is known as *isogamy*. After union of the gametes, quadriflagellated zygote is formed and after swimming for some time they come to rest.

In some species of *Chlamydomonas* like, *C. media* gametes from the same individual fuse to form zygotes i.e., *homothallic*.

Anisogamous Fertilization

In species like *C. braunii*, (Fig. 9.6) two types of gametes are formed *viz.*, large *macrogametes* 2-4 formed from each cell and small *microgametes*, 8-16 formed each cell. The macrogametes come to rest by withdrawing cilia while smaller microgametes continue to swim. These two gametes one micro and the other macro-gamete fuse by their anterior ends and the contents of the microgamete enter two macrogamete resulting in a zygote. This type of sexual union between dissimilar gametes is known as *anisogamy* or *heterogamy*.

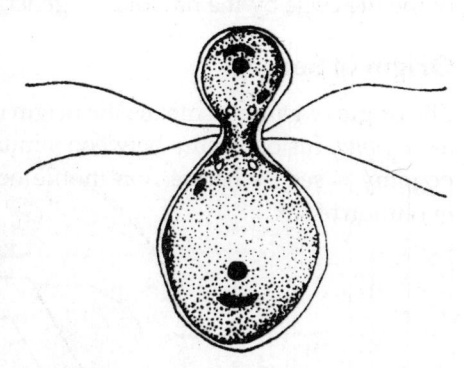

Fig. 9.6 *C. brauni*—Anisogamy.

Oogamous Fertilization

The fusion is more specialized in *C. coccifera* and *C. ooganum* (Fig. 9.7) in that the large female gamete loses its cilia and motility and functions as an immobile female gamete. The male gametes are small, biflagellated formed 16 from each cell. The fusion between such dissimilar gametes and one of them behaving as an immobile female gamete is known as *oogamy*.

Zygote Formation

Whatever may be the type of fusion from isogamy to oogamy through anisogamy the fusion product is known as zygote or zygospore. The zygote may retain its flagella or lose its flagella. After some motility it comes to rest and loses its flagella, rests for some time and secretes a thick cell wall. The thick wall is composed of cellulose and smooth or stellate outline. The zygote accumulates food material and its green colour is shadowed by hematochrome. It tides over unfavourable conditions.

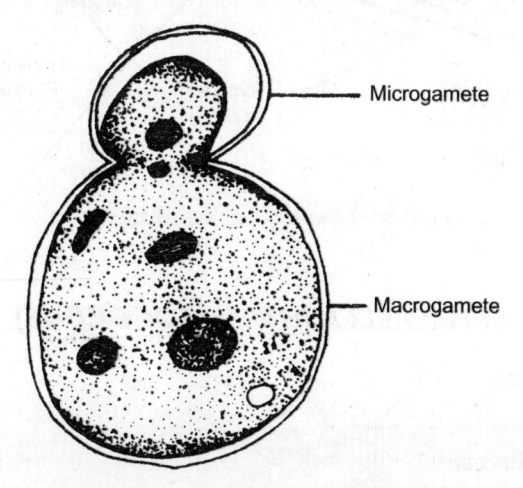

Fig. 9.7 *C. coccifera*—Oogamy.

Germination of the Zygote

After the advent of favourable conditions, zygote undergoes a meiotic division or reduction division followed by several mitotic divisions resulting in several *haploid* biflagellated zoospores. They escape by the rupture of the zygote and develop into separate individual cells.

In the life cycle of *Chlamydomonas* the diploid (2x) phase is represented by the zygote and the rest of the life cycle by the haploid (x) generation (Fig. 9.8).

Origin of Sex

The origin of gametes means the origin of sex. In the primitive form of sexual reproduction known as *isogamy*, fusion occurs between similar gametes. In the case of *C. oogamum* and *C. suboogamum*, oogamy is seen where a non-motile oogonium is formed which provides the final stage in the evolution of sex.

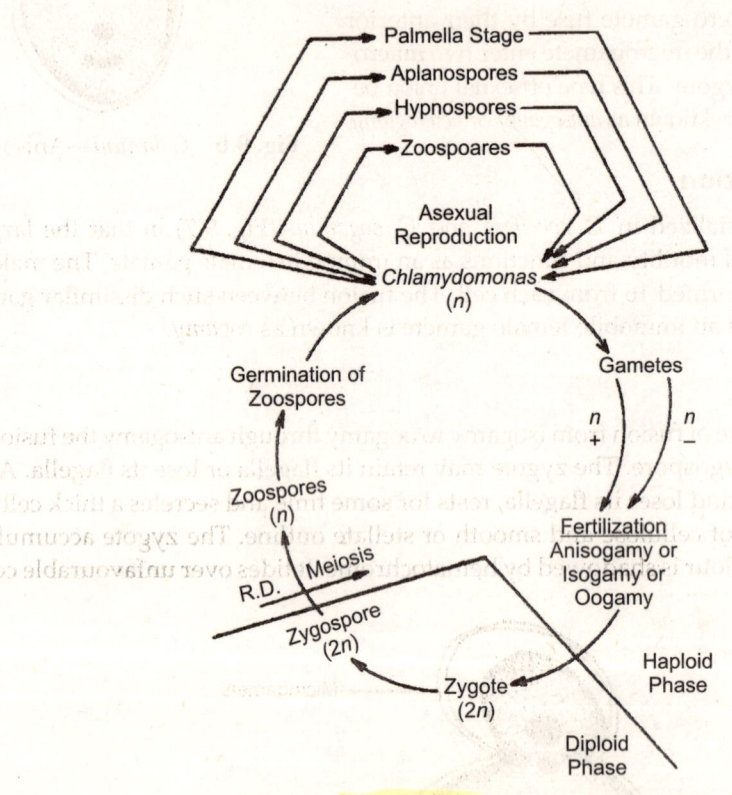

Fig. 9.8 *Chlamydomonas*—Life cycle.

SPHAERELLA (HAEMATOCOCCUS)

Class : Chlorophyceae

Order: Volvocales
Family: Chlamydomonadinaceae

Sphaerella is a freshwater alga and occurs in ponds and ditches. The active cell in this organism differs from that of *Chlamydomonas* chiefly in the separation of the protoplast from the cell wall. The cell wall itself is well defined but thin and is composed of cellulose (Fig. 9.9). It is separated from the cytoplasm by a thick mucilaginous inner wall which is penetrated by fine protoplasmic threads. Embedded in the central cytoplasm is a large hemispherical chloroplast which is hollowed out and resembles that in *Chlamydomonas*. In fact the structure of the whole protoplast is substantially similar to that of *Chlamydomonas*, except that there are numerous contractile vacuoles and several pyrenoids.

Haematococcus obtains its name from the fact that the cell is often red in colour. This is due to the presence of a carotenoid red pigment, haematochrome, which masks the green colour of the chlorophyll. The actual colour of any particular cell depends upon the relative proportion of haematochrome present which in turn depends on the supply of nitrogen compounds. The alga very readily forms cysts by the withdrawal of the flagella and the rounding up of the cell, in fact it is in this condition that the organism is generally seen.

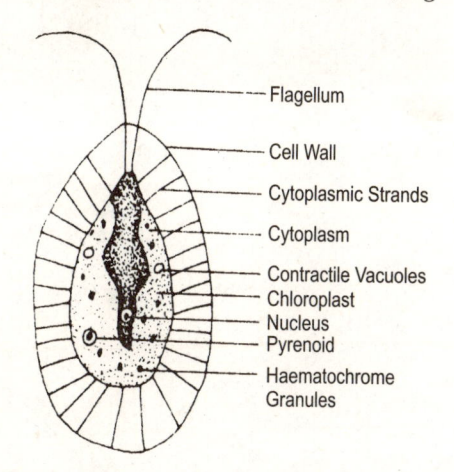

Fig. 9.9 *Sphaerella*—Structure of the thallus.

Haematococcus nivalis is closely similar to the type described above. The resting cells contain an abundance of red pigment and, since the alga is found on the snow in Alpine and Arctic regions, it gives the appearance of blood-red patches, from which it has gained the name of the "red-snow" plant. It also sometimes comes down in great quantities in rain, thus giving rise to stories about "rains of blood".

Reproduction

Reproduction is usually carried out by simple longitudinal division of the motile cell, later forming into zoospores. The contents of a cyst may sometimes divide to form an extensive palmella stage, and it is from such resting cells that the gametes are formed (Fig. 9.10). They are minute motile swarmers, with no cell wall, but otherwise like the normal cell.

Sexual reproduction is by the formation of gametes the plus(+) and minus(–). They fuse in pairs forming the zygote which divides meiotically to produce four daughter cells (Fig. 9.11).

Fig. 9.10 *Sphaerella*—Cyst formation.

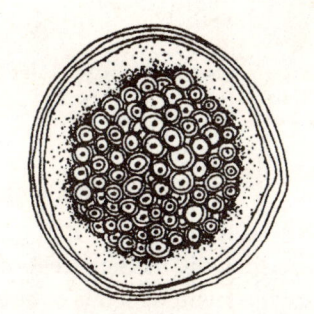

GONIUM

Class : Chlorophyceae
Order : Volvocales
Family : Volvocaceae

110 *A Textbook of Algae*

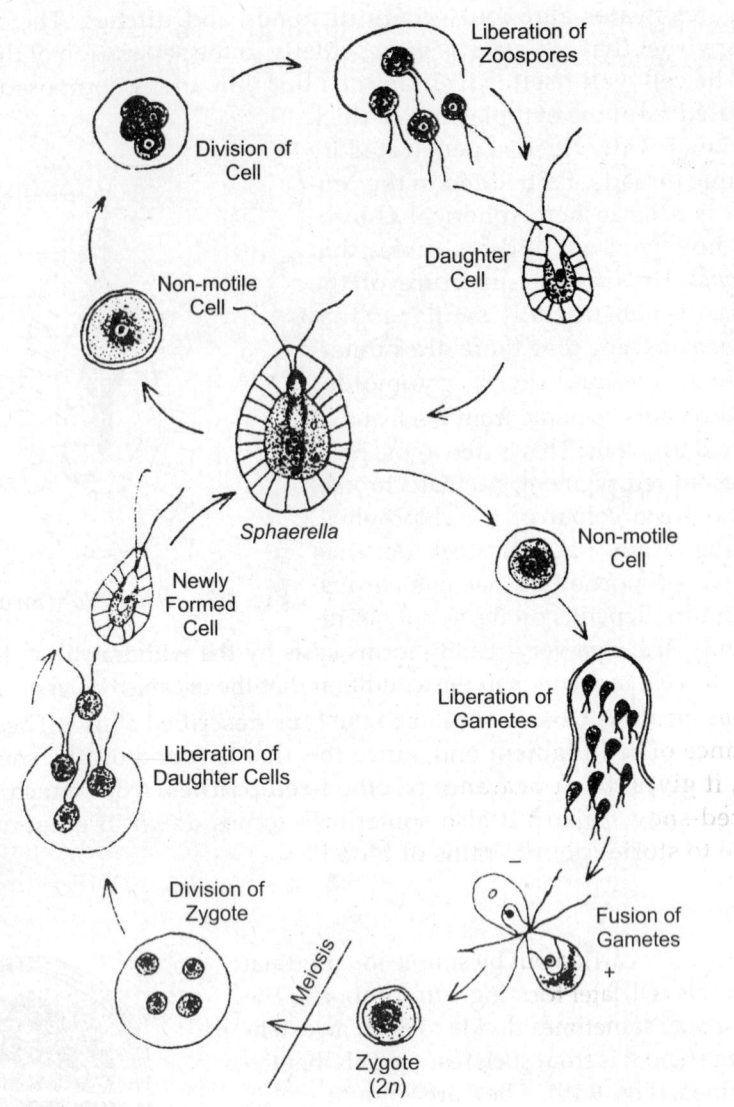

Fig. 9.11 *Sphaerella*—Asexual and sexual reproduction (Life cycle).

Gonium pectorale is the simplest of the colonial volvocales. The *colony* or *coenobium* consists of a sixteen-celled plate which is square in surface view but thin and slightly curved in side view (Fig. 9.12). It is made up of four central and twelve peripheral cells, each having the *Chlamydomonad* structure. The cells are connected by slight protrusions so that triangular spaces are formed between them, while the centre is made up of a squarish area of mucus. The whole coenobium is embedded in mucilage which possesses a firm boundary layer. The peripheral cells are inclined at an angle to the vertical axis of the plate, and each possesses a pair of long equal flagella which are directed towards the convex surface. When in motion the whole plate spins round on its axis and moves with the convex surface forwards.

In *asexual reproduction* all the sixteen cells of the colony divide simultaneously, each forming a daughter colony of sixteen cells. The divisions of each cell are all longitudinal. After division the daughter protoplasts in the mother cell in the form of a bowl or curved plesh, the *Plakea*. If a cell of a *Gonium* colony be artificially detached, it at once divides up so as to form a new colony of normal size.

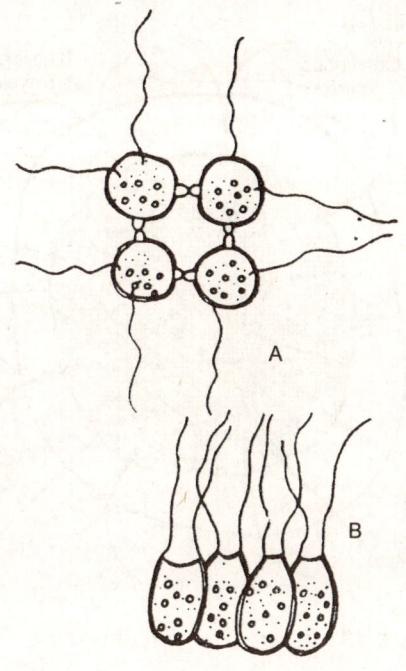

Fig. 9.12 *Gonium*—(A) Four celled colony, front view; (B) Same inside view.

Sexual reproduction is by means of naked isogametes which are produced in sexual coenobia of small size. *G. pectorale* is anisogamous. The gametes are liberated from the coenobium by the breakdown of the surrounding membrane. Gametes from the same colony do not fuse together, in fact there appears to be a physiological difference between these sexual coenobia the plus and minus strains, although they all look alike.

Gametes from one type of coenobium fuse only with those of another. The zygote so formed is at first quadriflagellate; it comes to rest and rounds off. A membrane is formed and a prolonged period of rest may follow. On germination the zygote divides to produce four cells which form a plate, and from each of these a fresh sixteen celled coenobium is produced. It is probable that the reduction division occurs during the first two divisions of the zygote, for it has been shown that there is a sexual segregation at this stage whereby two of the daughter coenobia are sexually distinct from the other two.

PANDORINA

Class :   Chlorophyceae
Order : Volvocales
Family : Volvocaceae

The colony consists of sixteen cells, each like a single *Chlamydomonas* cell. The cells are rather pyramidal in shape and are arranged in an oval group with their pointed ends inwards (Fig. 9.13). They are closely packed together, and the whole coenobium is surrounded by a mucous investment.

Each individual cell has two widely divergent flagella, by means of which the whole colony is propelled along, generally with a definite anterior end, in the cells of which the eye spots are larger and better developed.

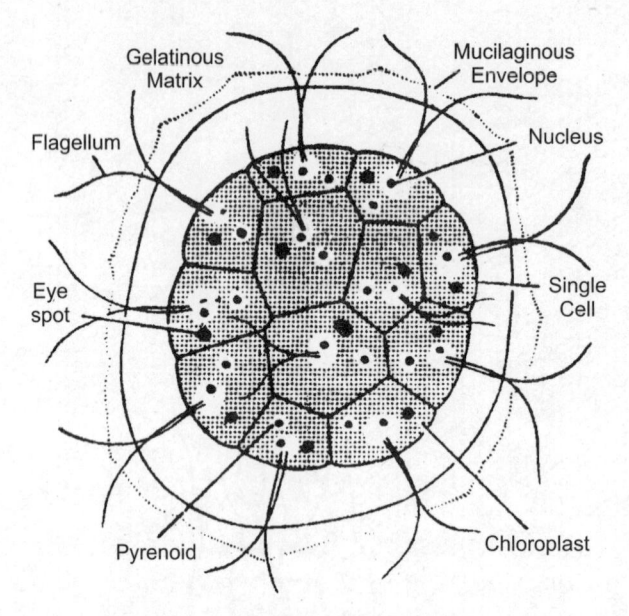

Fig. 9.13 *Pandorina*—Thallus structure.

Reproduction

Reproduction is both sexual and asexual. Asexual reproduction consists in the formation of daughter colonies within the cells of the parent. Each cell divides into sixteen cells, which at first form a curved plate. This later becomes flat and then curves in the reverse direction, the corners meeting to form a hollow coenobium. This is liberated as a new colony after the breakdown of the parent cell. This inversion of the daughter colonies during development should be compared with that described in *Eudorina* and *Volvox*. This liberation of independent zoospores rarely occurs.

Sexual reproduction consists in the liberation of the named protoplasts of the cells, which escape from the membranes and become the gametes. Conjugation is usually isogamous, but there is a tendency towards a preferential fusion of gametes of unequal size, thus showing an advance towards anisogamy, which becomes much more pronounced in *Eudorina* and *Volvox*.

The zygote germinates to liberate a single zoospore, which divides to form a new colony. Four cells are actually formed by the germinating zygote, but only one of these survives to form a zoospore, while the other three abort. This suggests that meiosis probably occurs at this stage, as in *Gonium*.

EUDORINA

Class : Chlorophyceae

Order : Vovocales
Family : Volvocaceae

In *Eudorina* the colony is larger than in *Pandorina*. It is oval in shape and consists generally of thirty-two globose cells loosely arranged on the periphery of a hollow mucilaginous matrix (Fig. 9.14). The

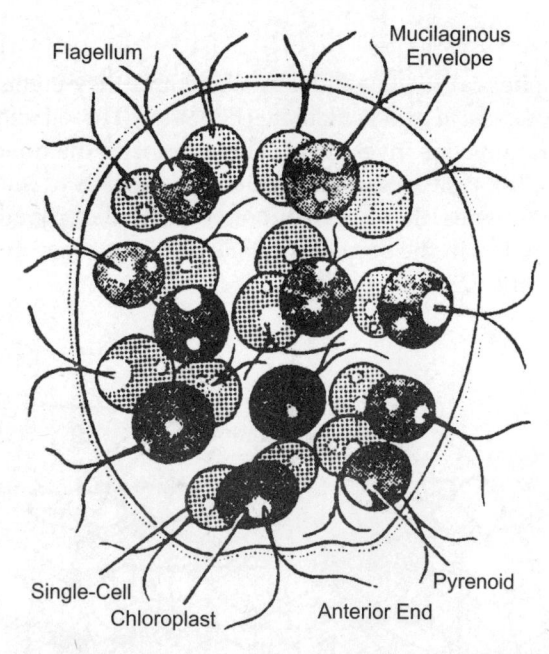

Fig. 9.14 *Eudorina*—Thallus structure.

biflagellate cells are of the typical *Chlamydomonad* type, but the chloroplast often has several pyrenoids and the cells show a progressive reduction in the size of the eye-spot as we pass from the anterior to the posterior end of the colony. Unlike *Pandorina*, the cells forming the colony are connected together by extremely fine protoplasmic strands, only visible after special staining. The flagella are long and pass outwards from the cells through minute open funnels in the mucilage investment.

Reproduction

Asexual reproduction consists in the formation of daughter colonies within the individual parent cells, whose protoplasts divide up into a number of parts equal to the number of the cells of the new colony. The young colony is at first flat, but finally curls over to form a hollow sphere. This new colony is liberated by the break-down of the parent cell wall. As all the cells reproduce simultaneously in this way the parent colony naturally disappears after reproduction.

Sexual reproduction is more complex than in *Pandorina* and shows a definite advance, to oogamy, the gametes being quite distinct in appearance and behaviour. The colonies are generally dioecious, that is to say, the male and female gametes are formed in different colonies. In the female colony the cells enlarge somewhat and become the female gametes, or oospheres, which, when mature, are fertilized *in situ*. In male colonies the cells divide to form groups of sixty-four (64) antherozoids, yellowish in colour, and each with two flagella. These are liberated as groups which swim to the female colonies, where the antherozoids separate. The zygotes germinate as in *Pandorina*.

VOLVOX

Class : Chlorophyceae
Order : Volvocales
Family : Volvocaceae

Structure

Colonies of *Volvox* are spherical to ellipsoidal in shape and they contain 500 to 50,000 cells (Fig. 9.15). Each cell is *Chalmydomonoid* and is biciliate (Fig. 9.16). These biciliate cells are embedded in the peripheral layer of the mucous investment. The interior of the hollow sphere is composed of watery nucilage. Each cell is provided with a gelatinous sheath of its own. The individual cells vary in shape in different species: they are ovoid or ellipsoidal or pyramidal. The number of such cells in a colony is determined in the very early stage itself, and they are all arranged in a specific manner. Hence, it is described as a *coenobium*.

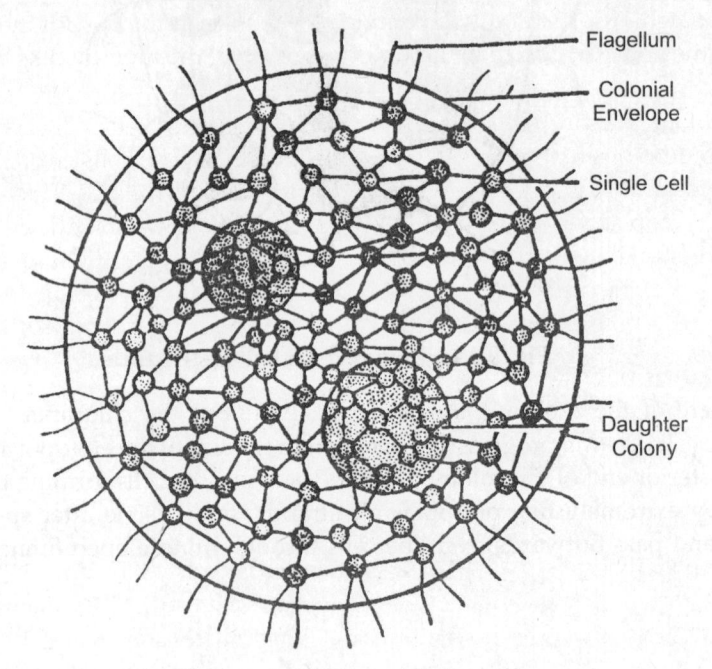

Fig. 9.15 *Volvox*—Vegetative colony.

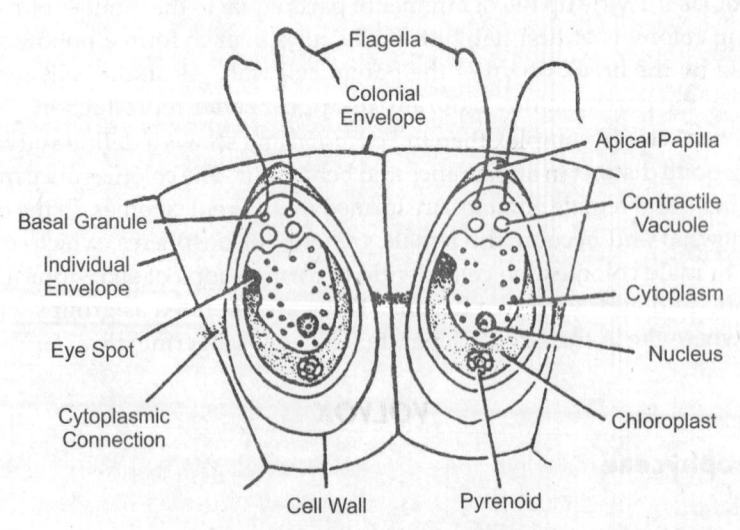

Fig. 9.16 *Volvox*—Cell structure.

The number of cells in a colony varies in different species. In *Volvox aureus*, the number varies from 500-1000 cells while in *V. globator*, the number is upto 20,000 cells. Most of the species, if not all, show delicate cytoplasmic connections known as plasmodesmata. But in *V. mononae* and *V. tertius*, there are no protoplasmic connections.

The colony is mostly composed of vegetative or somatic cells which are incapable of reproduction. A somatic cell of such a type is purely intended for assimilation of food material. The genus *Volvox* shows different types of cell structure. In *V. globator* the cell structure resembles *Sphaerella*. Its protoplast shows curved plate-like chloroplast with one or more scattered pyrenoids and 2-5 contractile vacuoles. In *V. aureus*, every vegetative cell resembles *Chlamydomonas* in its structure, in having a cup-shaped chloroplast and two contractile vacuoles at the base of the two flagella. The vegetative cells of the anterior side show larger eye spots than those of the cells towards the posterior end.

Young *Volvox* colonies are at first constituted purely by somatic cells which are all alike. But as the colony grows older, differentiation sets in. Some of the posterior cells enlarge very much, produce more pyrenoids and then develop into reproductive or *gonidial cells*. The gonidial cells have larger number of protoplasmic connections with the neighbouring somatic cells to get more of nourishment. The gonidial cells do not retain their flagella. These gonidial cells are superficially alike but differ in their behaviour. The gonidial cells that reproduce by asexual method are termed as *parthenogonidial* cells. The male reproductive cells are called androgonidial cells or *antheridia*, while the female reproductive cells are called *gynogonidia* or *oogonia*. In *V. rouseletii*, these three types of reproductive cells are found in different colonies.

Volvox exhibits asexual and sexual modes of reproduction. Reproduction is exclusively asexual at the beginning of growing season and exclusively sexual at the end of the season.

Asexual Reproduction

The asexually reproducing gonidial cells are 5 to 25 in number and they are confined to the posterior side. They are larger than the somatic cells and have no flagella.

All the divisions or a gonidial cell are longitudinal. The first division is longitudinal corresponding to the antero-posterior direction and this is followed by a simultaneous longitudinal division in a plane perpendicular to the first one. The four resulting cells divide lengthwise so that a curved plate-like, eight-celled *plakea* stage is formed. Divisions in longitudinal plane continue until daughter protoplasts of a definite number are fixed for the daughter coenobium. All these naked cells are arranged in the form of a hollow sphere with their anterior ends pointed towards the centre (Fig. 9.17). The hollow sphere has an opening, the *phialopore,* at the outer side. Now, there is a complete invagination (turning inside) through the phialopore so that the anterior ends of cells point outwards. Thus, the normal orientation is attained and then flagella develop at the anterior end. The cells acquire their cells walls. The process of turning inside out is termed as *inversion* and it was described by Powers and Pocock.

The daughter colonies remain inside the parent for a time. They finally get liberated into the surrounding waters when the parent colony ruptures. Thus, the liberation of daughter colonies results in the death of the parent colony itself.

Sexual Reproduction

It is oogamous (Fig. 9.18). The *antheridia* (male) and *oogonia* (female) are present in the posterior half of the colony. These sexually-reproducing cells are generally in small numbers. Species of *Volvox* may be dioecious *(V. aureus)* of monoecious *(V. globator)*. Monoecious coenobia are generally protandrous.

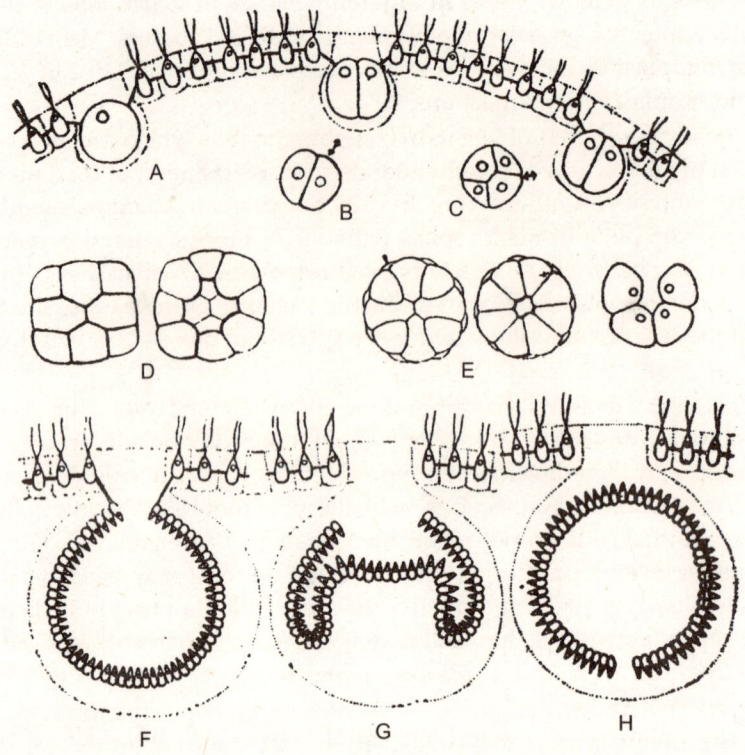

Fig. 9.17 *Volvox*—Diagrams showing stages in development of a gonidium into a daughter colony of *Volvox*—(A) gonidium; (B—E) two-, four, eight, and sixteen celled stages of development; (F) daughter colony just before inversion; (G) during inversion; (H) after inversion.

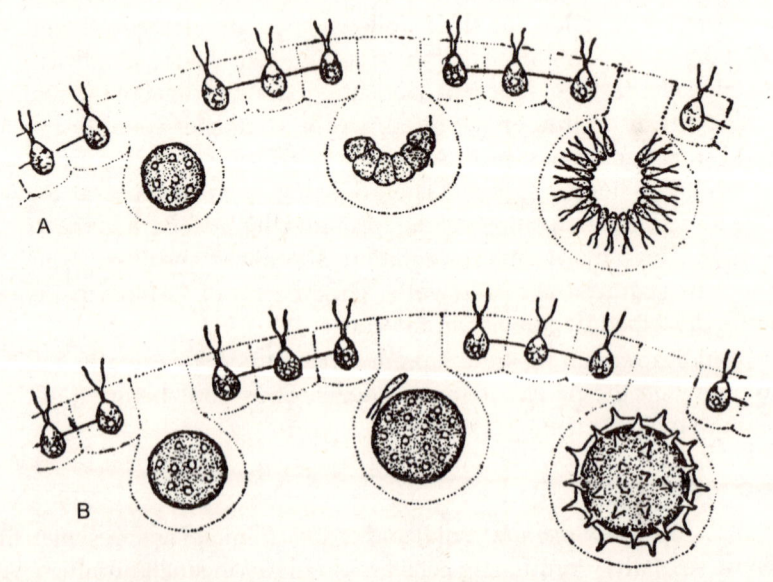

Fig. 9.18 *Volvox*—(A) Diagram or stages in development of a packet of antherozoids; (B) Diagram of a portion of a colony showing a developing egg, fertilization, and a mature zygote.

Antheridia

The male sex organs are termed by earlier phycologists as *androgonidia*. But now the term antheridia is widely used. The protoplast of the antheridium undergoes successive longitudinal divisions, in the same manner as in sexual reproduction. The number of permatozoids is 16, 32, 64, 128 or 256 depending upon the species. In some species, where 128 to 256 antherozoids undergo inversion, these elongate, bicilitate *spermatozoids* or *antherozoids* are all liberated together as a flat plate-like structure which swims about till it reaches the egg. The antherozoids later become separated, when fertilization is about to take place.

Oogonia

The female sex organs are, according to the earlier algologists, named as *gynogonidia*. But these are now known as *oogonia*. They resemble the asexually-reproducing gondidial cells. The oogonium is a flask-shaped structure opening to the outside by a narrow extension. The protoplast of oogonium rounds up into a spherical non-flagellated structure, the egg or the oosphere.

The oospore or zygote secretes a thick wall which is spinous or sometimes smooth. The green colour of the zygote is soon masked by the orange-red *hematocrome*. It accumulates sufficient food material to undergo a long period of rest for tiding over unfavourable conditions. The zygotes are liberated from the colony by disintegration of the gelatinous matrix.

During the process of germination, the zygote undergoes meiosis forming four haploid nuclei. Three of them degenerate so that its protoplast is uninucleate. The protoplast may become a bi-flagellate zoospore escaping into a vesicle formed by the innermost wall layer of the zygote. It undergoes longitudinal divisions and produces a few-celled colony (128 or 256 cells) in a manner similar to asexual reproduction. In some cases, after reduction division, the uninucleate protoplast does not develop into a zoospore. But it remains within the zygote and by the same sequence of divisions as in asexual reproduction, gives rise to a colony. Thus, Kirchner described the formation of a new colony direct from the zygote in *V. aureus*.

Colonies formed in sexual reproduction have 128 or 256 cells. These colonies multiply asexually and after six or more generations, mature colonies with the usual number of cells are formed.

EVOLUTIONARY TENDENCIES IN THE ORDER VOLVOCALES

Members of Volvocales occur in ditches of water, pools, ponds, and in large reservoirs. Unicellular Volvocales *(Chlamydomonas)* and colonial Volvocales *(Pandorina, Eudorina, Pleodorina)* are very common in temporary rain water pools. But *Volvox* is rather sporadic in occurrence and it is not found in temporary rain water pools. It is found commonly inhabiting permanent waters. These members are abundant in tropics.

Evolution of Thallus Organization

Chlamydomonas is the simplest member of Volvocales and it shows several variations in cell-shape, structure of chloroplasts, number of pyrenoids and their location.

In *Pandorina morum*, the colony is oblong, consisting of usually 16 to 32 closely packed pyriform cells whose broader ends are pointed outwards. A well-defined *mucilage* envelops the whole structure. Colonies with 32 cells look more or less like those of *Eudorina* and probably this is a connecting link between *Pandorina* and *Eudorina*. The coenobia are hollow in *Eudorina* and *Pleodorina*. In *Eudorina*, there are 32 cells usually arranged in definite transverse tiers (4, 8, 8, 4). But in *Pleodorina*, there are usually 128 cells or more and they do not show definite transverse arrangement. *Volvox* exhibits the highest type of colonial development and it consists of several thousands of cells. The colonies exhibit definite polarity with one end pointed forwards during movement. Sometimes, due to this marked polarity, the eye-spots are reduced in size in the posterior cells *(Pandorina)*.

Division of Labour and Origin of Soma

In *Pandorina*, all the cells are capable of reproduction and only an anterior cell may rarely fail to divide. In *Eudorina elegans*, of the 32 cells (4, 8, 8, 4) the anterior row of 4 cells lag behind others in dividing. In *E. illinoiensis*, these four cells are definitely smaller and purely vegetative or may divide and form only 16-celled colonies. In *E. indica*, coenobium is 64-celled and the anterior first two tiers of the cell are smaller. This tendency of forming vegetative cells culminates in *Pleodorina californica* where approximately, the cells of the anterior half of the coenobium are purely vegetative. *Pleodorina sphaerica* is further evolved in having 61 small somatic cells and 3 larger reproductive cells. In this species, the stages of formation of daughter colonies in asexual reproduction resembles that of *Volvox*. Perhaps it represents an interlink between *Pleodorina* and *Volvox*.

The coenobia of *Volvox* contain several thousands of cells and they are like hollow spheres. The *Chlamydomonas* like cells are embedded in the mucilage at the periphery, and the interior of the colony is filled up with mucilage. In *V. aureus*, the cells are 500-1,000 in number and in *V. globator* they are up to 20,000 in number. The cells may be interconnected together by fine cytoplasmic strands called plasmodesms: sometimes these are absent. Although there are thousands of cells, only a limited number are capable of reproduction. These are known as *gonidial* cells and they are confined to the posterior half of the colony. These are larger than the *somatic* cells which are assimilatory in function.

In unicellular *Chlamydomonas*, the entire protoplast of the parent cells is perpetuated in the offspring. So also, in *Pandorina*, since all the cells are capable of reproduction, the entire protoplast of each cell is distributed to the next generation. But in some species of *Eudorina*, *Pleodorina* and conspicuously in *Volvox*, some cells are purely vegetative and they perish after reproduction. Thus, this extreme specialisation, the division of labour, has brought in its wake, the death of the vegetative cells or soma.

Asexual Reproduction

This takes place in favourable conditions. In *Chlamydomonas*, the protoplast divides into four parts which escape out as zoospores. They directly enlarge into adult cells. Sometimes, a palmella stage may result under adverse conditions before the liberation of zoospores. In the colonial forms, the asexually reproducing cells vary in number in different genera. All divisions are longitudinal. After three divisions, the eight cells are cruciately arranged, forming a curved plate or *plakea*. In *Pandorina*, divisions stop at 16 or 32 celled stage: in *Eudorina* usually it stops at 32 and in some species at 64. In *Pleodorina*, it is 64 or 128. In *Volvox*, divisions continue until thousands of daughter cells are formed.

In *Pleodorina* and *Volvox*, all the daughter cells resulting from the plakea stage are arranged as a hollow sphere with a small opening, the *phialopore*. The whole structure turns inside out through the phialopore so that the anterior ends of the cells of the colony come towards the outer side. The daughter colonies get liberated by the decay of the parent individual.

Sexual Reproduction

In Volvocales, evolution of sexual reproduction can be traced from *isogamy* to *oogamy* through *anisogamy*. The evolution of sex is *polyphyletic* and it has taken place in algae along different lines.

A number of species of *Chlamydomonas* exhibit *isogamy*, where the fusing gametes are exactly similar in all respects. In *C. brauni*, the macrogametes are bigger than the microgametes and their union is known as *anisogamy*. In *Chlamydomonas, coccifera*, the vegetative cell withdraws cilia and acts as the female cell. The male gametes are small and spherical and they are formed by the division of the parent cell into 16 parts. The union between such an active biflagellate male gamete and a large round passive female cell is described as *oogamy*.

In the colonial forms, *Pandorina* shows *anisogamy* and the fusing gametes are unequal in size; the female is rather sluggish and the male is active. In *Eudorina,* colonies are dioecious or monoecious. The spermatozoids are liberated as a single bundle of 64 individuals. Each spermatozoid is elongate and biflagellate. These spermatozoids get separated when they approach the colony, where the ordinary vegetative cells act as female cells. The membranes of the female cells gelatinise before fertilisation. The zygotes remain within the female colony till its degeneration. In *Pleodorina*, the sexual reproduction is similar but oogonia and antheridia are confined to the posterior half of the colony. In *Volvox*, the colonies may be monoecious (*V. globator*) or dioecious (*V. aureus*).

Whatever may be the condition, the reproductive cells are confined to the posterior part of the coenobium. The oogonium produces a single rounded egg which is non-flagellate. The spermatozoids are narrow and elongated and so fusion can be described as oogamous. The zygote develops a thick spinous wall to enter a period of rest. In favourable conditions, it germinates to give rise to a colony. Thus, in *Volvox*, oogamy is more specialised than in *Eudorina* and *Pleodorina*.

The colonial development of plant body can be considered as a blind line of evolution that culminated in *Volvox* which represents the limit of evolution along this line.

CHLORELLA

Class : Chlorophyceae

Order : Chlorococcales
Family : Chlorococcaceae

Occurrence

Chlorella is unicellular and it occurs on damp soils, bark of trees, and in freshwater pools. Species inhabiting sewage may be colourless. The green cells of *Chlorella* found in aquatic animals like *Hydra, Stentor,* and freshwater sponges are described as *Zoochlorella*. These cells have a special capacity to escape digestion and live as symbionts in the tissues of the animals. The CO_2 released during the respiration of the animal tissues is utilised by *Zoochlorella* during phytosynthesis and oxygen is released.

Structure

Chlorella is unicellular and non-motile (Fig. 9.19). The cells are spherical or ellipsoidal having a cellulose cell wall. In each cell, the parietal cupshaped chloroplast is provided with or without a pyrenoid. The single nucleus is in the colourless central cytoplasm. Mitochondria and Golgi bodies are present. After completing the growth phase, the cells become multinucleate before the beginning of reproductive phase.

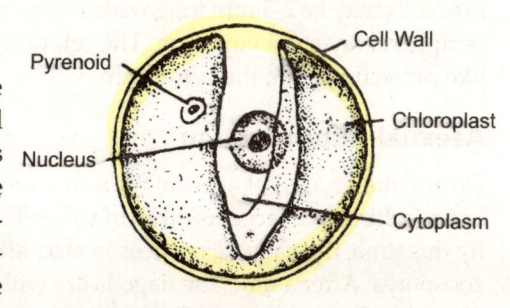

Fig. 9.19 *Chlorella*—Cell structure.

Reproduction

Chlorella is non-motile and it does not produce zoospores or gametes i.e., motile cells. Each cell produces by successive divisions, generally four autospores. Sometimes 8-16 autospores may be produced. They escape out when the parent cell wall breaks and later grow into new individuals.

Importance

The alga can grow in different conditions. Its photosynthetic pigments and reserve foods are like those of higher plants. Recent studies on the mechanism of photosynthesis are mainly based in *Chlorella* cultures. As *Chlorella* cells are rich in proteins, fats, carbohydrates and minerals, they may be used as food for human beings. But growing *Chlorella* for commercial purpose is not economically feasible. They may be used in submarines and space ships for regulating oxygen and carbon dioxide supply.

HYDRODICTYON

Class : Chlorophyceae

Order : Chlorococcales
Family : Chlorococcaceae

Occurrence

Hydrodictyon, commonly known as "water-net" occurs in fresh-water ponds and lakes. *H. reticulatum* is the common species and it may grow abundantly, covering the entire pond surface. *H. indicum* is also found in India and it was first described by M.O.P. Iyengar.

Structure

In *H. reticulatum,* the plant body is in the form of a hollow cylindric network that is closed at both the ends. This is known as a coenobium and a mature one may be as long as 20-30 cm. The network of the coenobium is formed of a number of meshes. Each mesh is hexagonal or pentagonal and the angles or corners are formed by the union of three large cylindric cells (Fig. 9.20). The nets are generally delicate and they easily break up into individual cells.

The coenobium of *H. africanum* is saucer-like and its cells are ellipsoidal.

Young cells of a coenobium are generally uninucleate having a parietal band-shaped chloroplast with a single pyrenoid. But as the cells grow and enlarge, they become multinucleate (coenocytic) and the band-shaped chloroplast becomes a complex reticulum forming many pyrenoids. The mature cells may be 2-3 mm long with a large central vacuole, so that the reticulate chloroplast is in the peripheral layer of cytoplasm. The cell wall is formed of cellulose. In *H. africanum* it produces knob like projections into the cytoplasm.

Asexual Reproduction

During the process of asexual reproduction, the protoplast of a mature coenocyte undergoes cleavage resulting in a large number of *biflagellate zoospores* (upto 20,000). Each zoospore is uninucleate. By this time, the vacuole reduces in size, affording sufficient space for the limited movement of the zoospores. After a time, the flagella are withdrawn and the zoospores develop into cylindrical cells. They develop separate walls and then orientate themselves into a net. The coenobium is released when cell wall of the parent cell ruptures.

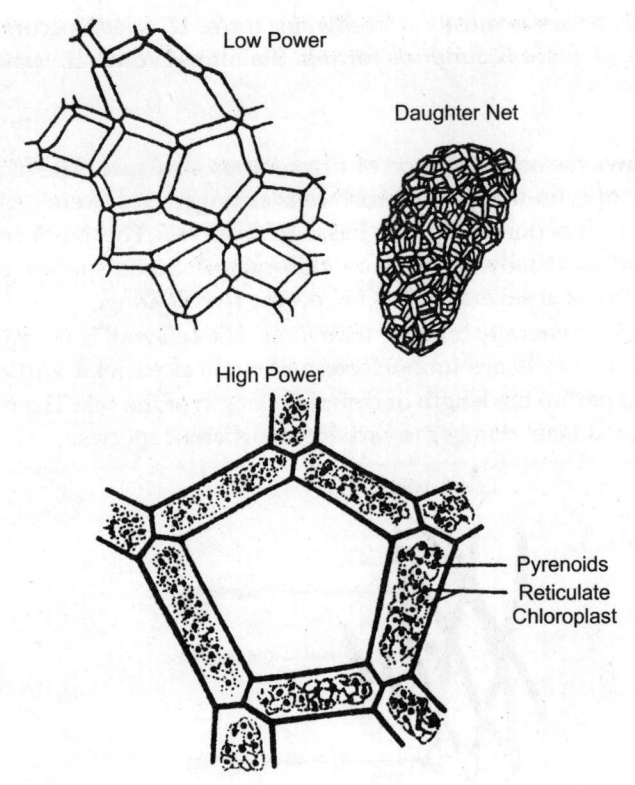

Fig. 9.20 *Hydrodictyon*—Colony and cell structure.

Sexual Reproduction

Hydrodictyon is monoecious or homothallic. During sexual reproduction, the protoplast of each cell by repeated cleavage, produces biflagellate gametes. The gametes are formed in large numbers than the zoospores and hence smaller in size. These gametes escape out through an opening of the parent cell wall. All the gametes are similar and two such structures may fuse together (isogamy). The fusion product withdraws the flagella and develops into a spherical zygote.

The zygote is thin-walled and green. It germinates immediately. It divides twice, during which reduction takes place, resulting in four biflagellate swarmers. The swarmers get liberated from the zygote. They swim, and on coming to rest, develop into thick-walled, non-motile *polyhedral cells* which are the usual means of perennation. These polyhedral cells resemble species of *Tetraedron*. After a time, the protoplast of each polyhedral cell undergoes mitotic divisions and forms many zoospores, as in asexual reproduction. These zoospores soon lose their flagella, develop their own cell walls and then arrange themselves into a net. This newly formed net escapes out when the cell wall of the polyhedral cell ruptures.

Sexual reproduction in *H. africanum* is not definite. The biflagellate gametes do not unite and they give rise to spherical resting spores which give rise to daughter nets.

ULOTHRIX

Class : Chlorophyceae
Order: Ulotrichales
Family: Ulotrichaceae

Ulothrix with about 30 species is mostly a freshwater form. *U. zonata* occurs in running water as green attached masses. *U. flacca* is a marine form in the littoral zones *U. implexa* is a lithophyte.

Structure

Ulothrix (Fig. 9.21) shows the simplest type of filamentous structure. The filament is unbranched and it consists of a row of cylindrical or squarish cells arranged end to end. All the cells are similar in structure as well as in function except the basal *rhizoidal cell*. This basal cell or holdfast acts as an attaching organ and is usually deficient in chlorophyll. Some species grow attached in the earlier stages but later break at several points to become free floating.

The cells of *Ulothrix* are generally broader than long. The cell wall is usually thin but in some, it is thick and stratified. The cells are uninucleate with typical parietal girdle-shaped chloroplast which extends to only a part of the length or complete length of the cell. The pyrenoids are embedded in the chloroplast and their number is variable in different species.

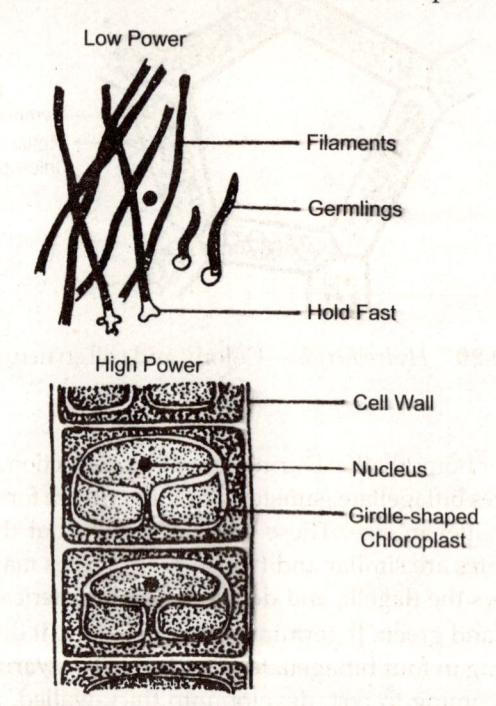

Fig. 9.21 *Ulothrix*—Filaments in low power and cell magnified.

Reproduction

Reproduction in *Ulothrix* takes place by three methods— Vegetative, asexual, and sexual.

Vegetative Reproduction. The filaments are found attached in the young stages. But as they grow in length, they break by accident into threads which will continue to float free in the water. Fragmentation of the main filament into several short pieces by disorganization of the cross-walls is a common method of vegetative multiplication.

Asexual Reproduction. All the cells of *Ulothrix* filament except the basal rhizoidal cells, are capable of giving rise to *zoospores*. (Fig. 9.22). The formation of these swarmers commences at the apex and then progresses downwards. In narrow species 1 or 2 *zoospores* are produced. In broader species, 4, 8, 16 and even 32 *zoospores* are produced.

The protoplast of the cell, just before undergoing cleavage, contracts and accumulates some reserve food material. The first division is at right angles to the long axis of the filament. The second division is at right angles to the first one. Further divisions result in 16-32 zoospores. The swarmers thus formed are liberated into a vesicle of mucilage through a lateral pore on the parent cell wall. The vesicle disappears after a couple of minutes and the zoospores are set free into the water.

In narrow-celled species all zoospores are similar. But in broad celled species like *U. zonata*, two kinds or swarmers are produced. (i) The larger zoospores are quadriflagellate and they are known as *macrozoospores*. In each cell 2 to 4 of them are produced. (ii) The smaller zoospores, called *microzoospores*, are either quadriflagellate or biflagellate. Each cell produces 8-32 micro-zoospores.

Both kinds of swarmers are approximately ovoid with a well-defined stigma. But the position of the eye-spot is variable (*U. zonata*). The macrozoospores have pointed posterior ends but the microzoospores have rounded posterior ends. In the former, stigma is anterior and in the latter it is in the middle of the body. The microzoospores move for a short time (about one day) while microzoospores move for three to six days.

The macrozoospore becomes broader than long before it settles on the substratum. It gets attached by the posterior end, produces a rhizoid and then develops into a new filament.

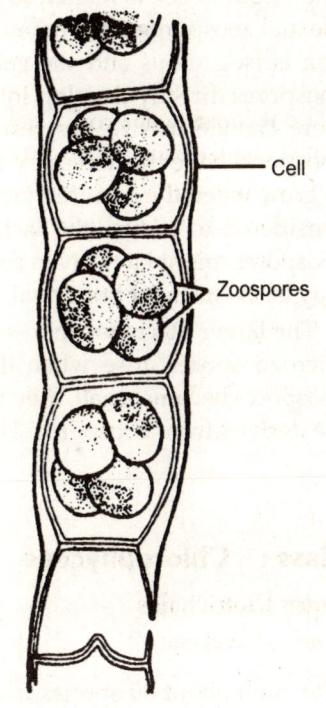

Fig. 9.22 *Ulothrix*—Asexual reproduction.

The microzoospore germinates at low temperatures in a similar way and it forms a narrow filament.

Asexual reproduction also takes place by *aplanospores*. The protoplast of a cell divides into 4-16 daughter protoplasts which develop into non-motile thin-walled aplanospores. They may be released or remain within the parent cell. When conditions are favourable, they germinate directly into new filaments.

In a few instances, food material is accumulated in each cell and the cell wall is very much thickened to form *akinetes (U. idiospora)*.

Sexual Reproduction. The sexual cells or *gametes* are formed in greater numbers than the zoospores in the same manner as in asexual reproduction. Their number is 16, 32 or 64. The gametes are usually liberated through a lateral opening in the mornings. Gametes are similar in all respects (*isogametes*) but they belong to + and − mating types. They are biflagellate and pyriform. Sexual union takes place between two gametes of + and − types, which belong to different filaments. The free swimming gametes come together by the anterior ends and then gradually fuse. Union between two similar gametes is described as *isogamy*. The fusion product is the quadriflagellate zygote. It swims for a short time and then comes to rest. Then it accumulates food material and develops thick wall to tide over the unfavourable period.

The first division of the zygote is the reduction division. By further divisions, it forms aplanospores or zoospores which are upto 16 in number. But the usual number is 4. During meiosis, sex factors segregate, so that half of the aplanospores or zoospores belong to + type and the other half to − type. The aplanospores or zoospores grow into new filaments.

Origin of Sex

The origin of sex is related to the origin of gametes. The gametes must have originated from the asexual zoospores that became too small to develop into new plants. In *Ulothrix*, sexual reproduction is isogamous and the gametes are like the biflagellate zoospores, but smaller in size. The zoospores directly develop into new plants but the gametes pair and fuse forming the zygote. A mere change in size has resulted in a fundamental difference in their behaviour. The larger zoospores have the necessary factors for growing into new plants. The smaller zoospores may fail to grow when the required factors are not sufficient. The chance fusion of such small zoospores is considered to restore the factors to optimum level for further development. Thus, the smaller zoospores might have given rise to the gametes. The actual nature of the factors is not known but it may be chemical or hormonal.

The larger macrozoospores arise when the parent protoplast of a cell divides into 2-4 bits. The microzoospores arise when the parent protoplast divides into a larger number of bits. If the zoospores become small, they fail to grow into new plants and behave like gametes. Thus, gametes are derived from zoospores. The origin of gametes has resulted in the origin of sex.

ULVA

Class : Chlorophyceae

Order: Ulotrichales
Family: Ulvaceae

Ulva with about 30 species is exclusively a marine form found attached to the rocks in the tidal zones of the oceans.

Structure

The thallus or plant body is a flat, broad, expanded leaf-like structure (Fig. 9.23). *Ulva* is frequently branched and the margin is often cut into serrations.

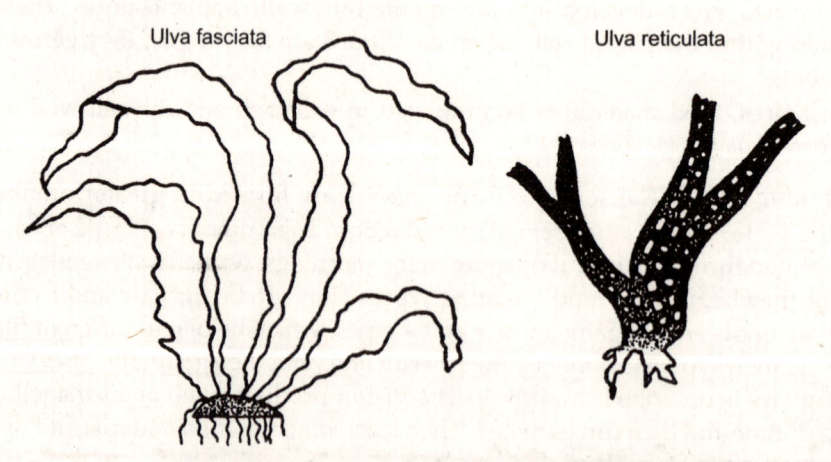

Fig. 9.23 *Ulva*—Habit. *U. fasciata* and *U. reticulata*.

Ulva is attached throughout its life time. The basal part is like a narrow stalk expanded into an attaching disc (holdfast). This holdfast is pseudoparenchymatous and it is formed by closely appressed rhizoids, produced by the lower portion of the thallus. A transverse section of mature thallus shows that it is two-layered in thickness. The two rows of cells have their long axes perpendicular to the surface of the thallus. Each cell has a single parietal chloroplast lying next to the outer face of the cell (Fig. 9.24). A single pyrenoid is embedded in it. The margin of the chloroplast is generally lobed. The cells are uninucleate. The nucleus lies towards the inner side of the cell. Cell division takes place always at right angles to the thallus surface, so that *Ulva* is only two-celled in thickness.

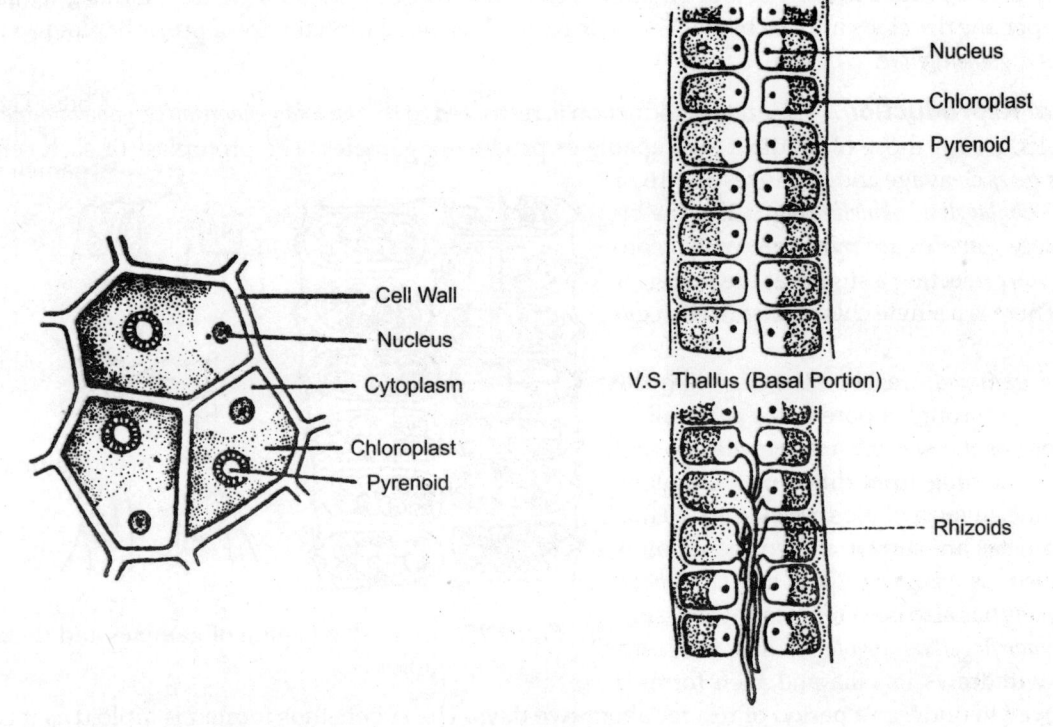

Fig. 9.24 *Ulva*—Cell structure. **Fig. 9.24(A)** *Ulva*—V.S. Thallus apical and basal portion.

In the lower region, certain cells of the thallus produce colourless tube-like outgrowths known as *rhizoids* (Fig. 9.24A). They grow between the two layers of cells and come out at the attaching portion to form the holdfast. The rhizoids are multinucleate produce *secondary thalli*. The plant body disorganises at the end of the season, but the basal holdfast persists and then produces new thalli.

Reproduction

Ulva reproduces by *vegetative, asexual* and *sexual* methods.

Vegetative Reproduction. Reproduction by fragmentation into pieces is unusual. But *Ulva* growing in estuaries may multiply by detached fragments of the thallus.

Asexual Reproduction. Asexual reproduction is confined to *Ulva* plants which are *diploid*. These *diploid plants* or *sporophytes* are exactly similar to the sexual plants which are haploid in chromosome number.

All cells of *Ulva* thallus, except those of the holdfast region, are capable of reproduction. Formation of zoospores commences in the marginal cells and then gradually extends to other cells. The protoplast of each cell divides and forms 4 or 8 *quadriflagellate zoospores*. During the formation of zoospores, *meiosis* takes place, so that they are haploid. The swarmers or zoospores are liberated through a pore in the cell wall.

In the process of zoospore formation, almost the entire contents of all cells of the blade are used up and only an empty mass of thin cell walls remain. So, the thallus collapses.

The zoospore, after a short period of activity comes to rest. It secretes a wall and then transversely divides into a lower cell and an upper cell. The lower cell gives rise to the holdfast, while the upper one develops into the blade. This new plant, developed from the zoospore, is haploid and so it is a *gametophyte*.

Sexual Reproduction. Sexual reproduction is restricted to the *haploid generation* or *gametophyte*. All cells, except those of holdfast are capable of producing gametes. The protoplast of each cell under goes cleavage and produces 8 or 16 or even 32 *biflagellate gametes* (Fig. 9.25). The biflagellate gametes are pyriform, with a conspicuous projecting a stigma at the posterior end. There is a single chloroplast in each gamete.

The gametes are liberated, just like the zoospores through a pore in the cell wall. A majority of the species are *heterothallic* and gemetes coming from different thalli alone will unite. In most of the species, the two fusing gametes are similar and so the union is described as *isogamy*. But in *U. lobata*, *anisogamy* has also been reported. The resulting *quadriflagellate zygote* swims for some time, withdraws its cilia and then forms a

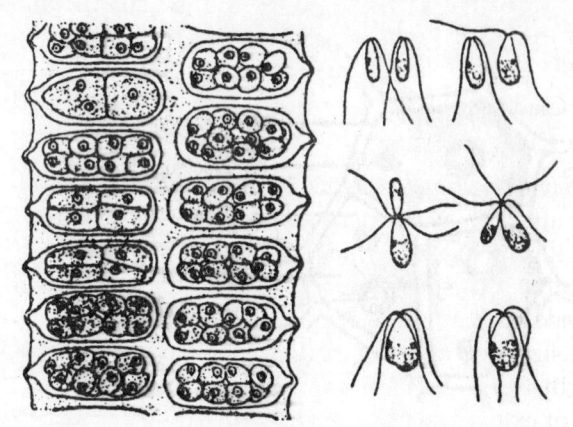

Fig. 9.25 *Ulva*—Formation of gametes and their union.

thick wall to undergo a period of rest for about two days. The zygote thus formed is diploid as it is formed by the union of two haploid gametes. The zygote undergoes repeated cell divisions and forms again the diploid plant of the asexual generation.

Alternation of Generations

In *Ulva*, there are two kinds of plants which are similar. One type of plant is diploid and it is the sporophytic generation. The *diploid* plant produces zoospores after reduction division. Hence, the zoospores are haploid and they develop directly into haploid *Ulva* plants.

The haploid plant reproduces by sexual method. The gametes are produced by simple cleavage and so they are haploid. Two of these gametes unite and form a diploid zygote, which develops into the *Ulva* plant of asexual diploid generation.

Thus, the life cycle involves a regular alternation of generations between an asexually reproducing diploid generation and sexually-reproducing haploid generation. These two generations are externally similar, and such an alternation of generations between two similar individuals was described by Fritsch as *isomorphic alternation* or homologous alternation. Smith G.M., on the other hand, is of opinion that *Ulva* has a *haplodiplontic* life cycle.

Evolution of Ulva

Ulva shows an advanced type of thallus organisation than that of *Ulothrix*. The rare tendency towards longitudinal division observed in *Ulothrix* (*schizomeris* stages) must have been made permanent so that division in several planes led to a thallus of *Ulva* type. In fact, the early stage of *Ulva* plant is filamentous and looks like that of *Ulothrix*. Thus, it reminds us of its probable ancestry from an *Ulothrix* type of plant.

The isomorphic life cycle can also be readily derived from the haploid life cycle of *Ulothrix* type. The zygote in *Ulothrix*, undergoes reduction division immediately during germination and so the diploid generation is represented only by the zygote. But in *Ulva*, the zygote undergoes equational divisions to form a diploid plant. Thus, the postponement of meiosis led to the intercalation of a new diploid asexual generation between two haploid sexual generations. This sudden change is effected in the life cycle by postponement of meiosis.

CLADOPHORA

Class : Chlorophyceae

Order: Cladophorales
Family: Cladophoraceae

Cladophora normally occurs in freshwater ponds and pools, attached to stones, snail-shells and other substrata. It grows as repeatedly-branched green filamentous tufts. But some species like *C. fructa* and *C. profounda* appear as compact cushions or hollow balls of the size of human head (Fig. 9.26). These species occur in freshwater lakes or as marine forms. Such forms of extraordinary growth are placed by some phycologists in a separate genus *Aegagropila*.

Fig. 9.26 *Cladophora*—Aegagrophilous habit.

Structure

Cladophora species are primarily attached but they may also be found free-floating in quiet waters of ponds. The thallus is attached to the substratum by means of *rhizoids* which arise from the base of the plant. The rhizoids are septate and they spread out as discs at the ends forming *stolons*. In some marine forms, even narrow aseptate rhizoids may arise from the branches to support better growth. Besides helping in attachment, the rhizoids which are rich in reserve foods persist through unfavourable season and produce the green filamentous branches again when conditions become favourable. Thus, they aid in perennation.

In *aegagropilous forms* where the whole plant is like a hollow ball, there will be several much branched individuals interlaced together. But Brand described certain special narrow branches called neutral branches which coil round other normal branches, thus making the whole structure more compact. The growth of the aegagropilous forms is slow.

Cladophora filaments are repeatedly branched (Fig. 9.27A). The branching is characteristic. Branches arise as lateral outgrowths at the upper end of a cell just below the septum. But during growth, the lateral branch is pushed aside (erection) so that branching looks apparently dichotomous. In some cases, two branches are produced on either side of the cell on the upper side giving a trichotomous appearance.

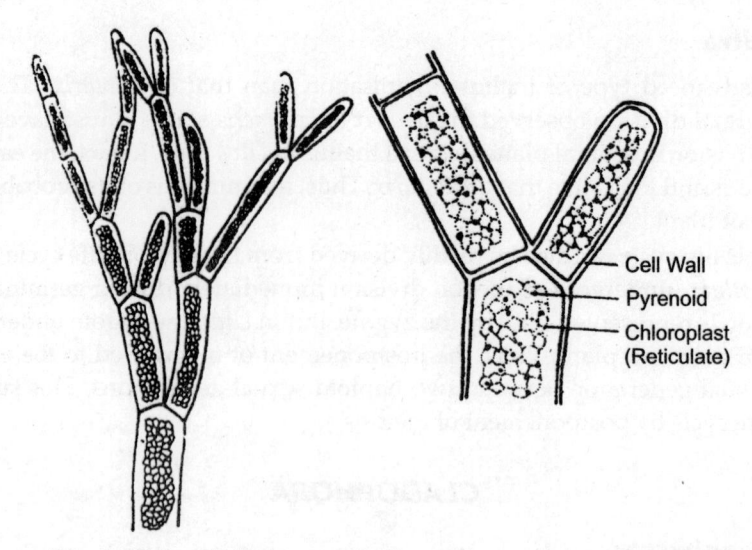

Fig. 9.27A *Cladophora*—Thallus and a portion enlarged.

The cells are cylindric and much longer than broad. They are multinucleate (Fig. 9.27B). Each cell has a big central vacuole and the cytoplasm forms a lining layer on the inner side of the cell wall. There are two views regarding the nature of the chloroplast, *viz*, (a) that the chloroplast is a single sheet of parietal reticulum lodged in the peripheral layer of cytoplasm and that the numerous pyrenoids are scattered at the intersections of the meshes of the chloroplast; (b) that the reticulum-like structure is formed of several discoid chloroplasts which are mostly parietal and a few lie in the cytoplasmic strands traversing the vacuole; each discoid chloroplast has, according to Schussing, a single pyrenoid embedded in it.

The cell wall is thick and stratified showing distinction into (a) inner lamella of cellulose (b) a middle lamella of pectic material (c) and outer layer of *chitinous* material.

Reproduction

In some *Cladophora* species, the cells of rhizoidal branches accumulate food material. The cells persist through the unfavourable period, although the thallus may itself disappear. They produce new thalli when the situation is favourable.

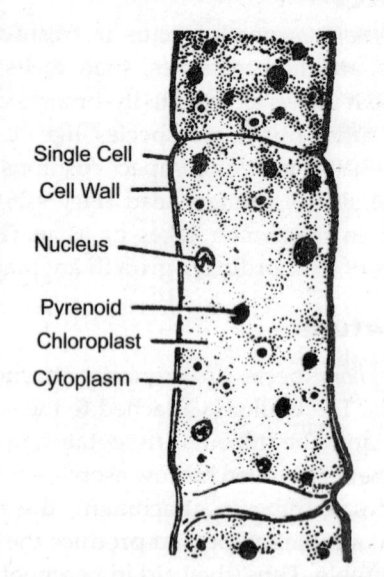

Fig. 9.27B *Cladophora*—Thallus, a portion enlarged.

In most of the species of *Cladophora,* there is alternation of generations between a diploid asexual plant and haploid sexual plant. Both the generations are exactly similar in external structure. Asexual plant produces zoospores and the sexual plant produces gametes.

Asexual Reproduction. It is found only in the *diploid plant* or *sporophytes*. Almost all the cells except the attaching cell are capable of forming zoospores. Zoospore formation commences at the terminal ends of the branches and then proceeds downward. Prior to the division, the protoplasm gradually encroaches on the vacuole and the nuclei divide actively. The divisions of the nuclei have been shown to be reductional. After the nuclear divisions, there is an aggregation of the cytoplasm around the nuclei, so as to form uninucleate protoplasts. Each uninucleate protoplast later transforms itself into a *quadriflagellate* zoospore. The zoospore is pear-shaped with a chloroplast and an eye-spot. The zoospores are liberated singly through an opening, at the upper end of the cell wall. Thus, several quadriflagellate zoospores are produced after meiosis, and hence they are haploid. The zoospore, on germination, gives rise to haploid gametophytic plants.

Sexual Reproduction. The sexual plant is produced by germination of the haploid zoospore. This haploid plant or gametophyte reproduces sexually. It produces biflagellate gametes, in the same manner as the zoospores. They escape out through a pore which is either terminal or lateral. All the gametes are exactly similar, but union takes place between two gametes of different plants, and so *Cladophora* is *heterothallic*. The union is described as *isogamy*, and as a result of this, a zygote is formed. This zygote is diploid, since it is formed by the union of two haploid gametes.

This zygote germinates directly without reduction division. The division of the nucleus is purely equational. The resulting plant is a diploid asexual plant i.e., the sporophyte.

Alternation of Generations

In most of the species of *Cladophora*, there are two generations, namely the diploid or asexual plant and the haploid or sexual plant. The diploid plant or the sporophyte undergoes meiosis and produces haploid zoospores which develop into the sexual plants or gametophytes. The gametophytes produce gametes. The gametes unite in pairs and form a diploid zygote which develops into the sporophyte. The two generations are exactly similar in external appearance and they differ only in the chromosome number. This type of alternation between two similar generations is described as *isomorphic* or *haplodiplontic*.

STIGEOCLONIUM

Occurrence. It is a common green alga found in well-aerated, fresh water. The genus comprises about 35 species. Of these 6 have been reported from India. These are *S. tenue, S. lubricum, S. lubricum formasalina, S. flagelliformis, S. nannum* and *S. attenuatum*. They occur in standing water of ponds, springs or flowing streams attached to submerged aquatic plants or stones.

Thallus (Figs. 9.27C-a, b, c). It is usually a heterotrichous filament frequently enclosed in a delicate gelatinous sheath which is of watery consistency and thus is not easily demonstrable. The filamentous thallus is differentiated into a **prostrate portion and an erect portion** (A). From the basal prostrate portion arise a number of upright filaments. The latter constitute the erect system. The upright filaments are sparingly branched in an alternate (*S. attenuatum*) or opposite. (*S. flagelliformis*) manner. The branches end in long, hyaline, multicellular hairs (C) or terminate in a point. The branch arises from the upper part of the parent cell. The cells of the main filament and of the laterals are of the same size. The projecting system shows diffuse growth which means growth by division of any of the cells in the filament. The prostrate system anchors the thallus to the substratum. The filaments forming the prostrate system exhibit apical growth and are made up of shorter cells.

There is great variability in the ratio of development of prostrate to erect system in the different species. In some the former and in others the latter is very much reduced. The more developed the prostrate system, the less developed is the projecting system and *vice versa*. In case the prostrate system is poorly developed, the basal cells of the projecting filaments put out rhizoids which serve as organelles of attachment. In form the prostrate system may be a cushion or a compact disc consisting of numerous branches packed together and coalescent to form a single layered stratum (*S. farctum*). In some species the prostrate system is a loosely and irregularly branched thread consisting of a few cells. In still other it is a richly branched compact expanse Fig. 9.27C(A).

Cell Structure. The vegetative cells are uninucleate, each with a single chloroplast which is parietal in position and girdle-shaped in form. It has one or more pyrenoids. Frequently the chloroplast occupies a part of the length of the cell.

Asexual Reproduction (Figs. 9.27D-a to f). Besides fragmentation, *Stigeoclonium* reproduces asexually by aplanospores and akinetes. The akinetes are produced in the prostrate system. Palmella stages have been reported by Livingstone (1900) in many species. In this state *Stigeoclonium* multiplies vegetatively. Exposed to desiccation, the component cells of the thallus separate, round off, become thick-walled and divide in any plane. This is the **Palmella stage** Fig. 9.27D(F). In this condition the cells may separate or remain in groups. Under favourable conditions they resume activity and again form a filamentous thallus. The normal method of asexual reproduction, however, is by zoospores.

Zoospore Formation (Fig. 9.27D). It is of frequent occurrence. The zoospores are quadriflagellate and are produced usually singly in the vegetative cells called zoosporangia. The zoosporangia, in the main, are confined to the projecting system (A). All the cells in the smaller branches (laterals) are often used up in the production of these swarmers. The mature swarmers are mostly liberated through a lateral aperture in the wall of the sporangium (B). In some species such as *S. amoenum*, the swarmers are of two kinds namely, **quadriflagellate macrozoospores** (C) and **quadriflagellate microzoospores** (D). According to Juller and Klebs, (1939) the smaller swarmers are also zoosporic in nature and may give rise directly to new plants. Godward (1942) reported that in *S. amoenum*, the quadriflagellate microzoospores behave as gametes. Besides the difference in size, the two kinds of zoospores differ in other respects as well. The microzoospores are narrower, have a lateral projecting stigma towards the back end of the body, are slightly paler than the macrozoospores, remain motile for a longer period and exhibit greater sensitiveness to light. The macrozoospores have a flat lateral stigma situated in the middle of the body.

Germination of Macrozoospores. After the usual swarming period, the macrozoospore comes to rest and attaches itself at its anterior end to the substratum. It withdraws its flagella and secretes a delicate wall around it. The body of the macrozoospore, in some species, undergoes a transverse division to grow directly into an erect thread. The cells of the new filament next to the substratum later develop into a prostrate system. In other species the one-celled germling first forms the prostrate system from which arise the upright branches constituting the projecting system.

Sexual Reproduction (Fig. 9.27E). It is isogamous and takes place by the fusion of biflagellate gametes produced singly in the vegetative cells called gametangia (A). The zygote formed is quadriflagellate. When fusion takes place between quadriflagellate swarmers functioning as gametes (mentioned above), the resultant zygote is octoflagellate (D). Copulation between gametes occurs during the swarming period. First the two gametes get entangled by their flagella. This is followed by lateral fusion beginning at the front end. The resultant motile zygote has two eye-sports and 4 or 8 flagella. After a period of motility the zygote resorbs its flagella, rounds up and secretes a wall around it (E). The eye-spots have disappeared. The green colour fades away. The zygote now becomes bright orange in colour.

Germination of Zygote (Fig. 9.27E). After the resting period the zygote germinates. The contents divide into four parts (F), each with an eye-spot. The division of the nucleus is presumed to be meiotic. Each uninucleate part develops 4 flagella to become a quadriflagellate meiozoospore (H). On liberation each meiozoospore germinates in the same way as the quadriflagellate macrozoospores to form a plant resembling the parent which produced the gametes. Occasionally the zygote undergoes no resting period. It germinates immediately in day or so as produces directly (without the intervention of spores) a small unbranched filament of diploid cells which represents the diploid stage in the life cycle. This stage later produces meiozoospores by meiosis. This particular species of *Stigeoclonium* (*S. subspinosum*) shows a tendency towards heteromorphic condition and thus exhibits alternation of generations which is heteromorphic.

Taxonomic Position:

Division	:	**Chlorophyta** (Chlorophycophyta)
Class	:	**Chlorophyceae**
Order	:	**Chaetophorales**
Family	:	Chaetophoraceae
Genus	:	*Stigeoclonium*
Species	:	*tenue*

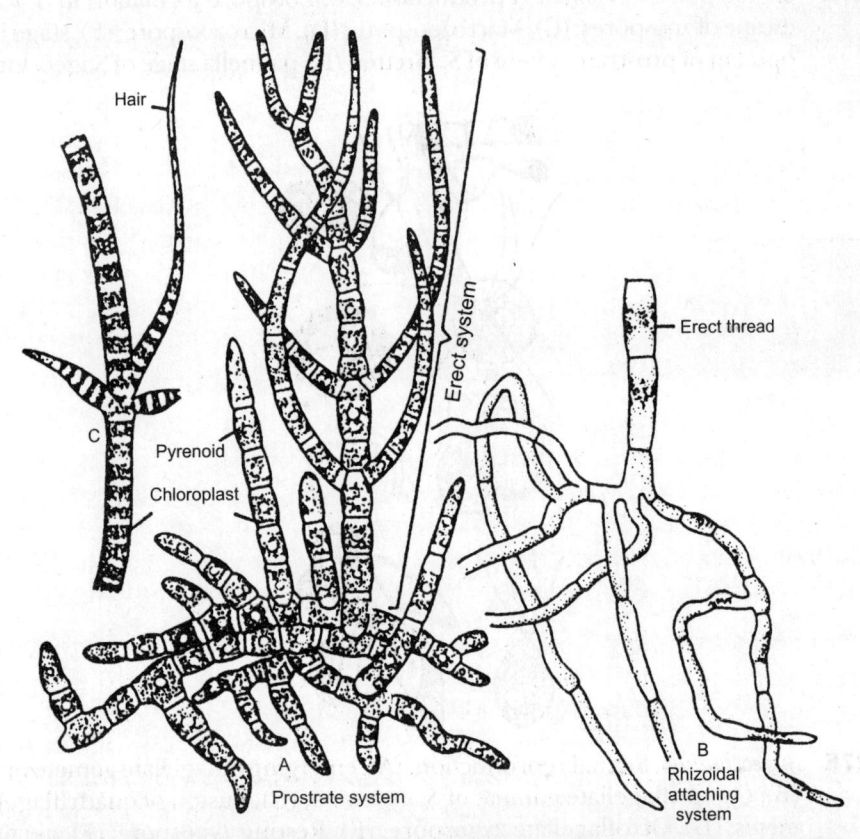

Fig. 9.27C *Stigeoclonium*. (A), Showing habit; (B), Rhizoidal attaching system of *S. amoenum* grown in culture; (C), part of an erect threads of *S. proteneum*.

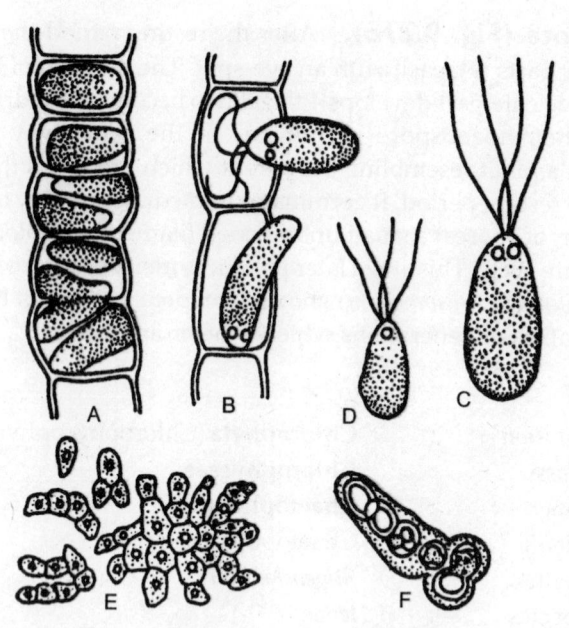

Fig. 9.27D *Stigeoclonium*—Asexual reproduction. (A), Zoospore formation in *S. tenue*; (B), escape of zoospores; (C), Macrozoospore; (D), Microzoospore; (E), stage in Development of prostrate system of S. farcum; (F), palmella stage of Stigeoclonium.

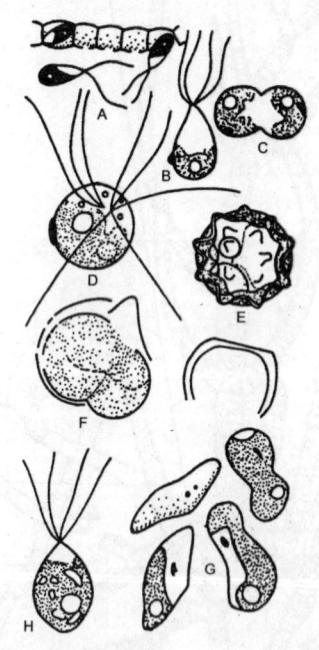

Fig. 9.27F. *Stigeoclonium*. Sexual reproduction. (A), emerging biflagellate gamete of *S. tenue*; (B), Quadriflagellate gamete of *S. amoenum*; (C), Fusion of quadriflagellate gametes; (D), Octoflagellate zygospore; (E), Resting zygospore; (F), germinating zygospore with the outer wall layer ruptured and the contents divided into four parts still enclosed by the inner layer, (G), Empty zygospore wall with the four meiozoospores having escaped from it; (H), liberated meiozoospore or gonozoospore.

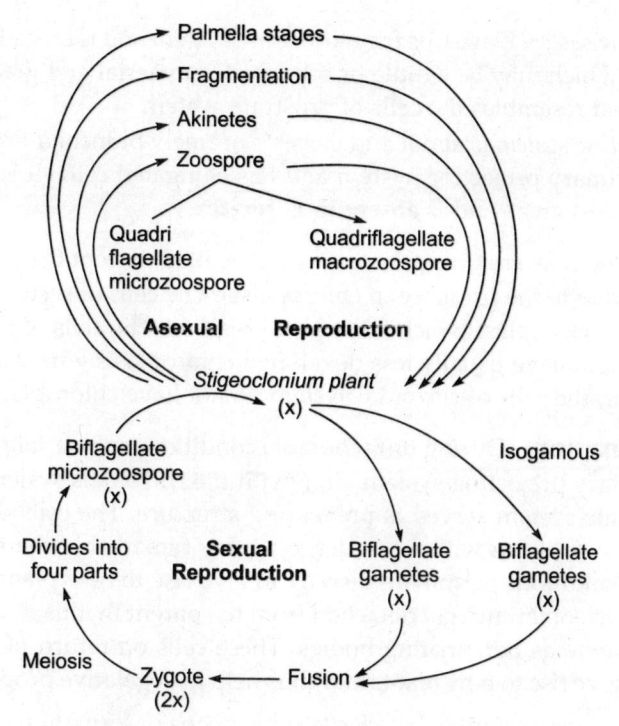

Fig. 9.27F *Stigeoclonium* life cycle.

FRITSCHIELLA

Occurrence A terrestrial alga which grows usually on moist alkaline soils was for the first time reported by M.O.P. Iyenger (1932) from South India and was named after his teacher and a great phycologist Prof. F.E. Fritsch. Later on it was also collected from other parts of India as well as from Nepal, Bangladesh, Burma, Sudan and Japan. *Fritschiella tuberosa* is the only reported species from India.

Organisation of Thallus The thallus of *Fritschiella* shows heterotrichous organisation and is differentiated into four systems (a) *Rhizoial system*, (b) *Prostrate system*, (c) *Primary projecting system* and (d) *Secondary projecting system*. Out of these, rhizoidal and prostrate systems are underground and buried in the soil whereas primary projecting system is subaerial. Only secondary projecting system is aerial and emerges out of the soil surface. (Fig. 9.27G).

The *rhizoidal system* consists of one or more septate rhizoid like elongated structure. It grows downwardly and does have colourless cells due to the absence of chloroplasts. However, it is absent in *Fritschiella simplex*, a species reported from Bangladesh.

The *prostrate system* is made up of rounded or irregularly swoller clustur of cells. This is branched filamentous, tuberous or parenchmatous and in mature thallus comprises of short congested branches which are differentiated into nodes and internodes. Certain nodal cells of prostrate system give rise to rhizoids towards lower side (which penetrate into the soil) and primary projecting system towards upper side. The *prostrate system* in young plants is not well developed but gradually it increases in size with the growth of the thallus and at maturity the thalli have very well developed prostrate system.

The primary projecting system Develops from prostrate system and is erect. It consists of uniseriate or biseriate filaments which may be simple or branched, sub-aerial and green. The cells are small and round in shape and resembles the cells of prostrate system.

The secondary projecting system is aerial and consists of freely branched uniseriate filaments. It is given out from the primary projecting system and has elongated cells. It is less developed in the thalli growing in exposed areas and is absent in *F. simplex*.

Cells Structure The cells are uninucleate and thin walled. Except the cells of rhizoidal system all the cells are green due to the presence of chloroplasts. The cells of secondary projecting system posses *curved plate like* chloroplast which have 2-8 pyrenoids. The cells of primary projecting system and prostrate system have usually less developed chloroplasts with 2-4 pyrenoids and dense cell contents. However, the cells of rhizoidal system do not have chloroplasts and are colourless.

Methods of Perennation During unfavourable conditions sub-aerial primary projecting system and aerial secondary projecting system alongwith the rhizoidals system usually degenerates and remaining prostrate system serves as preventing structure. The cells of the prostrate system perennate during dry conditions without undergoing any remarkable change in structure and on return of favourable conditions germinate directly to give rise to new plants. However, the nodal cells of prostrate system sometimes get detached from the parent thallus due to death and decay of internodal cells and serve as perennating bodies. These cells on return of favourable conditions germinate directly to give rise to new plants and thus help in vegetative propagation of alga as well.

Another method of perennation in *Fritschiella* is by means of formation of small (about 1mm in length) funnel shaped or club shaped tuber like bodies which are dark in colour and are ensheathed with the thick layer of cuticle. These structures germinate or return of favourable conditions to give rise to new plants.

Reproduction Genetically the plants in *Fritschiella* are of two type *haploid* and *diploid* but morphologically both the them are similar. The *haploid* plants are concerned with production of prostrate system of haploid plants. During the gametes formation the protoplast of gametangium divides repeatedly and large number of biflagellate gametes are produced. The gametes are morphologically similar to meiozoospores except that they are smaller in size. The gametic union results in the formation of diploid zygote which germinates immediately to give rise to diploid sporophytic plant which is morphologically similar to the haploid gametophyte.

The sporophyte or diploid plant produces haploid swarmer which are biflagellate or quadriflagellate. In *F. tuberosa* the meiozoospores are formed in the cells of prostrate system while in *F. simplex* they are formed in the cells of primary projecting system. The filaments, in the cells of which the meiozoo-spores are formed, gametes whereas *diploid* plants produce biflagellate and quadriflagellate meiozoospores by meiosis.

Asexual Reproduction The diploid plants are reported to multiply asexually by akinetes which are usually produced by the cells of ultimate branchlets. However, sometimes these structures are also found in any cell of the thallus.

Sexual Reproduction (Fig. 9.27G) The sexual reproduction is isogamous and biflagellate gametes develop in the cells of develop club shaped endings and get covered with a thick dark brown pellicle. The cells of these club shaped structures are similar to the other cells of the filament except that they are brick red in colour and have more homogenous cell contents. During the formation of swarmers the nucleus of these cells divide meiotically to form four haploid meiozoospores. Usually 2–4 biflagellate zoospores or a single quadriflagellate zoospore per cell is formed. Both the types of zoospores are formed simultaneously and are liberated through a lateral aperture in the sporangial wall. The meiozoospores on germination produce haploid plants.

Alternation of Generation *Fristchiella* exhibits isomophic *alternation of generation* in which haploid and diploid generations alternate with each other and have morphologically identical thalli. The sporophytic thallus produces through meiosis haploid meiozoospores which one germination give rise to the gametophyte. The gametophyte reproduces sexually and produces gametes of similar shape and size (isogametes) which after fusion form diploid zygote. The zygote after germination produces diploid sporophyte which is morphologically identical with the haploid gametophyte.

Taxonomic Position:

Division	:	**Chlorophyta** (Chlorophycophyta)
Class	:	**Chlorophyceae**
Order	:	**Chaetophorales**
Family	:	**Chaetophoraceae**
Genus	:	*Fritschiella*
Species	:	*tuberosa*

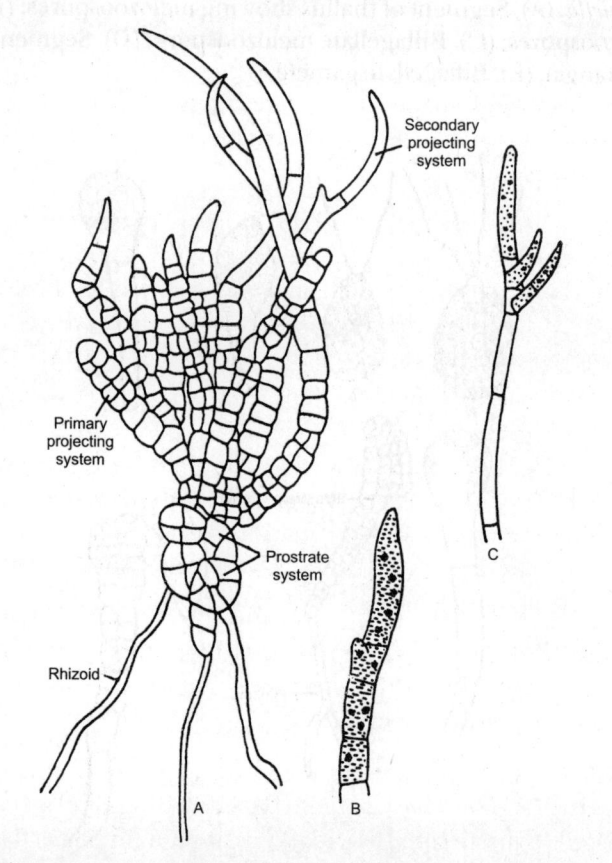

Fig. 9.27G *(A-C). Fritschiella.* (A), showing habit of plant; (B) and (C), segment of thallus showing secondary projecting system.

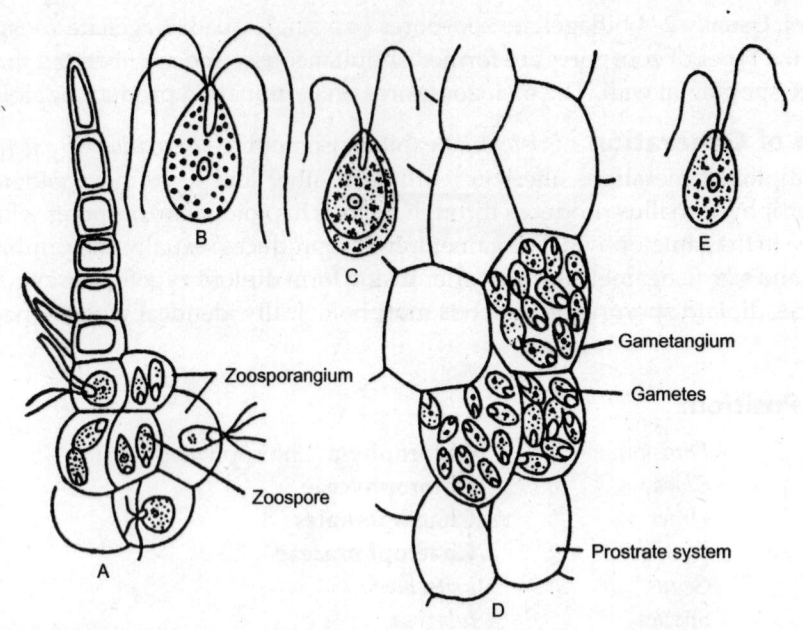

Fig. 9.27H *Fritschiella.* (A), Segment of thallus showing meiozoospores; (B), Quadriflagellate meiozoospores; (C). Biflagellate meiozoospore; (D), Segment of thallus showing gametangia; (E), Biflagellate gamete.

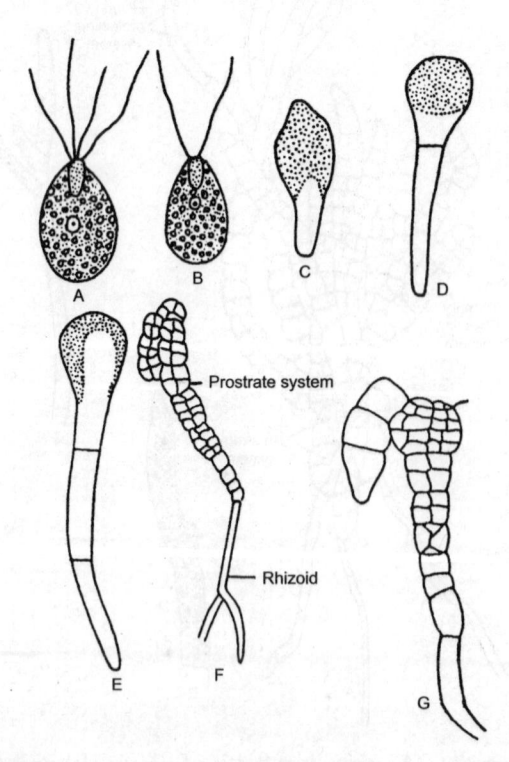

Fig. 9.27I *(A-G) Fristchiella.* Different stages of germination of meiozoospores and formation of thallus.

```
                    ┌─────────────────────┐
                    │    FRITSCHIELLA     │
                    │    GAMETOTHALLUS    │
                    └─────────────────────┘
                      ↑         ↑                          │
              Germination   Germination   Sexual           │
                              Reproduction                  ↓
              ┌──────────┐  ┌──────────┐              Isogametes
              │Biflagellate│ │Quadriflagellate│           (N)
              │meiozoospores│ │meiozoospores│
              │    (N)    │  │    (N)    │                 │
              └──────────┘  └──────────┘                   │
                    ↑         ↑                            ↓
                 ┌──────────────┐  Gametophytic Generation ┌──────┐
                 │  Sporogenic  │ ←──────────────────────→ │Fusion│
                 │    Meiosis   │                          └──────┘
                 └──────────────┘                              ↑
                        ↑        Sporophytic Generation        │
                 ┌──────────────┐                        ┌──────┐
                 │  Frischiella │ ← Germination ←        │Zygote│
                 │ sporothallus │                        │  2N  │
                 │     (2N)     │                        └──────┘
                 └──────────────┘
                      ↺ Asexual
                       reproduction
                         Akinete
```

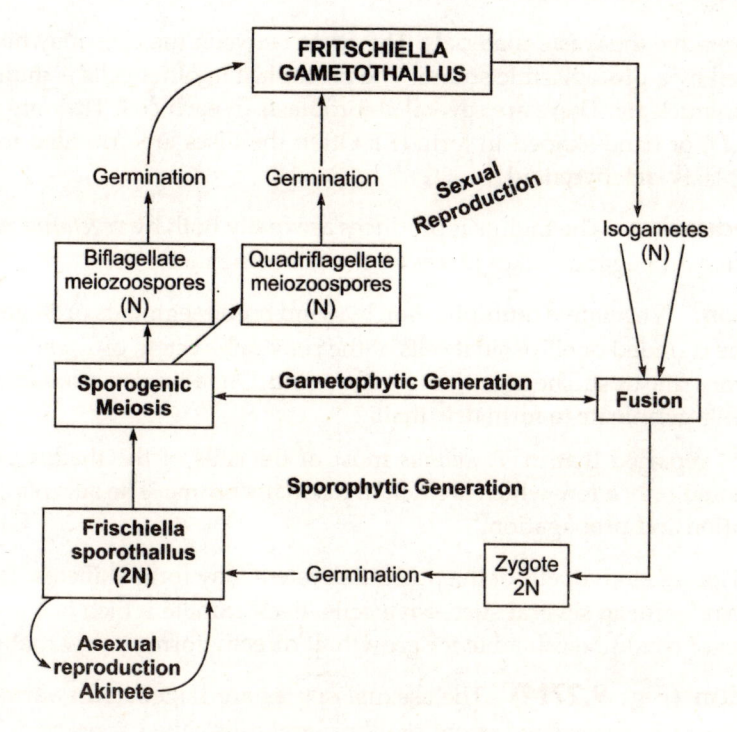

Fig. 9.27J *Fritschiella* life cycle.

TRENTEPOHLIA

The genus comprises about 50 species. Eleven species of this genus have been reported mostly from the North-Eastern region of India (Bruhl and Biswas, 1923). Jose and Chowdary (1980) also added 7 more to the list raising the number of Indian species of this genus to eighteen. All the species are strictly aerial and grow in diverse habitats but are especially abundant in the damp tropical and subtropical regions. They are found growing on moist soil in humid places or occur attached to rocks, walls of buildings, flower pots or grow on the bark of trees or leathery leaves. The yellow to range red colouring of the thallus, due to the presence in the cells of abundant haematochrome, makes it quite conspicuous. Some species occur as an algal component of many lichen thalli.

Organisation of Thallus (Fig. 9.27K) It is filamentous and branched. Usually it shows distinction into prostrate and erect systems (A). In many species both the systems are well developed and branched (*T. aurea*). The branching in the erect threads (B) may be alternate, opposite or unilateral. The laterals arise from the top, middle or subterminally from the parent cell. In some species (*T. umbrina*), the prostrate portion is well-developed but the erect threads are reduced. Some other species have the erect system more extensive than the prostrate system. The growth is apical.

Cell Structure (Fig. 9.27L) The cells may be cylindrical or barrel-shaped rarely moniliform (*T. monilia*). They usually have thick plainly stratified celluslose walls. The successive wall layers may be parallel or divergent. In the latter case the free end of the terminal cell of each branch bears a pectase cap (A) or a series of caps (B). The caps are periodically shed or pushed aside and replaced by new ones (C). West and Hood (1911) held that the caps are due to a secretion. Their function may be to reduce transpiration from the exposed tip or to afford protection. It may be a simple device to remove the waste materials. The septa between the cells may have each a single large pit penetrated by a protoplasmic strand. The protoplast in older cells is multinucleate but in young cells it is uninucleate. There are several chloroplasts in each cell. They are parietal in position and discoid (D) or band-shaped in form (E). Often the discs are arranged in bands (Geitler, 1923). The chloroplasts lack pyrenoid.

Asexual Reproduction The thallus reproduces asexually both by *vegetative methods* and *spore formation*. Vegetative propagation takes places by the following methods:

(i) Fragmentation. Vegetative multiplication by wind borne segments or fragments of the thallus is common. The rounded or ellipsoidal cells of the prostrate system of *T. umbrina* separate from one another to form a fine dust. The latter is blown by wind. On falling on a suitable substratum the one-celled fragments germinate to form new thalli.

Irgang (1927-28) reported that in *T. iolithus* most of the cells of the thallus perish in the dry season leaving behind only a few which are rich in haematochrome. The surviving cells serve as a means of perennation and propagation.

(ii) Akinetes. The vegetative cells of the prostrate system may form akinetes. The akinetes have thick walls and may occur in several successive cells. Each akinete which is a resting cell germinates on the return of conditions suitable for growth. It directly forms a new thallus.

Spore Formation (Fig. 9.27M) The asexual spores are flagellated swarmers (zoospores). They are differentiated in specialized ovoid or ellipsoidal cells called sporangia. The latter are of three types, pedicellate, funnel-shaped and sessile sporangia.

(a) Pedicellate sporangia (B). These are terminal or lateral in position on the branches of the erect system.

The pedicellate sporangium consists of a lower stalk cell terminating in a spherical body cell or the sporangium proper (B). The stalk cell frequently becomes bent in a typical knee-shaped manner. It is differentiated into a broad subspherical proximal portion and a narrow, cylindrical stalk-like distal portion. The latter ends in a sporangium. The pedicellate sporangium swarmers. Each sporangium produces numerous swarmers which in several species of *Trentepohlia* are quadriflagellate. In a few others they are reported to be biflagellate. The zoospores thus produced are liberated through a protruded terminal or lateral aperture. The liberated zoospores are somewhat flattened. Each germinates straightaway to produce a new individual.

Aplanospores In one species the contents of the sporangium before detachment are reported to divided to form aplanospores.

Sexual Reproduction (Fig. 9.27N)

It is isogamous. The isogametes are biflagellate (D) and produced in gametangia which are intercalary or terminal in position (A). They are formed like sessile sporangia by mere enlargement of the cell. Often they are produced on the basal part of the plant (Fig. 9.27I). The gametangia often differ from the sporangia of the same species. The protoplast of the gametangium produces a number of biflagellate gametes which are liberated through a somewhat protruded terminal aperture (D). The position of gametangia on the basal parts of the thallus favours inundation which admits of an easy sexual fusion between the gametes. Parthenogenetic development of gametes into filaments has been observed in some species (Meyer, 1936).

Taxonomic Position

Division	:	**Chlorophyta** (Chlorophycophyta)
Class	:	**Chlorophyceaes**
Order	:	**Chaetophorales**
Family	:	**Trentepohliaceae**
Genus	:	*Trentepohlia*
Species	:	*umbrina*

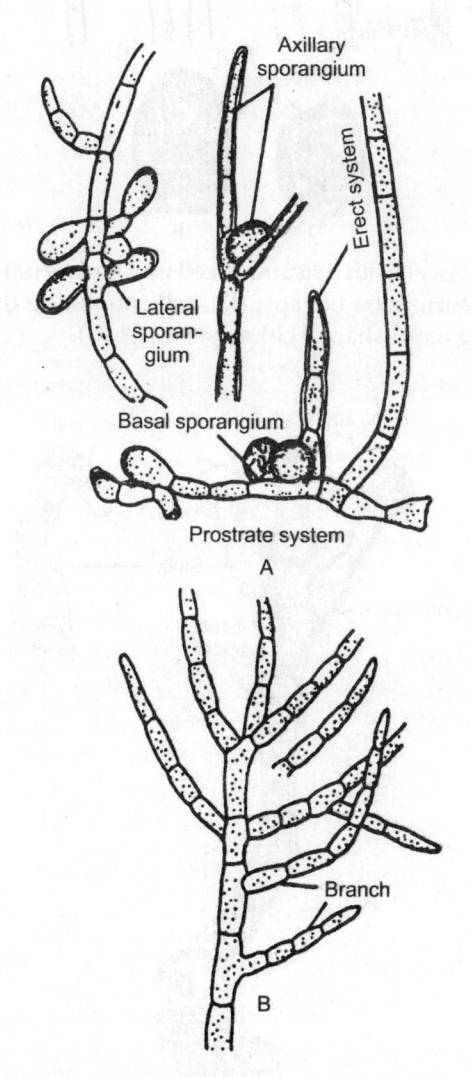

Fig. 9.27K *(A-B). Trentepohlia.* (A), Heterotrichous thallus of T. aurea with both the prostrate and erect systems well-developed; (B), Erect thread showing branching.

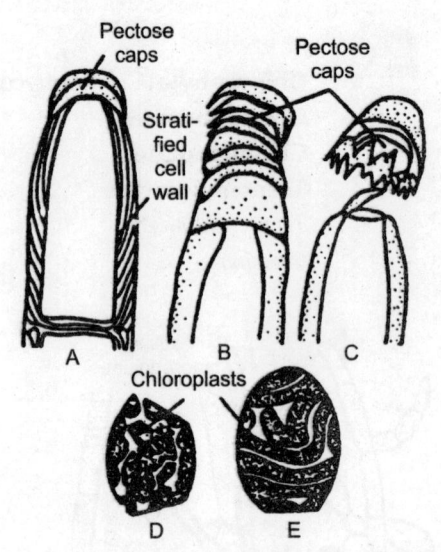

Fig. 9.27L *Trentepohlia.* (A), cell with a stratified cell wall and terminal pectose caps; (B) and (C), mode of formation of caps; (D), cell containing discoid chloroplasts; (E), cell containing band-shaped chloroplasts; (A–C).

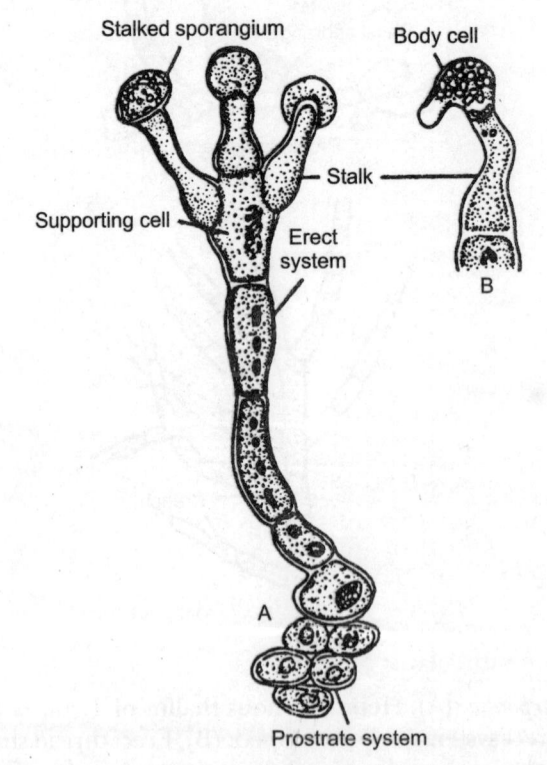

Fig. 9.27M *Trentepohlia. (A)*, Erect thread of T. umbrina showing the supporting cell bearing two lateral and one terminal stalked sporangia. (B), Pedicellate sporangium.

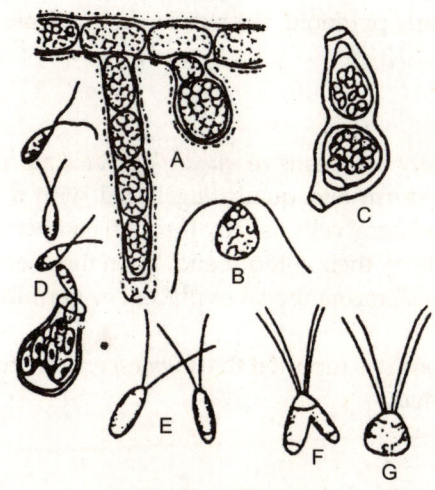

Fig. 9.27N *Trentepohlia*. Sexual reproduction. (A), T. aurea with a lateral branch bearing a terminal gametangium; (B), liberated gamete; (C), gametangia of T. umbrina; (E-F), stages in sexual fusion in T. bleischii; (G), Quadriflagellate zygote (G).

DRAPARNALDIA

Class : Chlorophyceae
Order: Chaetophorales
Family: Chaetophoraceae

Occurrence

It is a freshwater alga occurring in flowing waters attached to some substratum and is also found growing epiphytically on other plants.

Thallus Structure

The thallus (Fig. 9.28) shows heterotrichy having a distinct erect system and a poorly developed prostrate system. Holdfast represents the prostrate system and the projecting system is the dominant part of the thallus. The thallus is embedded in mucilage which is soft. The central axis bears branches of *laterals* and are profusely developed. These laterals serve in photosynthesis and in reproduction. The apices of the laterals usually end in pointed cells called *setae*. Numerous, multicellular, rhizoidal branches anchor the main axis to the substratum arising from the lower portion of the thallus. The cells of laterals are shorter with dense cytoplasmic contents and a single nucleus.

The main axis is composed of large barrel shaped cells with a vacuolate cytoplasm and a single chloroplast. The chloroplast has incised edges and forms a girdle in the

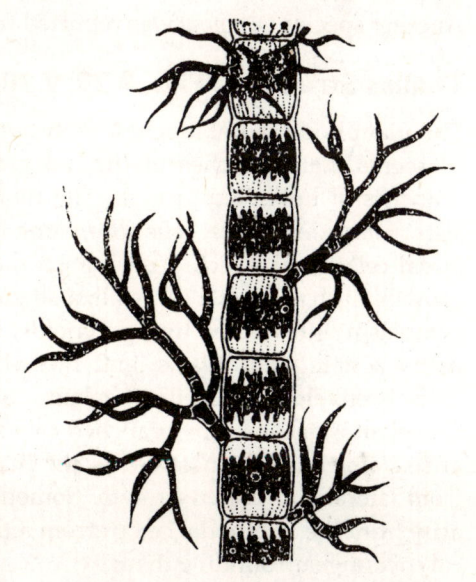

Fig. 9.28 *Draparnaldia*—Habit.

centre of the cell enclosing a single pyrenoid. Each cell is uninucleate occurring in the midst of the chloroplast.

Asexual Reproduction

Draparnaldia reproduces asexually by means of *quadriflagellated zoospores* produced mostly in the laterals. The zoospores are pyriform and quadriflagellated with a single chloroplast. After the zoospores are liberated from the lateral cells usually through an aperture, they swim for some time, rest for some time and then settle by their anterior end. Then they secrete a wall round them inside which the nucleus divides and wall formation takes place. The germling thus formed directly forms into the adult plant.

Under unfavourable conditions it is reported that *hypnospores* are formed singly in each cell with a thick wall and are later liberated.

Sexual Reproduction

It takes place by the formation of quadriflagellated gametes produced in lateral cells and are much smaller in size. It is of isogamous type and before union they withdraw their cilia and become amoeboid. These amoeboid gametes fuse in pairs to form zygotes. Zygotes secrete a wall around them and directly germinate into daughter plants.

DRAPARNALDIOPSIS

Class : Chlorophyceae
Order : Chaetophorales
Family : Chaetophoraceae

Occurrence

Draparnaldiopsis indica is a freshwater form growing on other aquatic plants in ponds and lakes. Another species *D. alpina* was reported from California by Smith and Klyver.

Thallus Structure (Figs. 9.29, 9.30 & 9.31)

The plant body is unbranched with two types of filaments, i.e., branches of limited growth and branches of unlimited growth. The filament consists of large internodal cells alternating with small nodal cells from which arise branches of limited growth. Each cell has a cell wall, a single nucleus, cytoplasm and one or two pyrenoids. There are many zonate chloroplasts, and the whole plant body is enveloped in a thick gelatinous material. The plant is anchored by branched rhizoids which arise from the lowermost cell of the main axis or from laterals of limited growth. Sometimes, they arise from the internodal cell or from a thick cortical covering surrounding them.

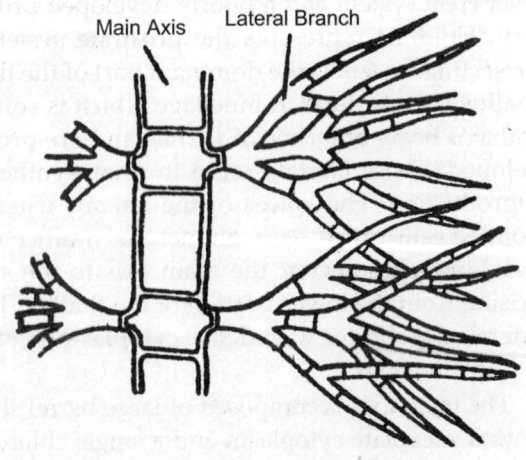

Fig. 9.29 *Draparnaldiopsis*—Habit.

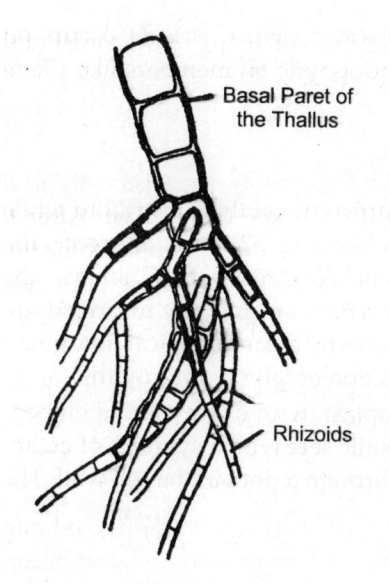

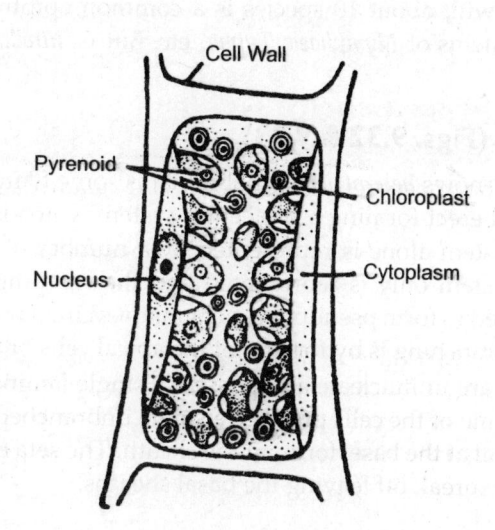

Fig. 9.30 *Draparnaldiopsis*—Basal portion.   **Fig. 9.31** *Draparnaldiopsis*—Cell structure.

Asexual Reproduction

It takes place by the formation of three types of zoospores, *viz.*, quadriflagellate macrozoospores, quadriflagellate microzoospores and biflagellate microzoospores. All the lateral cells except the nodal cells of the branches of unlimited growth and terminal cells of the branches of limited growth, are capable of forming swarmers. The cell divides into daughter cells and the contents recede from the cell wall and get converted into the zoospores. The zoospores may be ellipsoidal or ovoid or spherical in shape and have a single chloroplast, single nucleus, eye spot and a pyrenoid. After liberation they swim for some time, rest for some time and settle to bottom by their anterior end and lose their cilia. A thick wall forms and after rest, they divide forming a lower colourless rhizoidal cell and an upper cell which gives rise to the filament.

Sexual Reproduction

Gametes are produced on short laterals of the last order and are formed in the same way as zoospores from the cells. After gametes are formed they are liberated into the water by the gelatinization of their cell walls. They may be equal or unequal in size. They possess a single nucleus, with two equal cilia at the anterior end. There is a cup-shaped chloroplast, with a single pyrenoid in each cell.

Union of gametes takes place either isogamously or anisogamously by their anterior ends. After union the gametes lose their cilia, settles the zygote, encysts and then germinates after a period of rest. The zygote divides into two cells and the lower cell develops into the rhizoidal cell and the upper one into the filament proper.

COLEOCHAETE

Class : Chlorophyceae
Order: Chaetophorales
Family: Coleochaetaceae

Coleochaete, with about 10 species is a common epiphytic fresh-water alga. *C. scutata* occurs on leaves and stems of *Nymphaea*, *Typha*, etc. But *C. nitellarum* is endophytic on members like *Chara* and *Nitella*.

Structure (Figs. 9.32 & 9.33)

C. pulvinata shows *heterotrichous* habit that is some filaments lie prostrate on the substratum while others stand erect forming a projecting system which is cushion-like (Fig. 32). In *C. divergens*, the prostrate system alone is represented by a number of loose branched threads. In *C. scutata*, the prostrate system only is represented and the creeping threads which are radially arranged are laterally fused to form pseudoparerichymatous circular disc. It grows by means of apical (marginal) meristem. Branching is by forking of the apical cells or by production of lateral outgrowths.

The cells are uninucleate and possess a single laminate chloroplast, with one pyrenoid embedded in it. Some of the cells produce a single unbranched cytoplasmic seta with a cylinder of gelatinous material at the base, forming the sheath. The seta emerges through a pore in the cell wall. The setae readily break off leaving the basal sheaths.

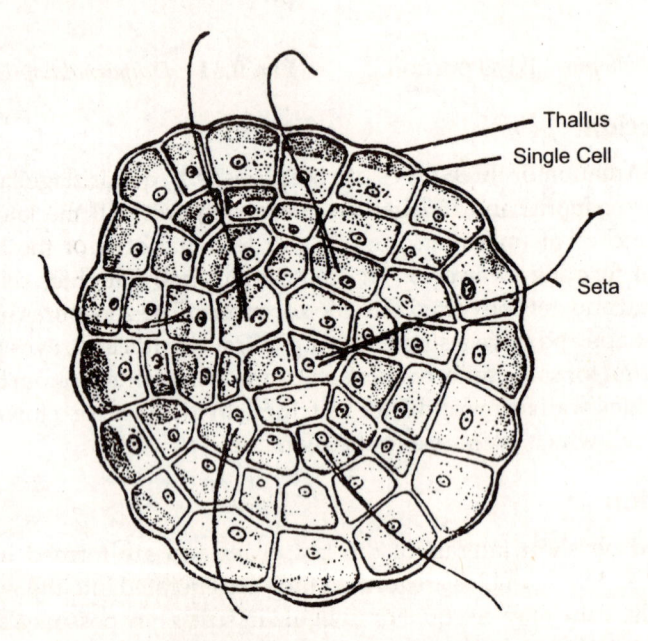

Fig. 9.32 *Coleochaete*—Habit.

Asexual Reproduction

It is by ovoid *biflagellate zoospores* produced singly from each cell. Each zoospore escapes out in an amoeboid manner through an apical opening formed on a papilla. The zoospore is peculiar in *lacking an eye spot* and the single chloroplast is laterally situated. The zoospore swims for some time, comes to rest and secretes a wall. It first divides transversely into two cells. The upper one forms a hair while the lower one divides in two directions, perpendicular to the substratum to form the disc (*C. scutata*). In heterotrichous forms, the zoospore first produces the branched prostrate system from which the erect system arises later (*C. pulvinata*). In *C. nitellarum*, the zoospore produces a tabular outgrowth which pierces the cuticle and then produces the filaments.

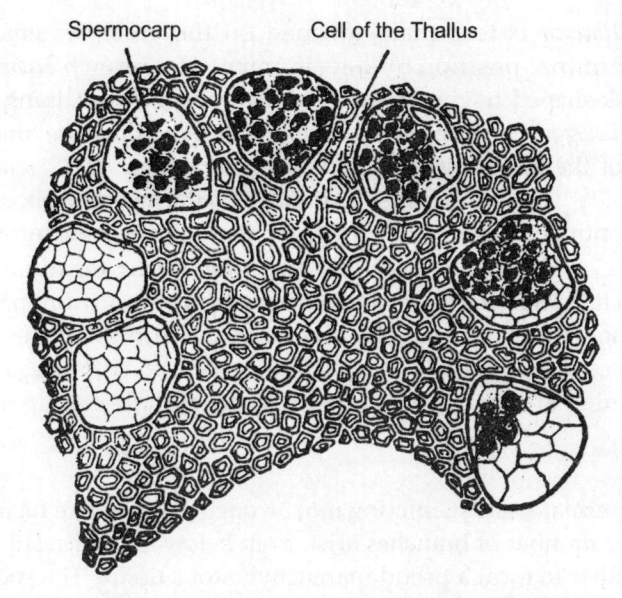

Fig. 9.33 *Coleochaete*—Thallus with spermocarp.

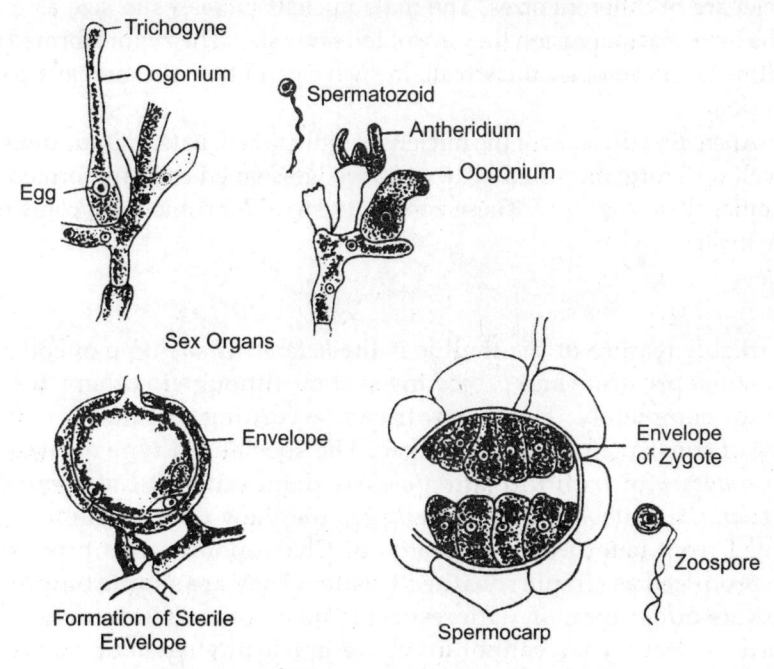

Fig. 9.34 *Coleochaete*—Sexual reproduction.

Sexual Reproduction (Fig. 9.33 & 9.34)

All species reproduce sexually but Lambart observed certain dwarf species reproducing purely by asexual method. The sex organs are antheridia and oogonia and sexual reproduction is by a specialised type of *oogamy*. The plants are heterothallic or homothallic. The oogonium, in *C. pulvinata* is terminally formed on the erect system. But it may be later displaced from its terminal position by development of a branch from the underlying cell. The oogonium is flask-shaped having a basal swollen protion containing the chloroplast and a narrow neck called *trichogyne* with colourless cytoplasm. When mature, the trichogyne breaks up and the protoplasm of the basal region rounds up into an egg. In *C. scutata*, the trichogyne is reduced to a short papilla. The oogonia arise terminally by the metamorphosis of the marginal cells. But the thallus may continue its marginal growth, so that they become intramarginal.

The *antheridia* of *C. pulvinata* are conical and they occur terminally in clusters on the projecting threads. The same filament which bears the antheridium may also bear the oogonium. In *C. scutata*, the antheridia are developed in the middle of the discoid thallus. Each antheridium produces a single oval or spherical green or colourless *biflagellate spermatozoid*. The spermatozoids are liberated when the antheridia break up apically.

Fertilization

This is affected by a spermatozoid swimming into an oogonium, where fusion takes place with the egg. Simultaneously, a number of branches arise from below and adjacent cells of the oogonium. They all coalesce together to form a pseudoparenchymatous tissue. The sheath thus formed takes up a reddish colour and the whole structure is termed by some algologists as *spermocarp* (Fig. 9.33). In discoid types, the investment is formed only on the side, away from the substratum. The male and female nuclei are of different sizes. The male nucleus increases in size as it approaches the female, but at the time of actual fusion they are of the same size. The zygote formed after the fusion, according to Oltmanns, develops a thick wall. In such a condition, the *spermocarp* crosses over the unfavourable period.

According to Allen, first division of the nucleus is reductional. Later mitotic divisions occur. This is followed by cell wall formation so that 16 or 32 wedge-shaped cells are formed. Each cell gives rise to a single biflagellate zoospore. These zoospores swim for some time, come to rest, and then give rise to new thalli.

Affinities

The most remarkable feature of the thallus is the *heterotrichous* type of construction. There is, in some, a distinct prostrate and projecting system although in others, the erect system is reduced partly or completely. This heterotrichy, according to Fritsch, is the most highly developed type of construction in green algae. The specialised type of *oogamous* reproduction and the *spermocarp* or fruit formation clearly point out that *Coleochaete* is much more evolved than *Ulothrix*. Fritsch, in fact, envisages the view that the higher land plants are probably derived from heterotrichous plants of Chaetophoraceous type. Hairs, found in *Coleochaete* are produced as simple sheathed bristles. They are simple outgrowths of the cell wall. Such hairs are quite common with respect to many other heterotrichous forms belonging to other orders. Hence, we cannot attribute much phylogenetic significance to these hairs.

The spermocarp formed in the life history of *Coleochaete* can be compared with the cystocarp of *Batrachospermum* and it represents the haploid phase. The life cycle of *Coleochaete* resembles the haplobiontic life cycle of *Batrachospermum*. Thus, there is a close resemblance between the sexual reproduction of *Coleochaete* and Red Algae (*Batrachospermum*). It may be an instance of parallel evolution along the *polyphyletic lines* (homoplasy).

OEDOGONIUM

Class : Chlorophyceae

Order : Oedogoniales
Family : Oedogoniaceae

Oedogonium is an unbranched filamentous form. Young filaments of *Oedogonium* are epiphytic on submerged leaves and stems of hydrophytes. In such cases, attachment is effected with the help of modified basal cell called *holdfast,* but older filaments get detached and float free on the surface of water.

Thallus Structure

The cylindrical cells are broader towards the upper end, and hence, there is a definite polarity between the base and apex. The cells of *Oedogonium* are uninucleate (Figs. 9.35 and 9.36) the chloroplast is a reticulate structure and the meshes or strands run parallel to each other along the axis of the cell. Pyrenoids are many and they are present at the intersecting points of the meshes. The pyrenoids accumulate starch in the form of plates. Sometimes, these starch plates may get distributed irregularly and obscure the structure of the chloroplast.

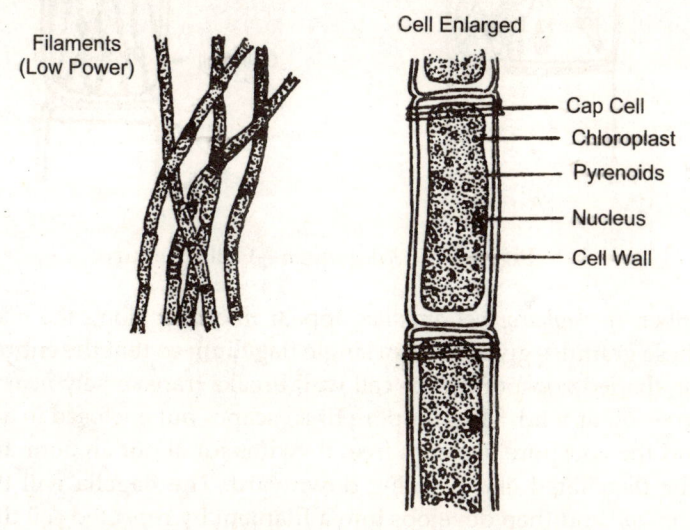

Fig. 9.35 *Oedogonium*—Thallus low and high power.

The cell wall except that of holdfast, consists of three portions. They are (*a*) an outer portion consisting of chitin, (*b*) a middle portion formed of pectin, and (*c*) an innermost portion consisting of cellulose.

Reproduction

It takes place by vegetative, asexual and sexual methods.

Vegetative. This takes place by accidental breaking of the filament into a number of portions which can grow into separate plants.

Asexual. It occurs by the formation of multiflagellate zoospore produced singly within a cap cell. Before zoospore formation, the whole protoplast contracts and, on one side, a hyaline region ap-

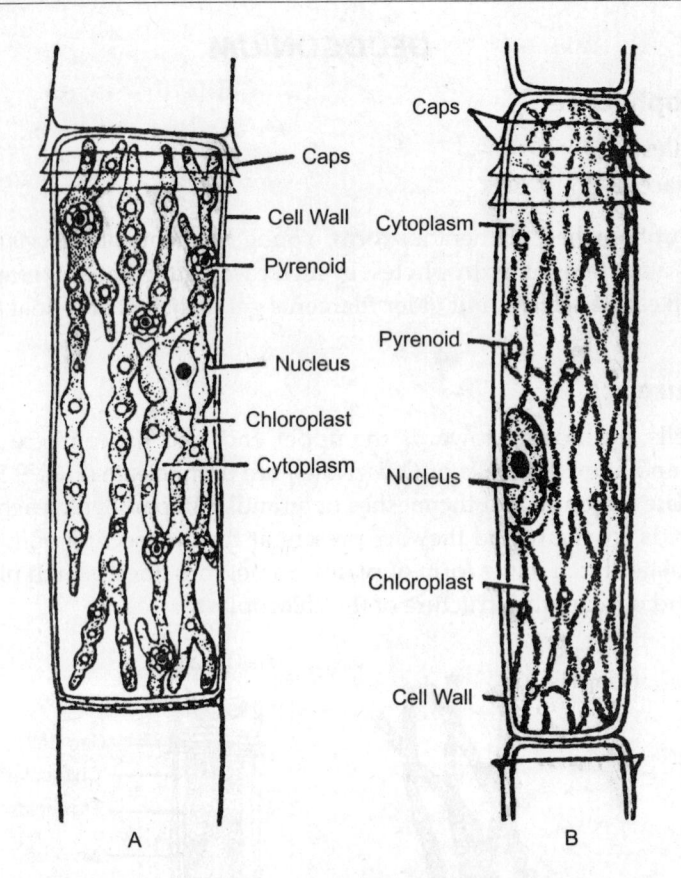

Fig. 9.36 *Oedogonium*—Cell structure.

pears. Then, a number of *blepharoplast granules* appear in a ring along the margin of the hyaline area. Each one of these granules gives rise to a single flagellum so that the entire structure becomes multiflagellate pear-shaped zoospores. The cell wall breaks transversely near the cap region and the upper parts comes off as a lid. The zoospore first escapes out enclosed in a vesicle. The vesicle soon disappears and the zoospore becomes free. It swims for about an hour and then settles on a substratum with the flagellated end pointing downward. The flagella will then be retracted. It develops a basal *hapteron* and then develops into a filament by repeated cell divisions.

In some species, *akinetes* are formed in chains: they germinate directly into new filaments.

Sexual reproduction The sex organs are *antheridia* (male) and *oogonia* (female) and they are unicellular structures. In *Oedogonium*, two kinds of species are distinguished depending upon the nature of the filaments bearing antheridia. In *macrandrous* species, the antheridia occur on filaments of normal size, but in the *nannandrous* species, the filaments bearing antheridia are few-celled and they are known as *dwarf* males (Fig. 9.37).

Macrandrous Species

In some species, filaments are *homothallic* bearing oogonia as well as antheridia. But in some, they are *heterothallic* having only antheridia or oogonia.

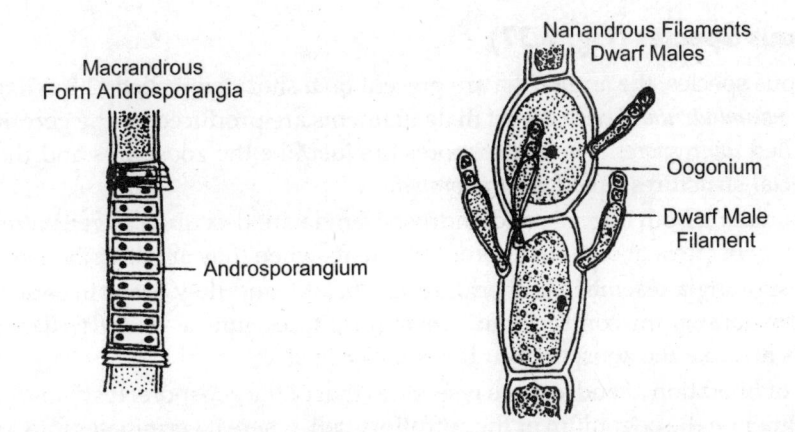

Fig. 9.37 *Oedogonium*—Sexual reproduction. Macrandrous and Nanandrous filaments.

Antheridia

The antheridia occur in series and their number varies from 2-40. They are terminal or intercalary in position. These are formed by successive divisions of an antheridial mother cell. The protoplast of each antheridium contracts and gets metamorphosed into an *antherozoid*. Sometimes, the protoplast may divide so the two antherozoids are formed. The wall of the antheridium breaks transversely and the spermatozoid first escapes into a vesicle. Soon, the vesicle disappears and the antherozoid becomes free. The antherozoid is multiflagellate and it is just like the zoospore but for its smaller size.

Oogonia (Fig. 9.38)

During the formation of *oogonium*, an oogonial mother cell divides transversely into two cells. The terminal cell remains as *suffultory* cell. In some species, oogonia may be formed in series, when the *suffultory* cell continues its activity as a oogonial mother cell (Fig. 9.38).

The oogonium is spherical and quite large and its protoplast rounds up into a single egg. Just before fertilization, the wall of the oogonium breaks and forms a pore on one side so that the antherozoids may enter. The egg develops a hyaline receptive spot before fertilization.

The antherozoid swims into the oogonium through the opening and enters the egg at the receptive spot. Thus, sexual union is *oogamous*. The male and female nuclei fuse to form a diploid zygote. It accumulates a reddish oil and secretes a thick wall. The wall of the zygote may be smooth or may show various designs. The zygote comes out, when the oogonial wall decays. It may undergo rest for a year or so, before it begins to germinate. The nucleus of the zygote undergoes meiosis resulting into four haploid nuclei which transform into *zoospores*. They are liberated into a vesicle when the wall of the zygote ruptures. This vesicle disappears and the zoospores become free.

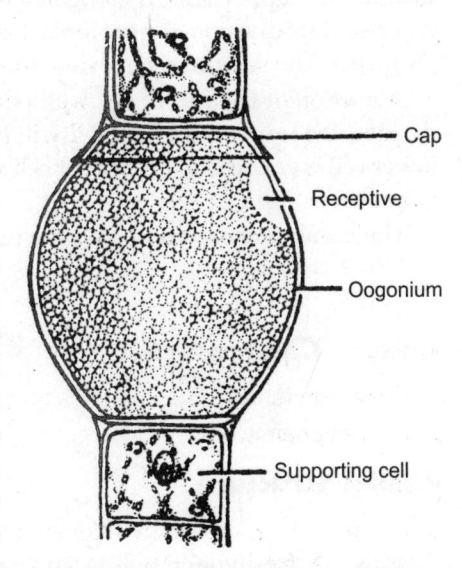

Fig. 9.38 *Oedogonium*—Oogonium.

In heterothallic species, two develop into the male filaments, and the other two develop into the female filaments on germination.

Nannandrous Species (Fig. 9.37)

In nannandrous species, the antheridia are present on a short few-celled (2-3 cells) filament, called *dwarf male* or *nannandrium*. These dwarf male filaments are produced by the germination of special structures called *androspores*. The androspores are just like the zoospores and they are produced singly in special structures called *androsporangia*.

Species producing both oogonia and androsporangia are described as *gynandrosporous*. In some others, they may be present on two different filaments when they are described as *idioandrosporous*.

The androsporangia resemble antheridia superficially and they occur in series. The protoplast of each androsporangium contracts and metamorphoses into a *multiflagellate androspore*. The androspore is just like the zoospore but it is smaller in size.

The mode of liberation of androspore resembles that of the zoospore. It swims for some time and then settles either on the oogonium or the suffultory cell, where it germinates into a few-celled *dwarf male* or *nanandrium*. This dwarf male consists of a lower stalk cell or stipe and one or two *antheridia*.

Each antheridium produces two multiflagellate antherozoids. The rest of the life history (fertilization and zygote germination) is same as in macrandrous species.

Algologists believe that nannandrous species have evolved from macrandrous species which are heterothallic and that the male filaments of heterothallic species got reduced in size, and gave rise to dwarf males. The *androsporangia* are supposed to have evolved from antheridia.

Cap-cell Formation

In *Oedogonium*, cell division is very characteristic and it may be intercalary or terminal. Only the holdfast is incapable of cell division.

Before cell division, a *transverse ring* of *thickening* appears just below the septum, at the upper end. It consists of *cellulose* and *amyloid*. The ring develops by intussusception. The nucleus migrates towards the upper end. It elongates first, and then divides mitotically into two daughter nuclei. After nuclear division, the protoplast also divides by transverse furrowing between the two daughter nuclei. The septum clearly demarcates the two daughter cells.

A portion of the parent cell wall persists as a cap on the upper daughter cell. Thus, when a cell divides, the upper daughter cell will have an *apical cap* and the lower one will be without it. The lower cell is encased in the parent cell wall which forms a *sheath*. The upper cell with the apical cap is known as *cap cell*.

The number of caps indicates the number of times a cell has divided.

ZYGNEMA

Class : Chlorophyceae

Order: Zygnemales
Family: Zygnemaceae

Thallus Structure

The unbranched, uniseriate filaments of *Zygnema* are found free-floating, in freshwater ponds and streams. However, in the young stages, they may be attached (Fig. 9.39). All the cells of a filament are essentially alike, and there is no differentiation into base or apex. The cells are longer than broad, cylindrical and continuously arranged end to end. The cell wall is thin and it is

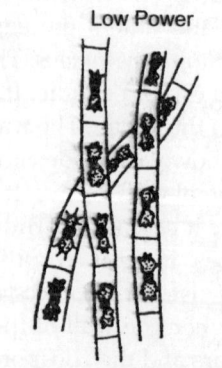

Fig. 9.39 *Zygnema*—Thallus in low power.

made up of *cellulose;* usually, on the outer side, it is provided with a mucilage (pectose) sheath, due to which the alga is slimy to touch. This mucilage sheath, shows a fibrillar structure.

The protoplast of each cell shows two *axile stellate* chloroplasts, lying on the longitudinal axis (Fig. 9.40). Each chloroplast has a *central pyrenoid*, and the peripheral portion is drawn into radiating processes. The nucleus is in the middle of the cell between the two chloroplasts, embedded in a massive strand of cytoplasm.

Cell Division

All the cells of *Zygnema* are capable of cell division and hence growth is not localised. In this process, the nucleus first divides into two; then the cell gets divided into two daughter cells such that each cell is having one daughter nucleus and one chloroplast of the parent cell. The nucleus, at this stage, is at the side of the chloroplast. The chloroplast then divides and the pyrenoid also accomplishes division. When the two daughter chloroplasts are thus well formed, the nucleus takes up its usual position midway between the two axile stellate chloroplasts.

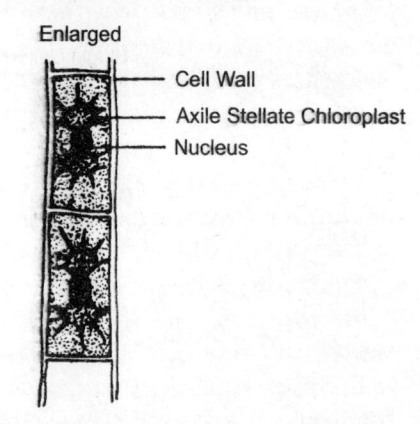

Fig. 9.40 *Zygnema*—Cell structure.

Reproduction

The cell division is only useful in the growth of the filament, but it is not helpful in multiplication of the filaments. *Zygnema* reproduces very commonly by vegetative and sexual methods, and asexual reproduction is absent.

Vegetative reproduction. *Zygnema* reproduces in a prolific manner by fragmentation. Fragmentation involves the breaking of a filament into several pieces, and each piece by cell division and enlargement, develops into a separate filament.

In cultures, some species are reported to multiply by akinetes or aplanospores. In *Z. sterile*, akinetes are formed. In *Z. giganteum* akinetes are formed in a chain. Aplanospores are formed in Z. *spontaneum*.

Sexual reproduction or conjugation. Sexual reproduction takes place readily in *Zygnema*, and it occurs in certain seasons only. The process of conjugation takes place usually after a period of vegetative activity. All the cells are potentially capable of taking part in reproduction and the vegetative cell itself acts as gamete-producing cell (gametangium).

Scalariform conjugation (Fig. 9.41). In scalariform conjugation (ladder-like), tubular connections are established between the opposite filaments. At the commencement of the process, two filaments of *Zygnema* come together and lie parallel to each other althrougt the length. Then, the opposite cells produce papilla-like outgrowths which grow into cylindric peg-like structures. They come into contact by growing towards each other. After, this the end walls at the point of mutual contact dissolve, and an open passage is formed from one cell to the other by establishing a conjugation tube.

The protoplast of each cell rounds up into an *amoeboid gamete* which is non-motile (aplanogamete). In some species (Z. *pectinatum)*, the opposite gametes are exactly similar, both in structure (morphological) and activity (physiological). The two gametes fuse together in the conjugation canal, midway between the opposite cells taking part in conjugation. The zygote results by *isogamy*.

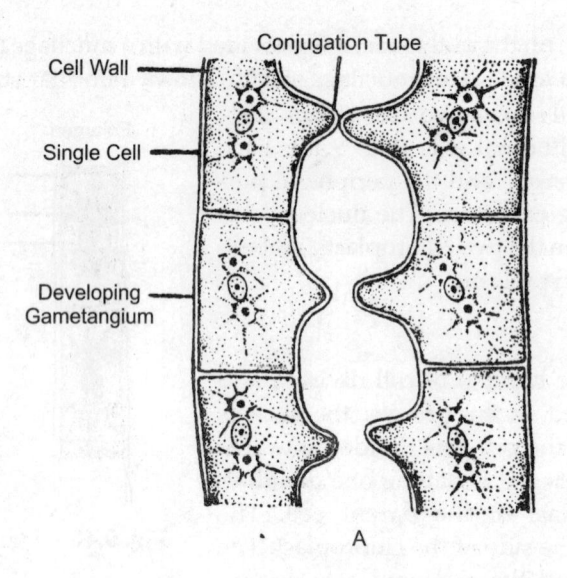

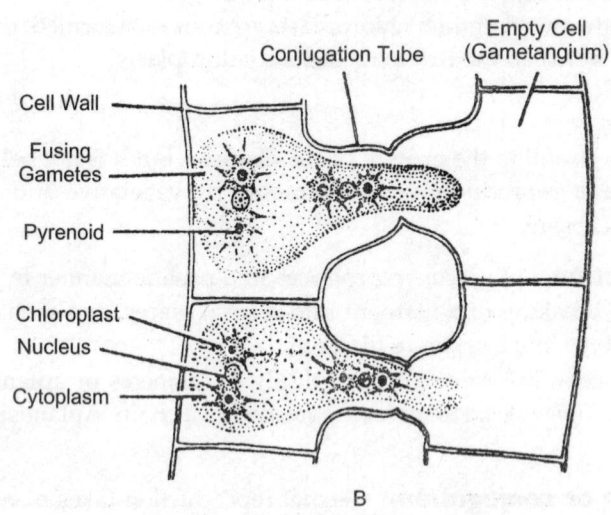

Fig. 9.41 *Zygnema*—Conjugation.

Some species show physiological anisogamy in which the male gametes are active and the female ones are passive. The male gamete slowly migrates into the opposite cell and fuses with the female gamete. The zygote in such a case, is formed in the femal cell. The highest type of sexual differentiation is reported in *Z. stellinum*, where the female cells which are longer than the male ones show bigger chloroplasts and pyrenoids.

Lateral conjugation. Lateral conjugation is not so common as scalariform conjugation.

In *Z. pectinatum,* the same filament shows scalariform conjugation as well as lateral conjugation. Lateral conjugation takes place by the establishment of a conjugation tube between two cells of the same filament, lying immediately one above the other. The two gametes fuse in the middle of the passage, to form the zygote.

Zygospores. The zygospores are either spherical or ellipsoidal in shape, having a thick cellulose wall distinguished into three layers. It accumulates a large amount of fat as reserve food. In isogamous forms, the chloroplasts of the two gametes retain their identity in the zygote, but in others (anisogamous forms), the chloroplasts of male gamete usually disintegrate. The union of the two nuclei is often delayed, even after cytoplasmic union. The zygotes are liberated by the decay of the cells, in which they are lodged.

Zygote germinates after a period of rest. The diploid zygote nucleus undergoes *reduction division* and forms four *haploid* nuclei. Three of them degenerate and only one remains withing the zygote. The outermost layers or the zygote wall burst at one end and the inner layer grows out as a long tube which undergoes division into two cells. These cells, by repeated divisions, give rise to a filament.

Azygospores or parthenospores. The contents of a cell of *Zygnema* may get rounded up and get surrounded by a thick wall giving rise to the azygospores of parthenospores. These parthenospores bear a close resemblance to the zygote. *Z. oudhensis* (Indian species) shows in the same filament, formation of zygotes as well as azygospores. The azygospores are formed from the gametes, which fail to fuse together.

SPIROGYRA

Class : Chlorophyceae
Order: Zygnemales
Family: Zygnemaceae

Spirogyra is strictly a freshwater form and the filaments are found as free-floating tangled masses.

Structure (Fig. 9.42)

Spirogyra filaments are simple and unbranched consisting of a row of cylindrical cells arranged end to end (Fig. 9.42). The cells are longer than broad. All the cells are alike and so, there is no distinction into base or apex.

The cells have cellulose cell wall (Fig. 9.43), cell wall is covered with mucilage sheath which makes the filament slimy to touch. The protoplast of each cell contains cytoplasm, a single nucleus and one or more chloroplasts. Streaming movements are seen in some species. The nucleus is

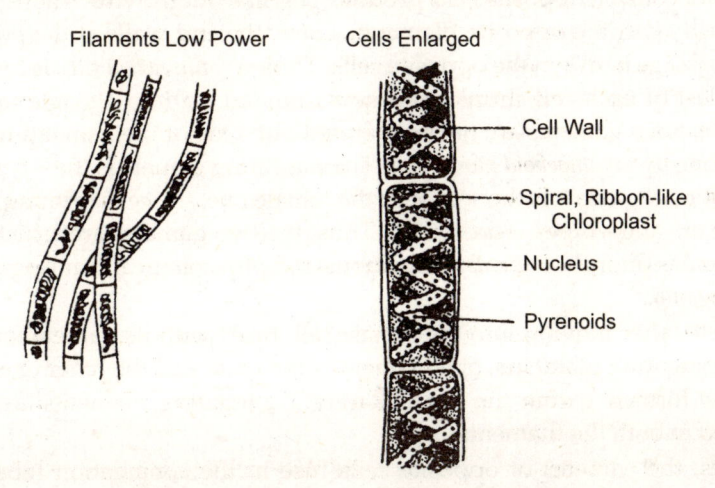

Fig. 9.42 *Spirogyra*—Thallus structure.

suspended at the centre by cytoplasmic strands. The chloroplasts are ribbonshaped and spirally twisted, and they show several pyrenoids embedded in them, along their length. The chloroplasts are narrow with a smooth margin or broad with serrated margins with scattered pyrenoids. The cell sap is rich in tannins. All cells of the filament are capable of cell division. Cell division is followed by elongation of daughter cells resulting in the growth of the filament.

Reproduction

In *Spirogyra*, multiplication takes place by vegetative or sexual methods of reproduction. *Asexual reproduction is completely absent.*

Vegetative reproduction. Long filaments of *Spirogyra*, while floating, get broken up into small parts or fragments by accident. Each piece grows into a separate *Spirogyra* filament. This is known as fragmentation. Species with replicate septa readily multiply by fragmentation.

Sexual reproduction. Sexual reproduction takes place by means of *conjugation*. The conjugation, taking place between two cells of opposite filaments, is known as scalariform conjugation. Lateral conjugation takes place between two adjacent cells of the same filament. The ladder-like or scalariform conjugation is very common in most of the species of *Spirogyra*. Lateral conjugation is rather not so very common, but in *S. affinis* and *S. tenuissima* it is almost a rule. Sometimes, the same filament may show both types of conjugation.

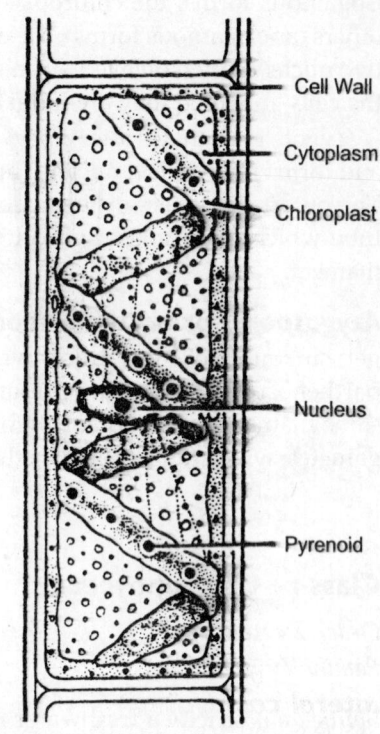

Fig. 9.43 *Spirogyra*—Cell structure.

Scalariform Conjugation (Fig. 9.44)

During scalariform conjugation the filaments lie parallel to each other in a common mucilage (Fig. 9.44). At the point of contact, opposite cells produce peg-like outgrowths. They grow into tube-like structures and push apart the opposite filaments. Later, the end walls at the point of contact dissolve, forming a passage between the opposite cells. Thus, a *conjugation tube* is formed. Meanwhile, the entire protoplast of each cell shrinks and gets rounded up into a single motionless gamete, called *aplanogamete*. Such gametes are never liberated out. One of the gametes moves from its cell into the opposite one by an *amoeboid movement*. The migrating gamete is the – type or the male one and the other stationary gamete is the + type or the female one, the cells forming them being called as male or female or –and +types respectively. Thus, the two gametes are structurally similar but functionally dissimilar (morphologically isogamous and physiologically anisogamous). The union is described as *isogamy*.

The male gamete, after moving into the female cell, fuses with its gamete forming the zygote. Usually, in the conjugating filaments, one filament is the male and the other one is the female and all the zygotes are formed within the latter. Rarely, conjugating filaments are homothallic and zygotes are formed in both the filaments.

In some species, the gametes of opposite cells fuse in the conjugation tube and so they are morphologically and physiologically isogamous.

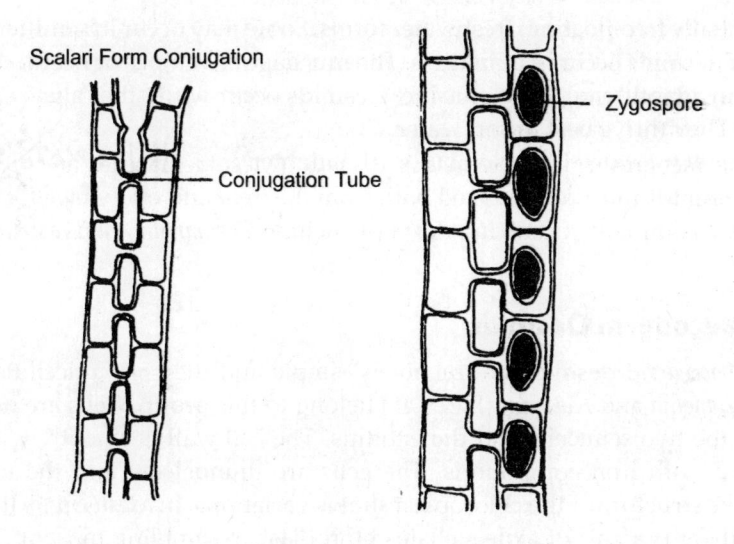

Fig. 9.44 *Spirogyra*—Scalariform conjugation.

Lateral conjugation. At the septum, a lateral protrusion is produced by cell of a filament, which are immediately present one above the other. The septum dissolves and a passage is established between adjacent cells. Male gamete moves into the female cell through the conjugation tube and fuses with its gamete, forming the zygote. After lateral conjugation, a female cell having a zygote is always found to be adjacent to an empty male cell.

Zygote. In a *zygote*, cytoplasm of both gametes first mingle together and this is followed by delayed nuclear fusion. Zygote stores food and develops a thick wall to cross over the unfavourable period. The zygote accumulates a large amount of fat.

Germination of zygote. Zygotes are released by the decay of cell walls. It germinates when conditions are good. The diploid nucleus (2x) of zygote undergoes divisions twice, of which the first is reductional. This results in four haploid nuclei, but only one remains and the others degenerate. The outer thick wall of the zygote bursts and the inner layer grows into a long tube which undergoes transverse division into two cells. The lower cell contains scanty cblorophyll and it develops into a rhizoid which remains in the zygospore-membrane for some time. The upper cell continues to divide so that a filament is formed.

Sometimes, conjugation may not lead to fusion of gametes. In such cases, the gamete rounds off and secretes a membrane resulting in azygospore (*S. mirabilis*).

DESMIDS

Class : Chlorophyceae
Order: Conjugales
Family: Desmidiaceae

Occurrence

Desmids are essentially free-floating, freshwater forms. Some may occur in semi-terrestrial habitats. A good number of desmids occur in planktons. The mucilage envelopes or needle-like processes on the cell wall help in adapting to planktonic life. Desmids occur with other algae on the margins of ponds, pools, etc. They thrive best in soft water.

Desmids can be recognised into *Saccoderm desmids* (Mesotaeniaceae) and *placoderm desmids* (Desmidiaceae). Desmids must be observed both from the *front* and end views for determining the species. Some of the commonly occurring desmids include *Cosmarium, Staurastrum,* and *Euastrum* etc. (Fig. 9.45).

Structure of Saecoderm Desmids

The members of *Saccoderm* desmids are relatively simple and they are unicellular. *Mesotaenium, Cylindrocystis, Spirotaenia* and *Netrium* (Fig. 9.46) belong to this group. Cells are normally rod-like and do not show the two semicells and the isthmus. The cell wall is smooth, without pores and never impregnated with iron compounds. The cells are uninucleate and the nucleus is in the middle of a cell. The structure of the chloroplast shows variations. In *Mestoenium* it is an axile plate; in *Cylindrocystis,* there is a pair of axile stellate chloroplasts resembling those of *Zygnema,* and in *Spirotenia,* it is a spirally-twisted, parietal ribbon-shaped structure like that of *Spirogyra.* Pyrenoids are variable in number.

Amorphous colonies may be formed by species of *Mesotaenium* and *Cylindrocystis* growing on terrestrial substrate due to a common envelope of mucilage formed by gelatinisation of the outer layer of the wall.

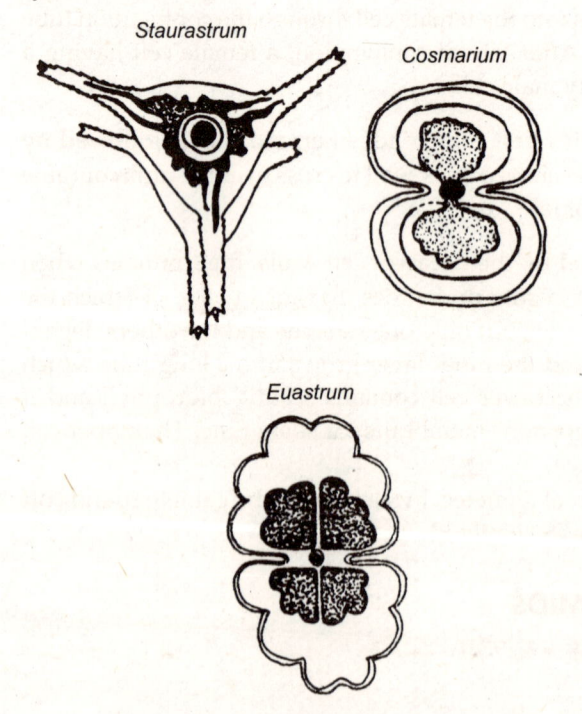

Fig. 9.45 *Desmids*—Structure.

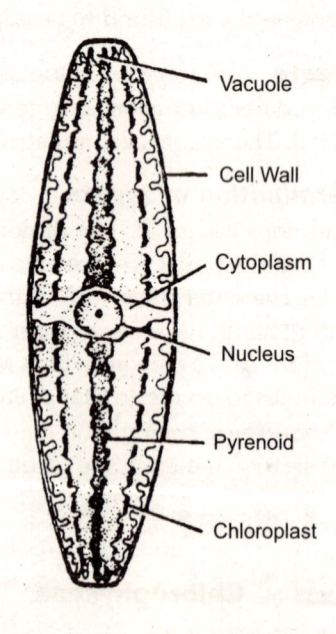

Fig. 9.46 *Netrium* structure.

Structure of Placoderm Desmids

Placoderm desmids constitue the true desmids and they are in all, about, 2,500 species. The bewildering diversity of form observed in these species and the remarkable variations of the chloroplasts are the unique features. Most of them are called *semicells* which are mirror images of each other. The gap between the two semicells is known as the *sinus*. In *Closterium*, it is semilunar in shape with pointed ends and in *Penium*, cells are cylindric.

Some are cylindrical *(Pleurotaenium)* and others are disciform *(Micrasterias)*. Unicellular desmids of this group show great variation in structure and ornamentation of wall e.g., *Cosmarium, Closterium, Xanthidium, Microasterias*, etc (Fig. 9.47). The wall may be smooth or may bear granules, warts, spines or other protruberances in a characteristic manner. In *Xanthidium* forked spines are present at the angles.

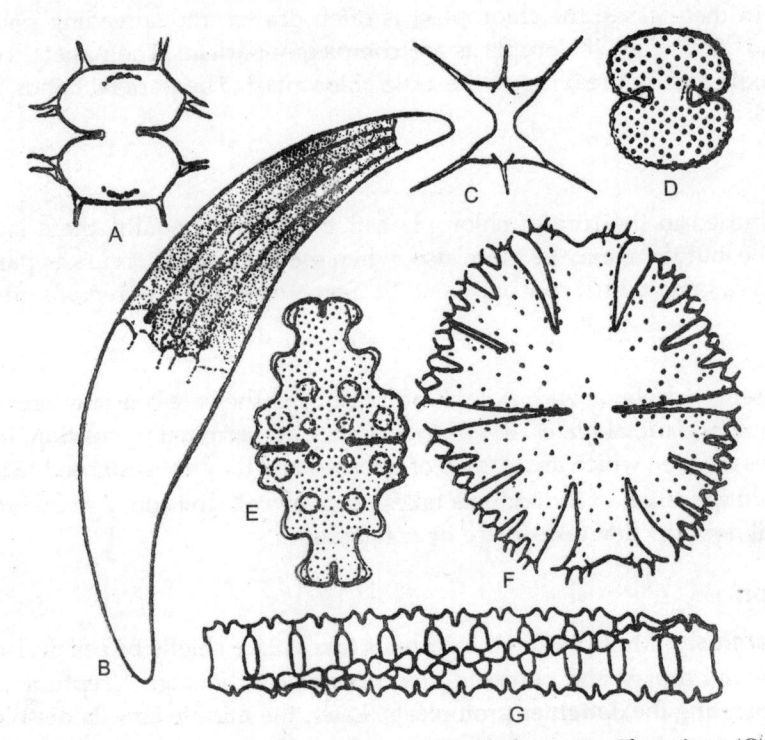

Fig. 9.47 Range of structure in Desmids—*(A) Xanthidium, (B) Closterium, (C) Staurastrum, (D) Cosmarium, (E) Euastrum, (F) Micrasterias, (G) Desmidium.*

Some *placoderm* desmids are *colonial*. The colonial forms may be filamentous as in *Spondylosium* where a number of *Cosmarium*-like cells are united into a chain. *Desmidium* also is filamentous. In *Cosmocladium*, it is an amorphous colony in which a number of irregularly placed *Cosmarium*-like cells are united together by mucilage threads.

Cell Wall

In *Saccoderm* desmids, it is formed of a single piece. But in *placoderm* forms the wall consists of two halves which overlap and fit in together in the middle region of the cell. These halves come apart during conjugation. The cell wall is formed of cellulose impregnated with iron, often with an outer gelatinous sheath. In amorphous colonial forms, the mucilage is present in abundance. The mucilage is secreted through mucilage pores in the cell wall. In Saccoderm desmids, the cell wall is not impregnated with iron and mucilage pores are absent. In most of the placoderm desmids pores are present, except in the isthmus region. The mucilage pore is a narrow canal in the inner layer of the wall, but it enlarges into a cylindrical collar in the outer layer.

The cells are uninucleate and the nucleus is cradled in a mass of cytoplasm in the isthmus region.

Chloroplast Structure

Chloroplasts show variation in number and structure. In placoderm desmids, generally, a single chloroplast is present in each semicell e.g., *Cosmarium* spp. Typically, the chloroplast is axile. It may extend into radiating arms into the angles of semicell.

Several species of *Cosmarium*, show a tendency of development of the axile chloroplast towards parietal nature. In these cases, the chloroplast is often drawn into spreading parietal plates. In *Pleurotaenium* and *Xanthidium*, chloroplasts are completely parietal. The parietal condition is derived from the axile one by breaking of the axile chloroplast. The parietal bands may be few or several.

Pyrenoids

The *pyrenoid* is buried in the core of chloroplast at the centre. Usually, there is one when the chloroplast is axile, but there may be more also, when the chloroplast occurs as parietal plates. In *Closterium*, there is a longitudinal row of pyrenoids. Sometimes, these pyrenoids are scattered.

Vacuoles

These may be absent in desmids with axile chloroplasts. But they are centrally present when chloroplasts are present as parietal plates. In *Closterium*, they are terminal in position. In the vacuoles, vibrating granules are seen which are formed of gypsum and they are supposed to act as *statoliths*.

In desmids, multiplication of individuals takes place by cell division. Asexual reproduction is absent and sexual reproduction takes place by conjugation.

I. Cell Division

Saccoderm desmids. Multiplication of desmids takes place chiefly by cell division. Generally, the division of the chloroplast and pyrenoid precedes nuclear division. A septum arises as a ring-like ingrowth separating the daughter protoplasts. Later, the middle lamella dissolves so that the two daughter cells separate.

Placoderm desmids. In these desmids, cell division is more complicated than in Saccoderm desmids. Several forms like *Cosmarium, Euastrum* having a marked constriction, show a different pattern. The isthmus region between the two semi-cells elongates, so that the two semi-cells are slightly moved apart. Then the nucleus divides into two. A septum develops in the middle of the isthmus and then splits up into two. This results in the formation of two individuals from the parent, each having an old semicell and a portion of the isthmus attached to it which enlarges into the other semicell. The chloroplast now enlarges and extends from the old semicell into the new one. Then it divides followed by the division of pyrenoids. As the new semicell fully develops, structural peculiarities of the cell wall develop, characteristic of the species. Often the cells formed after division, fail to separate giving rise to a *pseudo-filamentous* structure (*Cosmarium moniliforma*). If growth conditions are very favourable, cell division may take place even before the complete formation of, semicells.

Asexual reproduction is absent as in all conjugales. But in *Hyalotheca* and *Spondylosium*, aplanospore formation is recorded.

II. Sexual Reproduction

In desmids, sexual reproduction takes place by *conjugation* as in all other conjugales.

Sacoderm desmids. It takes place typically by *conjugation*. During this process, two individuals lie in a parallel manner in a common mucilage. They produce peg-like processes which meet end to end. A conjugation tube will be established when the separating end-walls dissolve away. In some cases (*Netrium*), the conjugation processes arise at the point of contact and the conjugating cells are pushed apart as the process elongates. By this time, the entire protoplast of each individual contracts into a non-flagellate *aplanogamete*.

The two aplanogametes will fuse in the conjugation tube (isogamy). As the *zygote* enlarges, the conjugation tube may widen and lose its identity (*Cylindrocystis*).

In *Spirotaenia*, two individuals lie in contact. Each individual produces two aplanogametes. This is followed by the gelatinisation of the cell wall and fusion of aplanogametes of opposite cells in pairs. Due to this, two zygotes will be found in a common mucilage. There is no formation of conjugation tube.

Placoderm desmids. In *Placoderm* desmids, only in a few genera, conjugation takes place by the formation of distinct tubes (*Closterium, Hyalotheca*). However, in genera where there is a marked constriction between the semicells, (*Staurastrum*) two individuals come together (parallel or inclined) and lie together in a common envelope of mucilage. The angle of approach, and the method of contacting varies according to the genus. Each individual breaks at the isthmus region and the two semicells move apart. The two protoplasts meet midway between the individuals and fuse together to form the zygote.

In *filamentous* forms, usually, there is dissociation into separate cells and conjugation takes place between cells derived from two different filaments. Some of the filamentous forms show definite conjugation canals. In *Desmidium cylindricum* zygote is formed in the female cell. Rarely, in some filamentous forms, lateral conjugation between two adjacent cells is observed (*Spondylosium*). Double zygospores are seen in *Closterium lineatum*, in which two gametes are formed from each individual.

Zygote. The fusion of the gametes is not immediately followed by the union of the two nuclei. The two nuclei fuse just before germination. During the time of fusion, the chloroplasts are not clearly distinct.

In some forms, during the formation of zygote, of the two chloroplasts contributed by the two gametes, one may disintegrate; in others, where four chloroplasts are contributed by the two gametes, two disintegrate during maturation of zygote.

In Saccoderm desmids, the zygote has usually a smooth thick wall of three layers. In placoderm desmids, zygote is usually ornamented. It undergoes a period of rest. During germination, it undergoes *meiosis* and forms 4 haploid nuclei. Usually, in *Saccoderm desmids* all the four nuclei are functional so that four individuals are formed. But in *Netrium*, of the 4 nuclei, two degenerate and so, only two cells are formed.

In *Placoderm* desmids, usually two individuals are produced from a zygote. Rarely four individuals are produced (*Staurastrum* sp.) and in a species of *Hyalotheca* only one individual is formed around a functional nucleus, and the other three nuclei disintegrate.

CHARA

Class : Chlorophyeae
Order : Charales
Family : Characeae

Occurrence

Chara with about 90 species, is widely distributed in standing waters. It occurs in submerged ponds and pools forming an extensive growth. The plant is attached to the muddy bottom of the pond or pool by means of rhizoids. They occur in waters which may be deficient in oxygen. Mostly, they are freshwater forms; however, a few are marine forms.

The meadows of *Chara* plants, growing under the waters often form calcareous deposits, since the plants are abundantly encrustated with Calcium Carbonate.

Thallus Structure

The plant is much branched and it is anchored to the substratum by means of branched uniseriate filamentous *rhizoids*. The erect axis grows indefinitely by means of a dome-shaped apical cell. The plant is distinguished into internodes and nodes (Fig. 9.48). The internode is constituted by a long cylindric cell encircled by a single layer of *cortical* cells which are much smaller in diameter. At the node, a pair of central cells are surrounded by a peripheral group of 6-20 cells (Fig. 49). From the node, short branches of limited growth, otherwise called primary laterals or 'leaves' (6 to 20) arise in whorls giving the plant an *Equisetoid* look. The short branches or 'leaves' are also constituted by nodes and internodes (about 3 to 6 nodes) but its internodal cells are comparatively short. Branches of this type do not grow indefinitely as the apical cell ceases its activity by assuming the form of an

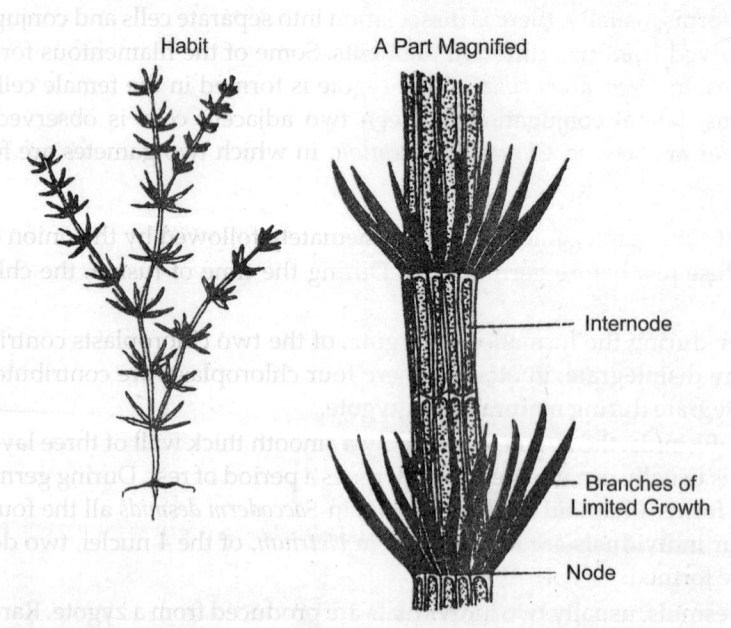

Fig. 9.48 *Chara*—Habit and part of thallus magnified to show nodes and internodes and branches of limited growth.

elongated pointed structure. Branches of unlimited growth may also arise from the axils of these 'leaves'.

Generally, a branch arises in the axil of the oldest lateral belonging to a whorl.

Apical Growth of Thallus

Branches of unlimited growth grow by means of an *apical domeshaped* cell (Fig. 9.50). If the apical cell is damaged, growth is continued by the uppermost short lateral which becomes continuous with the main axis. The apical call cuts off a derivative cell (D.C.) towards the posterior side. The derivative cell divides transversely into a lower internodal initial and an upper nodal initial. The lower internodal initial without division, elongates and develops into the internode.

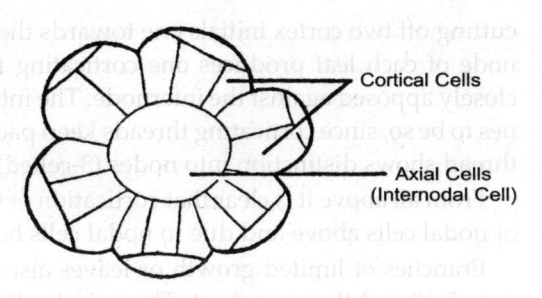

Fig. 9.49 *Chara*—T.S. of internode.

In the nodal initial, the first division is vertical so that two semicircular cells are formed. Subsequent divisions are in curved planes and intersecting so that 6-20 peripheral cells are formed surrounding a pair of central cells. Each peripheral initial divides into an apical initial that gives rise to the 'leaf' and a basal nodal cell. So, the number of leaves produced is equal to the number of peripheral cells.

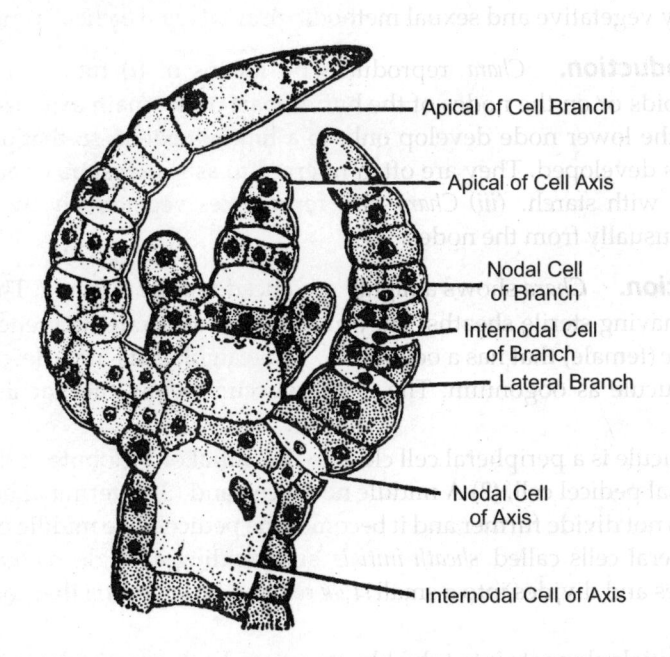

Fig. 9.50 *Chara*—Vertical section of thallus apex.

Cortication

The internodal cells of the main axis, as well as those of leaves are ensheathed by a single layer of cortical cells and is known as *cortication* (Fig. 9.49). Each developing leaf of a node has its basal cell cutting off two cortex initials one towards the apex of thallus and another towards base. The basal node of each leaf produces one corticating thread upwards and another downwards which lie closely apposed against the internode. The internode is corticated from the beginning and it continues to be so, since corticating threads keep pace in growth along with the internode. The corticating thread shows distinction into nodes (3-celled) and internodes (1-celled).

From all above it is clear that cortication of the upper half of internode is due to the development of nodal cells above and due to nodal cells below.

Branches of limited growth or leaves also grow just like the main axis. But the internodes are only 5-10 and they are short. The apical cell becomes pointed and stops its activity so that these primary laterals or leaves are limited in growth. At the nodes of the leaves, single-celled branches occur and they are known as *secondary laterals* or *stipules*.

Cell Structure

The apical cell contains dense cytoplasm and a central nucleus. In the nodal cell, a single nucleus and several chloroplasts are present in the dense cytoplasm. But in the internodal cell, there is a large central vacuole and several nuclei and many discoid chloroplasts occur in the peripheral layer of cytoplasm. The chloroplast lacks pyrenoid. The cytoplasm shows rotation or streaming around the vacuole.

Reproduction

Chara reproduces by vegetative and sexual methods. *Asexual reproduction* is entirely absent.

Vegetative reproduction. *Chara* reproduces by means of (*i*) tuberous structures (bulbils) formed on the rhizoids or on the nodes of the buried part of the main axes. (*ii*) In some cases, the peripheral cells of the lower node develop only to a limited extent, so that an aggregate of cells looking like a star is developed. They are often referred to as starch stars or amylum stars as they are copiously filled with starch. (*iii*) *Chara* also reproduces vegetatively by *protonema*-like outgrowths, produced usually from the nodes.

Sexual reproduction. *Chara* shows a highly advanced type of oogamy. The sex organs have a complex structure having sterile sheaths. They are the *globule* (male) that enclose the antheridial filaments and *nucule* (female) that has a oogonium. Some algologists have described the globule as antheridium and nucule as oogonium. The globule occurs below the nucule at the base of the laterals (Fig. 9.51).

The initial of a nucule is a peripheral cell close to the initial of a globule. It divides into a row of three cells: (1) A basal pedicel cell, (2) A middle nodal cell and (3) A terminal oogonial mother cell. The lowest one does not divide further and it becomes the pedicel. The middle cell or the nodal cells cuts off five peripheral cells called *sheath initials* surrounding a single *central cell*. The oogonial mother cell elongates and divides into a small *stalk cell* and an *oogonium* that contains a single large egg (Fig. 9.54).

The five sheath initials elongate into tube-like structure. Each one divides transversely forming a small terminal *coronal cell* and elongate *tube cell* (Fig. 9.52). Thus, the crown will have five coronal cells. The five tube cells coil spirally around the oogonium forming a sterile sheath. The oogonium accumulates a large amount of reserve foods. At the time of fertilization, tube cells separate so that five slits will be formed in between the coronal cells for the entry of sperms. Smith prefers the term *nucule* due to the presence of sterile sheath.

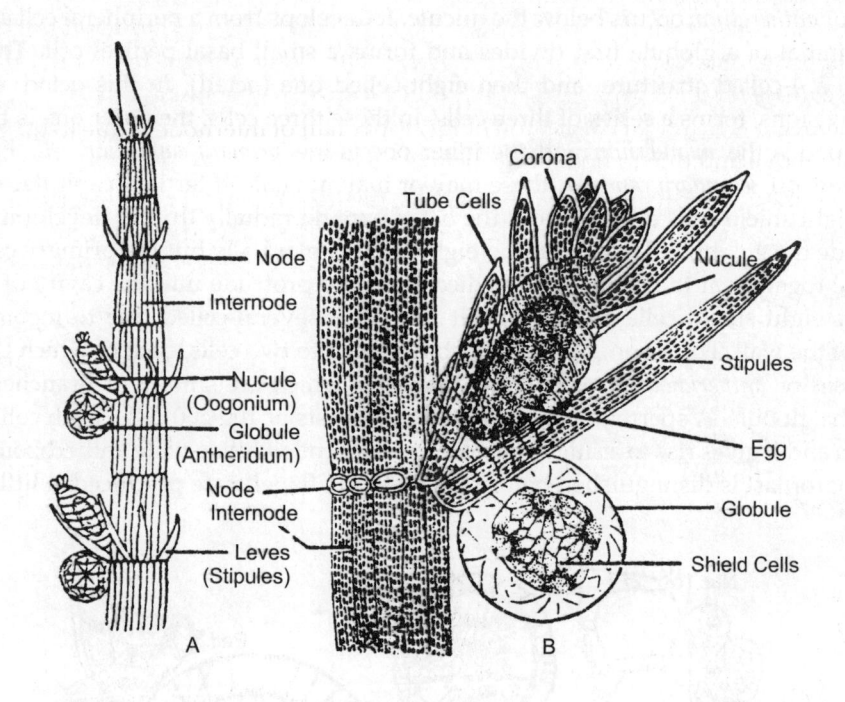

Fig. 9.51 *Chara*—Sex organs detailed structure.

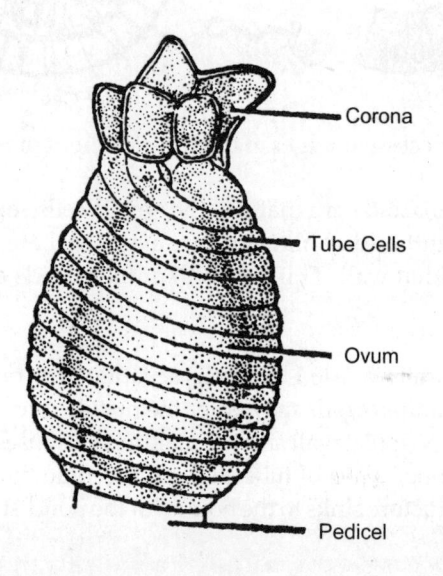

Fig. 9.52 *Chara*—Oogonium (nucule) structure.

Globule Structure (Figs. 9.53, 9.54 & 9.55)

The *globule* or *antheridium* occurs below the nucule. It develops from a peripheral cell at the node of a leaf. The initial of a globule first divides and forms a small basal pedicel cell. The upper cell divides into a 4-celled structure, and then eight-celled one (octad). In this octad, each cell, by periclinal divisions, forms a series of three cells. In these three cells, the outer one is the *shield cell*, the middle one is the *manubrium* and the inner one is the *primary capitulum cell*. Each primary capitulum cuts off *secondary capitula* (these may or may not cut off tertiary capitular cells). Meanwhile, the eight shield cells derived from the octad expand radially during development forming cavities inside the golobule. The manubria (eight) elongate radially but the primary capitular cells remain close together at the centre. The pedicel cell may protrude into the cavity of globule. Although only eight shield cells are present, it may look several-celled due to incomplete radial ingrowths of the wall. The secondary capitula bifurcate into two cells, each of which gives rise to a *spermatogenous* or *antheridial filament*. These filaments which are simple or branched fill up the cavities of the globule. A spermatogenous filament consists of 100-200 cells. Each cell is known as antheridium and it gives rise to a single, narrow, spirally coiled biflagellate antherozoid. A spirally-coiled blepharoplast is distinguished before the two long flagella are produced, a little behind the

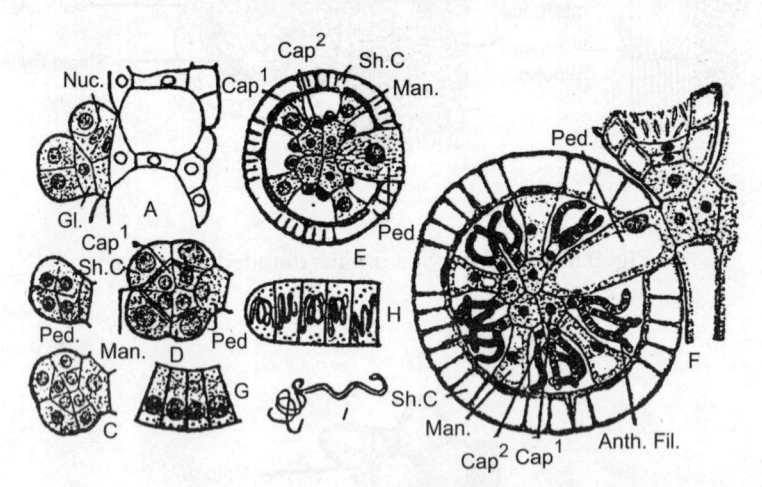

Fig. 9.53 *Chara*—Successive stages in the development of antheridium, (globule).

anterior end. When the antherozoids are mature, the shield cells separate from each other and the manubria, capitulae and the antheridial filaments are exposed. Later, each antherozoid escapes out through a pore in the antheridial wall. Thus, several antherozoids are released in the morning.

Fertilization

In a mature nucule, the five spirally-twisted tube cells separate from one another just below the corona forming narrow slits. Several antherozoids may enter the nucule through these slits. One of them penetrates through the gelatinised oogonial wall and the male nucleus fuses with that of the egg. The zygote secretes a thick wall and the inner walls of tube cells also become thickened. The enveloping threads become silicified. The whole structure sinks to the bottom of the pond. It germinates after a period of rest.

Germination of Zygote

Prior to germination, the zygote nucleus migrates towards the apex. It divides twice, of which the first is reductional forming four daughter nuclei. A cross wall is formed asymmetrically resulting in

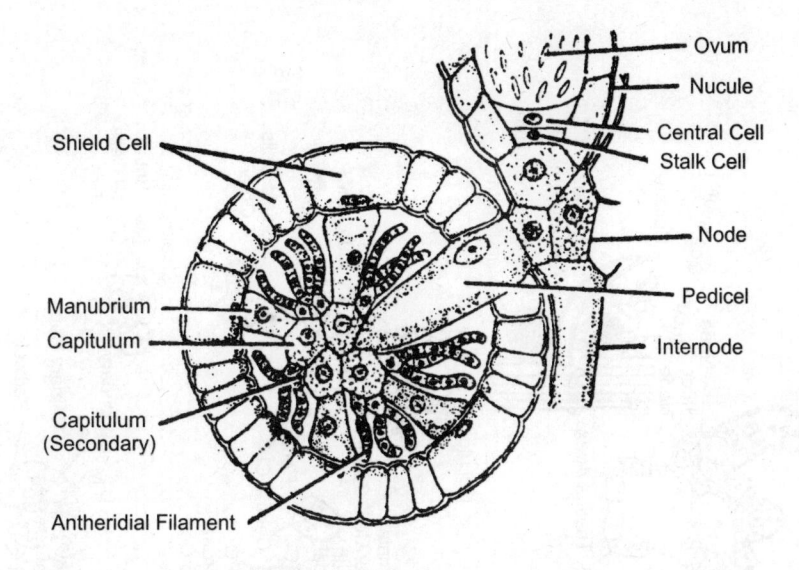

Fig. 9.54 *Chara*—Antheridium (globule) structure.

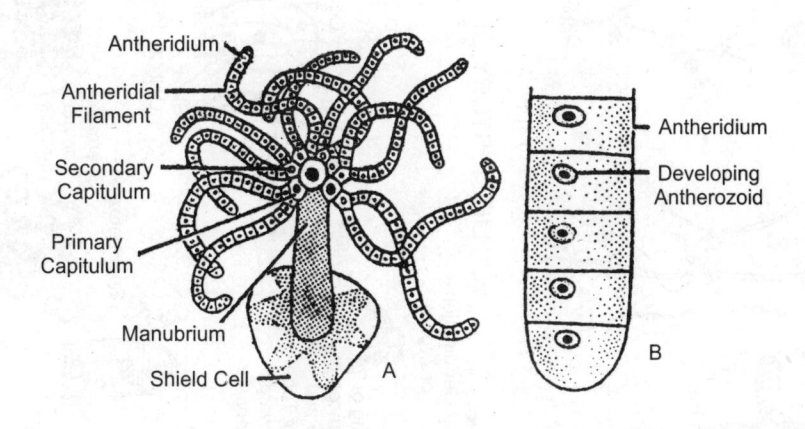

Fig. 9.55 *Chara*—Antheridium enlarged.

a small distal lenticular cell and a large basal cell. The distal cell contains one nucleus and the basal cell contains three nuclei. These three nuclei degenerate later. The oogonial envelope breaks open and the lenticular cell gets exposed. It divides vertically into two cells namely the *rhizoidal initial* and *protonemal initial*. The rhizoidal initial develops into the primary rhizoid distinguished into nodes and internodes. Secondary rhizoids arise in a whorl at the nodes. The protonemal initial gives rise to the green filamentous primary protonema which is also differentiated into nodes and internodes. The lowermost node of primary protonema gives rise to colourless *secondary rhizoids* and green secondary protonema. The second node of primary protonema produces a whorl of appendages one of which develops into the *Chara* plant. The entire life cycle of *Chara* is depicted in Fig. 9.56.

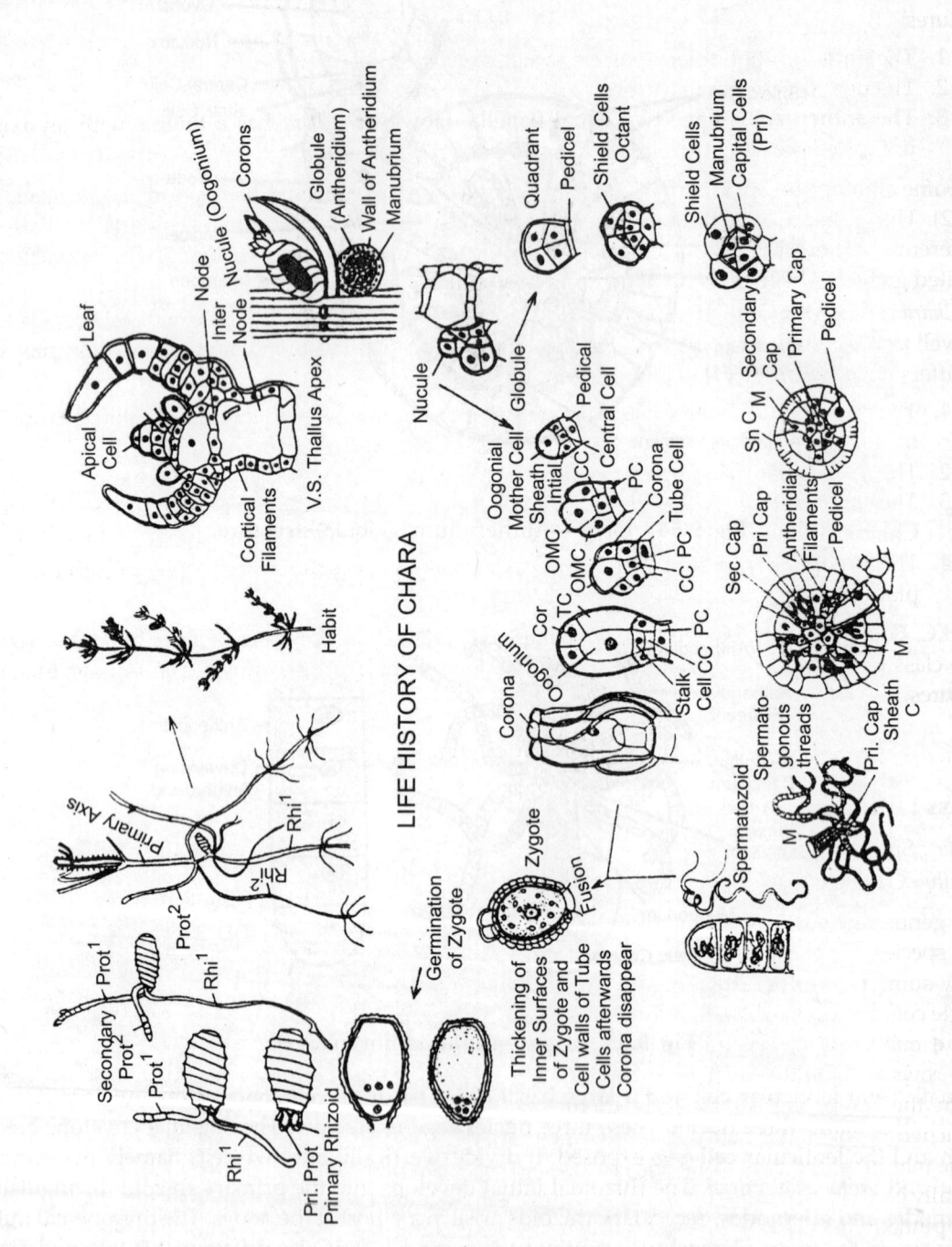

Fig. 9.56 *Chara*—Life history.

Affinities

Fritsch included *Chara* under Charales of class Chlorophyceae because of the following common features:

1. The photosynthetic pigments are same.
2. The products of photosynthesis are carbohydrates.
3. The antherozoids have two equal flagella. However, *Chara* has a thallus with an axis of unlimited growth bearing leaf-like short laterals at the nodes.

Some algologists consider that Charales might have evolved from Chaetophorales (Desikachary, 1962). This is because of the resemblance *Chara* bears to *Draparnadiopsis*. They are: 1. Thallus is differentiated into nodes and internodes. 2. The axis has cortical covering. 3. The laterals have limited growth. 4. The reproductive organs are restricted to laterals.

Chara is placed under Charophyceae of Chlorophyta (Morris, 1968). The structural organization as well as the complex sex organs point out to a greater specialisation than that of Chlorophyceae. It differs from Chlorophyceae in the following points:

1. The erect axis is differentiated into nodes and internodes with laterals of limited growth at the nodes.
2. The sex organs are complex being provided with sterile sheaths.
3. The antheroloid is an elongated biflagellate spiral structure unlike any found in Chloropbyceae.
4. The zygote germinates and produces a protonemal structure which gives rise to the adult plant.

H.C. Bold (1967) segregated *Charales* completely from Chlorophyceae of algae and gave a separate class by itself as *Charophyta* on par with Algae, Fungi and Bryophyta due to its specialised features.

BRYOPSIS PLUMOSA

Class : Chlorophyceae

Order : Siphonales
Family : Caulerpaceae

The genus *Bryopsis* is distributed mainly in tropical seas, but one species, *B. plumosa*, is commonly found in spring and early summer on the British coasts. The thallus consists of a single coenocyte, but this enormous cell shows differentiation into a main axis, from which arise, towards the upper end, two rows of lateral branches or pinnae (Figs. 9.57 & 9.58). From the lower end of the axis there is formed a little-branched horizontal rhizome anchored by rhizoids. This rhizome may produce numerous upright axes, so that each plant actually consists of a little tuft of vertically growing filaments. The pinnae vary in length, those nearest the base being the longest and decreasing regularly towards the apex. Each pinna is an elongated sac, and there is a constriction at the base where it joins the main axis. No true septa are formed in the coenocyte prior to the reproductive phase and the entire plant has one continuous vacuole lined by cytoplasm and containing numerous minute, round chloroplasts and nuclei.

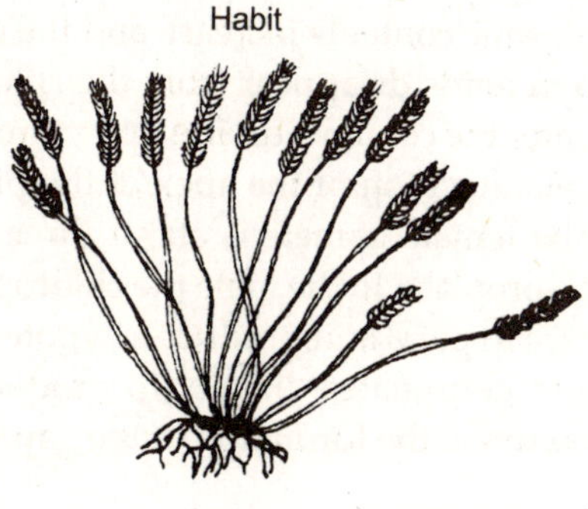

Fig. 9.57   *Bryopsis*—Habit.

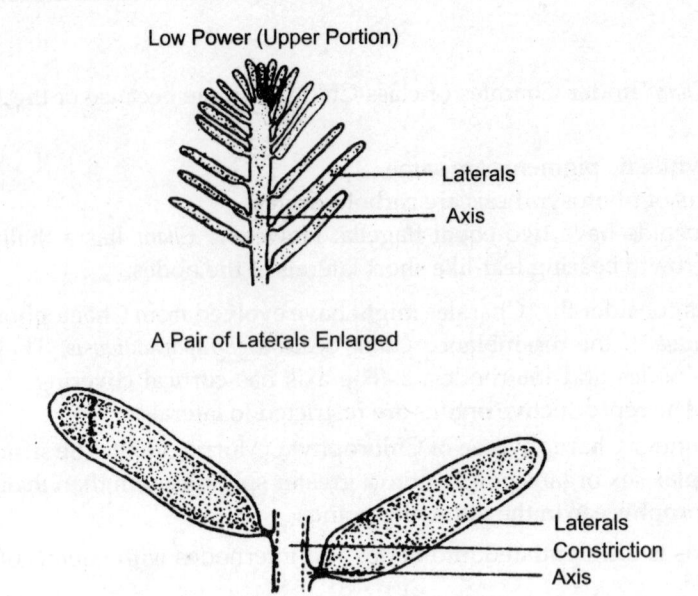

Fig. 9.58 *Bryopsis*—Structure in low power and high power.

Vegetative propagation may be effected by the detachment of pinnae which become plugged at the point of constriction. These are able to develop new rhizoids and grow into fresh plants. It is interesting to note that in conditions of dull light or when plants are placed upside down the apices of the pinnae develop rhizoids.

Reproduction

The only known method of reproduction is by means of gametes. There is no asexual method of reproduction.

Sexual reproduction. The gametes are not alike and are usually produced on separate plants. The first stage consists of the cutting off of a pinna from the main axis by means of a septum which arises as a ring-like thickening. The whole of the pinna thus forms a gametangium whose protoplasmic contents increase and the chloroplasts multiply by division. In the male gametangium the pyrenoids disappear from the chloroplasts; but they remain in those of the female gametangium. Later the contents divide up by simultaneous cleavages to form gametes, which are liberated by the gelatinization of the apex of the pinna. The gametes are pyriform with two equal, apical flagella. The female gamete is about three times as large as the male and is provided with a deep-green chloroplast. In the male the chloroplast is yellowish in colour and probably not functional. Gametes fuse in pairs and produce a zygote which may, for a time, retain all four flagella. Later it rounds off and germinates directly to produce a new plant. The plants are diploid and reduction division occurs in the formation of the gametes.

10 Class: Xanthophyceae

FAMILY I. BOTRYDIACEAE

The family includes the most primitive siphoneous bladder-like coenocytic forms anchored to the substratum by a rhizoidal system. Asexual reproduction takes place by biflagellate zoospores and sexual reproduction is either isogamous or anisogamous. The family includes a monotypic genus *Botrydium*.

BOTRYDIUM (Fig. 10.1A to 10.1H)

Occurrence. *Botrydium* is a commn terrestrial alga found growing on the mud or damp soil on the bank of pools, ponds and streams. (Fig. 10.1A). It has six species. Of these *B. granulatum* is of wide

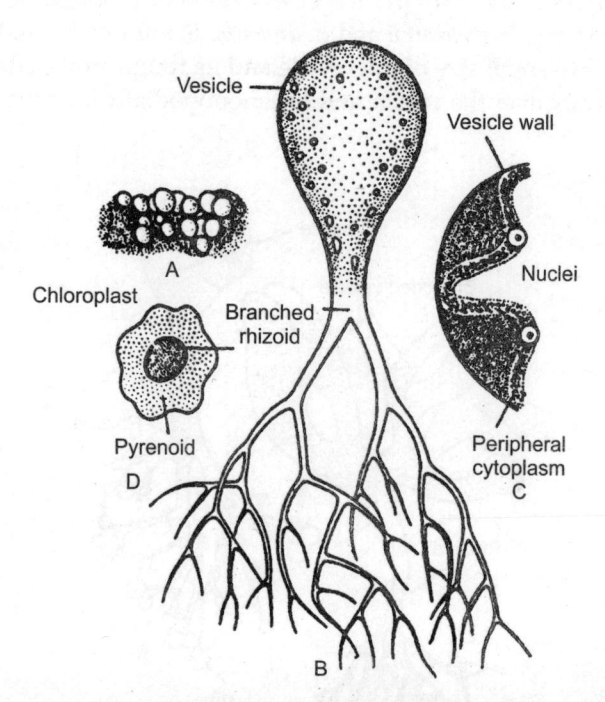

Fig. 10.1A (A-D). *Botrydium granulatum*. (A), several individuals showing habit; (B), thallus showing differentiation into balloon shaped overground part and colourless dichotomously branched underground part; (C), portion of young vesicle; (D), chromatophore of young thallus with a pyrenoid.

occurrence. It is found on bare damp soil or exposed mud of drying ponds and pools. Under favourable conditions it appears in countless numbers. *Botrydium* frequently grows intermingled with a green alga *Protosiphon* to which it bears a superficial resemblance. However, it differs from Protosiphon in its branched rhizoid, inability to divide vegetatively and many discoid chromatophores besides pigmentation, food reserves and flagellar morphology.

Thallus (Fig. 10.1A) The bladder-like thallus of *Botrydium* consists of a yellow-green pear-shaped or spherical, aerial portion which is anchored to the substratum by a colourless, branched rhizoidal portion (B). The subterranean rhizoid is dichotomously branched. The tiny, balloon-like overground portion, which is 1-2 mm. in diameter, is called the **vesicle**. It has a relatively tough wall chiefly cellulose in nature. The wall is, sometimes, encrusted with carbonate of lime. Internal to the wall is a lining layer of **cytoplasm** surrounding a large, central **vacuole** (C). The cytoplasm contains numerous small scattered nuclei and several discoid or fusiform **chromatophores** which are distributed evenly in one or more layers within the cell membrane wall external to the nuclei. They are often connected with one another by strands of dense cytoplasm. Each chromatophore has a naked pyrenoid located in the centre (D). There is never any starch in the protoplast. The photosynthetic reserves accumulate as oil and fat (Lipid). Smith (1955) reported the occurrence of **leucosin**. Within the lining layer of cytoplasm is the **central vacuole**. The central sap-vacuole and the lining layer of cytoplasm are continuous throughout the plant. There are no chloroplasts in the rhizoidal portion but the latter contains numerous nuclei scattered throughout the cytoplasm. The coenocytic siphoneous thallus-like plant body of *Botrydium* has all the essentials of a multicellular organism but it is not divided into cells. It is more appropriate to call it **acellular** rather than **unicellular**.

Besides *B. granulatum* three other important species have been reported from different parts of the world. These are *B. tuberosum, B. wallrothii* and *B. divisum. B. wallrothii* (Fig. 10.1D) differes from *B. granulatum* (Fig 10.1B) in (i) small size of the vesicle and its rough stratified wall, (ii) deposition of carbonate of lime uniformly over the vesicle and (iii) monopodially branched underground rhizoidal system.

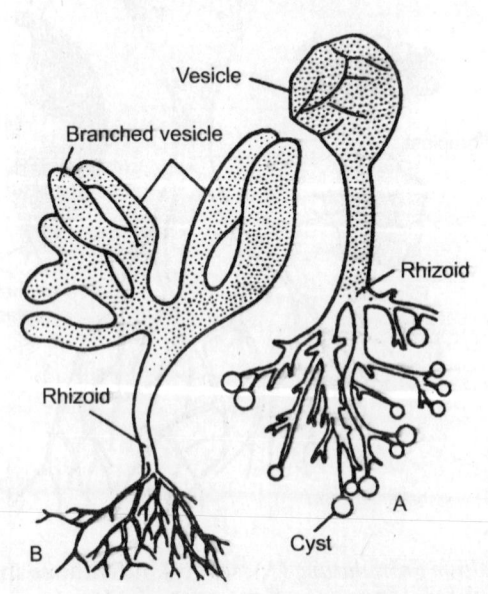

Fig. 10.1B (A-B). *Botrydium.* A, B, tuberosum with rhizoidal branches bearing terminal tuber-like cysts; B, B. divisum with a branched vesicle.

Class: Xanthophyceae **171**

B. tuberosum (Fig. 10.1B) was reported from India by Iyengar (1925) and from Russia by Miller (1927). The thallus consists of a small aerial vesicle attached to the substratum by a dichotomously branched rhizoidal portion. There is no deposition of lime on the vesicle. The rhizoidal branches bear small cyst-like structures at their tips. Iyengar (1925) described a new species of *Botrydium* from India. It is *B. divisum* (Fig 10.1B). It differs from all other species of *Botrydium* in having a branched aerial vesicle which is not covered with lime.

Reproduction *Botrydium* reproduces both asexually as well as sexually. Vegetative reproduction is not known.

(i) Asexual reproduction. It takes place by the formation of diverse types of spores. They may be zoospores, aplanospores or resting spores depending upon the conditions in which the alga grows.

Zoospores (Fig. 10.1C) The zoospores are formed when the plants are submerged under water. The protoplast of the vesicle divides into innumerable small uninucleate parts (A). Each uninucleate daughter protoplast metamorphoses into an ovoid or pyriform zoospore furnished with two unequal flagella inserted a little to one side of its anterior end (C). The longer flagellum is of tinsel (or **pantonematic**) type. It bears a double row of numerous fine lashes along its entire length giving the flagellum a feathery appearance. The shorter flagellum is of **whiplash** (or **acronematic**) type. It lacks the lashes and has a smooth surface. The zoospore has two, sometimes more chromatophores which are lateral in position. There in no eye-spot. They are liberated through an apical aperture formed by the gelatinisation of the wall of the vesicle (B). Each zoospore after the usual swarming period comes to rest, resorbs its flagella and secretes a wall around it. It then germinates by giving out a tubular colourless rhizoid at its attached end.

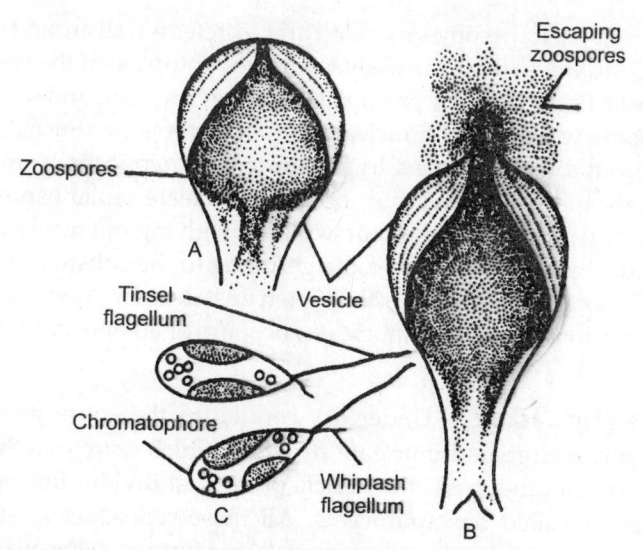

Fig. 10.1C (A-C). *Botrydium*. Zoospore formation. (A), vesicle protoplast forming numerous zoospores; (B), vesicle discharging zoospores; (C), two liberated zoospores.

(ii) Aplanospores (Fig. 10.1D) The aplanospores are formed under certain conditions particularly when the plants grow in the damp air on a wet soil but are not submerged. The protoplast of the vesicle divides repeatedly to form uninucleate daughter protoplasts. These fail to develop fla-

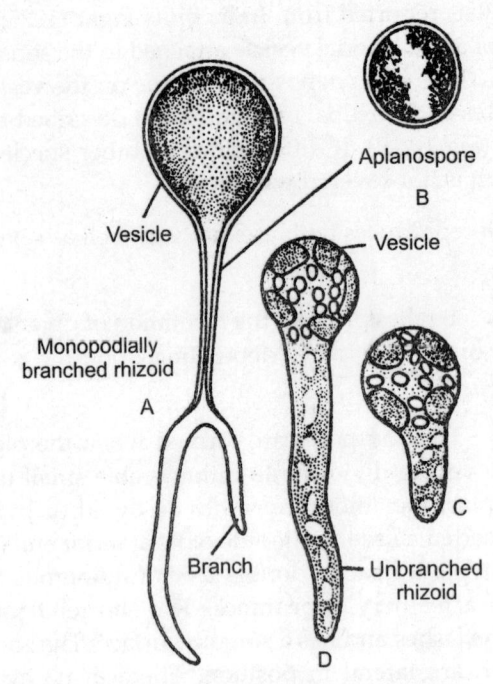

Fig. 10.1D (A-D). *Botrydium*. (A), A thallus of B. wallrothi with a monopodially branched rhizoidal system; (B), Aplanospore; (C-D), Stages in germination of aplanospore.

gella. Each daughter protoplasts becomes rounded and secretes a wall around it before liberation. It is an aplanospore. In *B. wallrothii* there is a cleavage of the protoplast of the vesicle into multinucleate daughter protoplasts. Each daughter protoplast becomes rounded and secretes a wall around it to become an aplanospore which is multinucleate. Under adverse circumstances each aplanospore secretes a thick wall around it to become a **hypnospore**. The hypnospores may be uninucleate or multinucleate. The zoospore, uninucleate or multinucleate aplanospore and uninucleate hypnospore, each germinates directly into a new plant by giving out a colourless, simple tubular rhizoid towards the attached end (C). It fixes the germling to the substratum. The young plant of *Botrydium* (D) with an unbranched rhizoid is thus often mistaken for *Protosiphon*. The multinucleate hypnospores on germination, produce uninucleate zoospore or aplanospores which give rise to the new thalli.

(iii) **Resting bodies (Fig. 10.1E)** Under dry conditions the entire protoplast of the vesicle rounds off to form a single large multinucleate structure which secretes a thick wall around it to become a **macrocyst**. In still other cases the vesicle protoplast divides into several, multinucleate thick-walled hypnospores called the **sporocysts**. All these reproductive structures (zoospores, aplanospores, hypnospores, macrocysts and sporocysts) are formed in the overground vesicle of the plant. Sometimes particularly during periods of scarcity of water the protoplast of the vesicle in *B. granulatum* passes down into the rhizoidal portion of the plant. There it divides to form several, thick-walled multinucleate resting cysts called the **rhizocysts** (A). They are globose or ellipsoidal in form and are serially arranged in the rhizoids. With the return of condition favourable for growth the protoplast of the rhizocyst divides to form uninucleate zoospores (B) or aplanospores each of which gives rise to a new plant. Sometimes the rhizocyst directly germinates into a new plant (C).

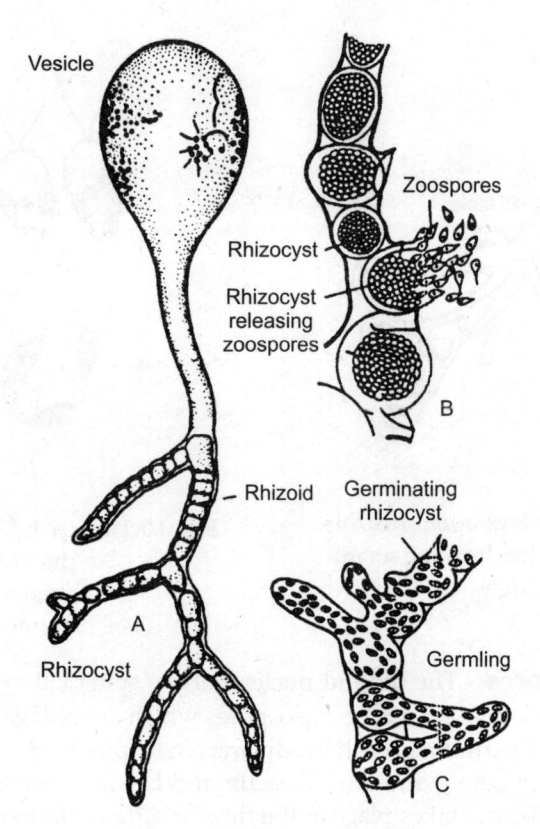

Fig. 10.1E (A-C). *Botrydium*. (A), formation of rhizocysts in B. granulatum; (B), rhizocyst releasing zoospores on germination; (C), rhizocysts directly germinating into new plants.

The protoplast of the aerial vesicle in *B. tuberosum* migrates into the rhizoidal portion. There it collects at the tips of the rhizoids which get inflated and swollen into round, thick-walled multinucleate hypnospores or cysts often erroneously called the tubers (Fig. 10.1F).

Sexual Reproduction (Fig. 10.1G) *B. granulatum* is monoecious (homothallic) and sexual reproduction is isogamous. Some species are reported to be dioecious (heterothallic) and may be anisogamous (Moewus, 1940). The gametes are formed during the rainy season. The protoplast of the vesicle divides into uninucleate parts each of which metamorphoses into a biflagellate gamete. The gametes are obpyriform to broadly ellipsoidal in form (A). Each gamete has 1 to 3 or 4 chromatophores. The eyespote may be present or lacking. The two unequal flagella are attached to the broader anterior end. The longer flagellum is of tinsel-type and shorter is of whiplash type. The gametes may escape through an apical aperture formed by gelatinisation of the vesicle apex. Usually isogametes conjugate before they are set free. They meet and fuse by their non-flagellate posterior ends (B, Rosenberg, 1930). The gametes which fail to fuse may develop parthenogenetically. The spherical zygote secretes a wall around it (E) and germinates. It undergoes no resting period. The division of the zygote nucleus is meiotic. According to Moewus (1940) the heterothallic species produce the gametes which are of unequal size (anisogametes) and are liberated through an apical pore formed by the gelatinisation of the vesicle apex. The liberated gametes become opposed in pairs at their anterior ends and then fuse laterally.

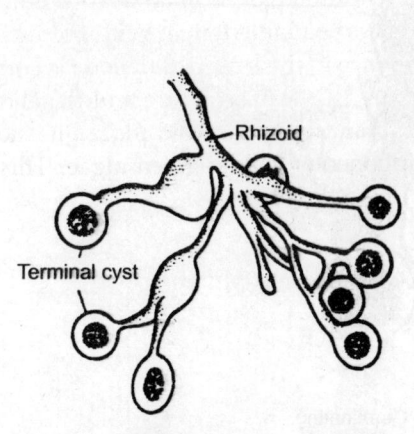

Fig. 10.1F *Botrydium tuberosum*. Rhizoidal branches bearing terminal cysts singly.

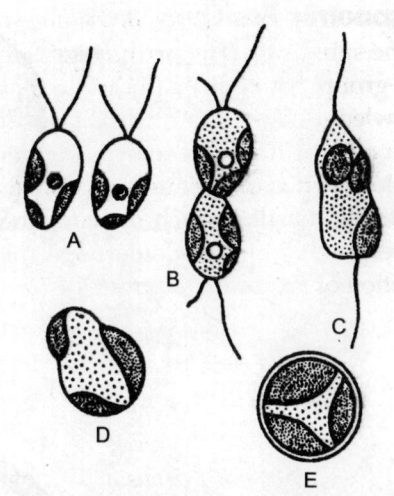

Fig. 10.1G (A-E). *Botrydium*. Sexual reproduction. (A), isogametes; (B-D), stages in gametic union; (E), zygote.

Formation of Meiospores The diploid nucleus of the spherical zygote undergoes meiosis. The protoplast divides to form 4 or 8 motile meiospores which are biflagellate. The meiozoospores escape from the zygote and germinate, each producing a new plant. On the other hand Rosenberg (1930) suggested that the zygote germinates directly and by mitosis produces a new *Botrydium* thallus which is diploid. Meiosis takes place at the time of differentiation of gametes.

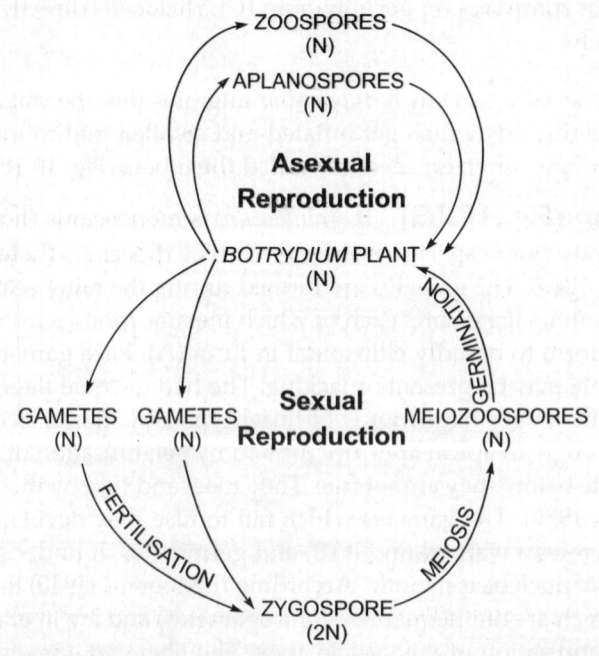

Fig. 10.1H Graphic representation of the life cycle of *Botrydium*.

Class: Xanthophyceae

Taxonomic Position *Botrydium* was formerly placed in class Chlorophyceae and was included in the sub-group 'Heterokonatae' (meaning motile cells with unequal flagella) in contradiction to sub-group "Isokontae" (meaning motile cells with flagella of equal lengths). With the increasing knowledge about its structure it was found that the difference in the length of flagella is correlated with other differences such as pigmentation, food reserves, cell wall structure and flagellar morphology. Thus the genus was shifted to a separate class Xanthophyceae and placed in the order Heterosiphonales which includes the multinucleate siphoneous yellow-green algae. This order corresponds to order Caulerpales (old Siphonales) of class Chlorophyceae. Thus the taxonomic position of *Botrydium* at present is:

Division	:	**Xanthophyta** (Xanthophycophyta)
Class	:	**Xanthophyceae**
Order	:	**Heterosiphonales**
Family	:	**Botrydiaceae**
Genus	:	*Botrydium*
Species	:	*granulatum*

VAUCHERIA

Class : Xanthophyceae

Order: Vaucheriales
Family: Vaucheriaceae

Vaucheria, with about 40 species is mostly a freshwater form and only a few are marine. They form green felty mats on damp soils.

Structure

The tube-like filamentous body of *Vaucheria* is very often fixed to the substratum by means of rhizoid-like branches.

The filaments of *Vaucheria* attain a length of several centimetres and they show monopodial branches sparingly (Fig. 10.2). Terrestril forms get attached to the soils by means of rhizoid-like branches with few chromatophores. The siphoneous filament grows in length by elongation of terminal portion and they lack septa. But septa are formed when sex organs are formed.

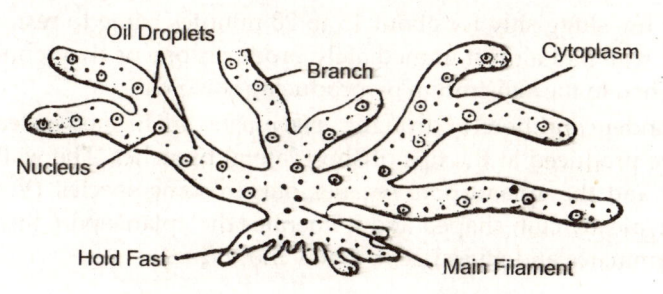

Fig. 10.2 *Vaucheria*—Thallus.

The cell wall is thin and it is composed of cellulose, often impregnated with pectic substances. A single central vacuole runs throughout the length of the filament and the cytoplasm is confined to a layer, on the inner side of the cell wall. This layer of cytoplasm contains numerous nuclei (coenocytic) towards the inner side and several discoid chloroplasts towards the outer side. Chloroplasts are devoid of *pyrenoid*. In *Vaucheria*, oil is stored in droplets in the cytoplasm. But under constant illumination, partial or complete formation of starch may take place.

Reproduction

Vegetative reproduction: *Vaucheria* multiplies in a prolific manner vegetatively by fragmentation. The detached portions grow into separate individuals.

Asexual reproduction (Fig. 10.3).
Typically, asexual reproduction takes place by means of large *multiflagellated zoospores*. In terrestrial forms they are produced when flooded with water. The swollen distal ends of branches develop into *sporangia*. In each sporangium, which is club-shaped, a single multiflagellate zoospore is formed. During the development of sporangia, some of the branches, terminally swell up and become club-like, as cytoplasm with nuclei and chloroplasts stream into the distal portion. The tips of these branches become deep green. The protoplast divides transversely, a little below the swollen distal part and a transverse wall is laid so that the terminal part is marked off as a sporangium.

In the sporangium, originally the nuclei are towards the inner side and chloroplasts towards the periphery. But later, the positions are reversed such that the nuclei come to lie in the peripheral layer of protoplasm. The protoplast of the sporangium contracts followed by the formation of a pair of unequal flagella, opposite to each nucleus. Thus, a multinucleate, multiflagellate syn-zoospore (compound zoospore) is formed. The sporangium forms a small pore at the apex by gelatinisation through which the naked compound zoospore squeezes out and escapes into the surrounding water. Zoospore formation can be induced by removing the filaments from nutritive solutions to pure water. In nature, they are generally liberated out in the mornings.

The multiflagellate zoospore is interpreted to be a compound zoospore which has failed to divide into uninucleate biflagellate zoospores.

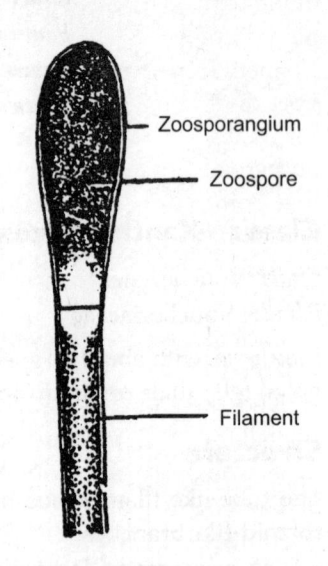

Fig. 10.3 *Vaucheria*—Asexual reproduction.

The zoospores swim, sluggishly for about 15 to 20 minutes come to rest, and withdraw their flagella. It secretes a wall and almost immediately produces one or two tubular outgrowths. The outgrowth gets attached to the substratum by producing rhizoids.

In some species, under conditions of drought, *aplanospores* are formed instead of zoospores. The aplano-sporangia are produced at the tips of short lateral branches. The wall of the sporangium ruptures irregularly and the aplanospore escapes out. In some species (*V. pilobloides*) the sporangium is usually typically club-shaped and it liberates the aplanospore through an apical pore. The aplanospore germinates and gives rise to a new individual.

Gongrosira stage.
In terrestrial forms, under conditions of desiccation, filaments of *V. germinata* undergo transverse septation of the filaments into a series of short segments separated by thick gelatinous walls. The segments secrete thick walls and they are known as thick-walled *akinetes* or *hypnaspores* (Fig. 10.4). These stages are referred to as *gongrosira* stages. The hypnospores or cysts may separate and germinate directly into new filaments. Sometimes, the contents of the hypnospore

may divide into a number of amoeboid masses. These amoeboid protoplasts escape through a pore in the wall, come to rest, secrete a wall, and later germinate into new filaments.

Sexual reproduction. The sex organs of *Vaucheria* are *antheridia* (male) and *oogonia* (female) (Fig. 10.5). In monoecious or homothallic species, the two sex organs are produced adjacent to each other. In dioecious or heterothallic species, the male and female sex organs occur on different plants. There is considerable difference in the arrangement of sex organs in species of *Vaucheria*.

The *antheridia* may develop slightly earlier or simultaneously along with oogonia. They are produced at the ends of short laterals, as curved structures. The antheridium is cut off from the filament by a septum. The protoplast of the antheridium undergoes divisions into several uninucleate bits, each of which transforms itself into a *biflagellate spermatozoid* or *antherozoid*. The antherozoids are oval and the two flagella are laterally inserted, one pointing forward and another backward. One of them is acronematic and another is pantonematic. The antherozoids are liberated out through one or more apical pores.

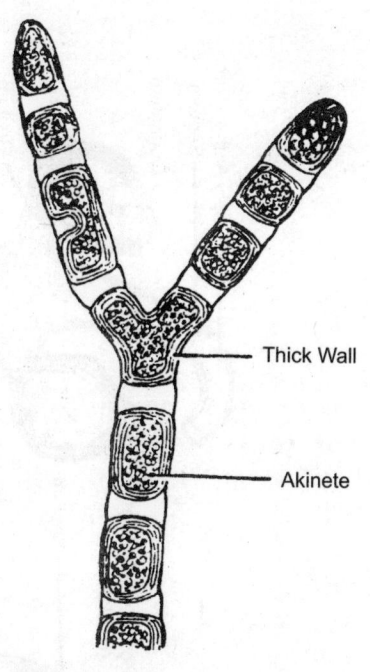

Fig. 10.4 *Vaucheria*—Akinete formation.

The *oogonia* are sessile or stalked and rounded. There is a septum at the base, making it off from the filament. It is rich in reserved food (oil) and contains several chloroplasts. It is typically uninucleate. It has been described, that this condition is attained by degeneration of all other nuclei, except one. Some consider that all the nuclei except one migrate with part of the cytoplasm back into the filament and that the cross wall is laid only very late in the development of the oogonium.

The oogonium, when fully developed produces a beak towards one side. At this region, an opening is formed by gelatinization of the beak. Sometimes, the aperture is apical in position. The cell sap is extruded out from the central vacuole, and as a result, the *ovum* contracts.

Fertilization. During fertilization, several spermatozoids enter the oogonium through the aperture formed at the beak region. But only one of them fuses with the ovum. The male and female nuclei do not fuse immediately. The male nucleus becomes as big as the female one and then they fuse together.

The zygote secretes a several-layered thick envelope, and it is filled with oil globules. The chloroplasts almost disappear. The zygote takes a prolonged period of rest for several months, before it germinates into a new filament. It is probable that the division of zygote is *meiotic*.

Affinities. *Vaucheria* was placed under Chlorophyceae in Siphonales by Fritsch for the following reasons:

1. Siphoneous structure of the filaments.
2. Formation of starch as reserve food under induced conditions.

Vaucheria is placed under *Vaucheriales* as an order in class Xanthophyceae by Ian Morris, due to the following reasons:

1. Chlorophyll *b* is absent and chlorophyll *e* is identified in the zoospores.

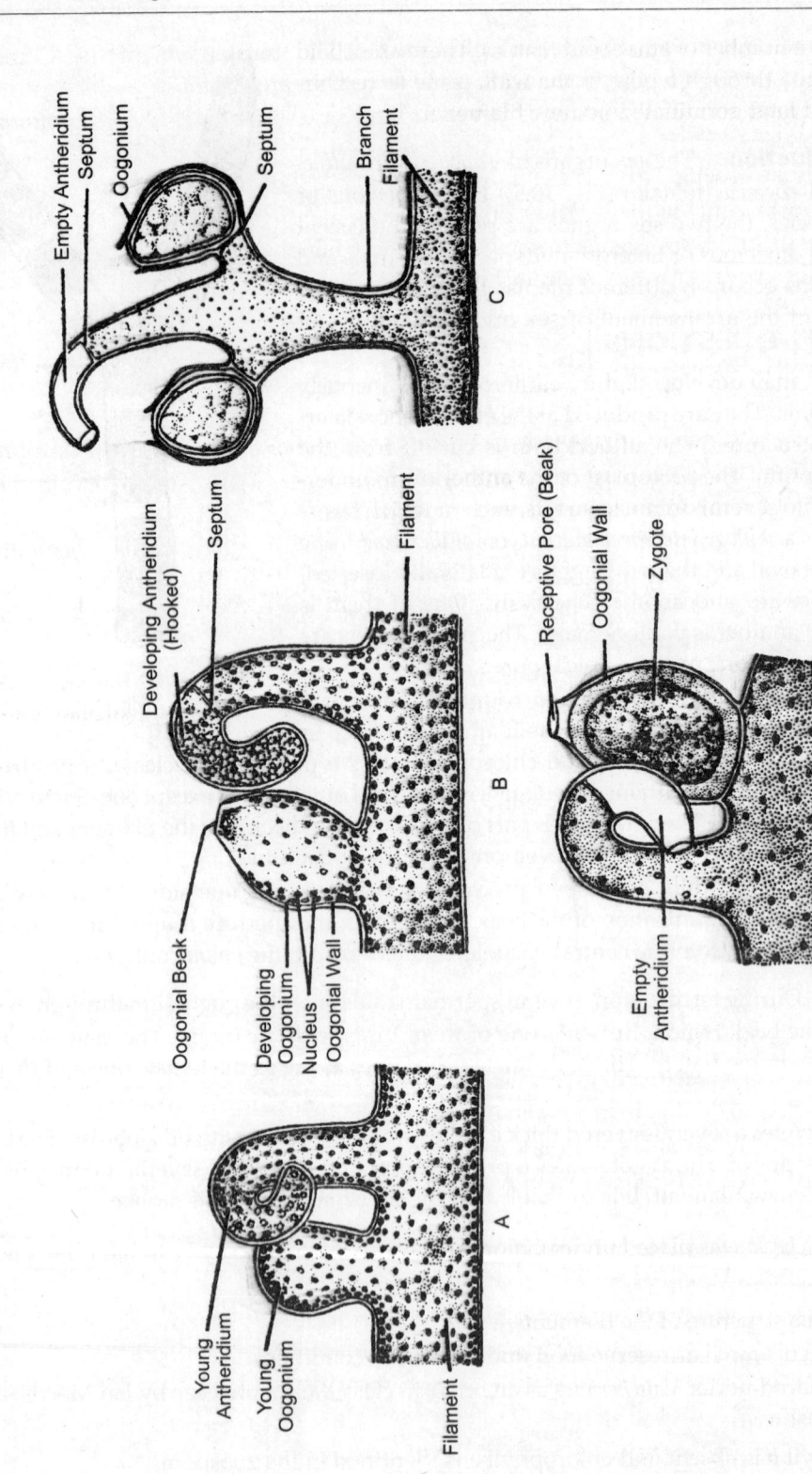

Fig. 10.5 *Vaucheria.* Sexual reproduction.

2. Carotenoids are in excess of the chlorophylls. The carotenoids of the Caulerpales (Chlorophyceae) namely siphonein and siphonoxanthin are absent.
3. Under natural conditions, starch is never formed as reserve food. Oil is accumulated as reserve food.
4. Pyrenoids are absent.
5. Flagella are unequal in the spermatozoid.

The resemblance of the sex organs to some members of Fungi (Oomycetes) is a case of parallel evolution and does not signify any relationship.

IMPORTANT QUESTIONS

Essay Type

1. With the help of suitable diagrams describe the range of thallus structure in Chlorophyceae.
2. Give an account of cell structure in green algae.
3. Describe in detail the various modes of reproduction in green algae.
4. Describe in detail the various types of chloroplasts found in different genera belonging to Chlorophyceae.
5. Give an account of different life cycles met within green algae.
6. Give an account of various types of chloroplast in green algae.
7. Describe the various modes of perennation in green algae.
8. Describe the various modes of sexual reproduction in green algae studied by you. Give suitable examples as well.

Short Answer Type

9. Write short notes on:
   (a) Distribution of green algae
   (b) Chlorophyceae
   (c) Pigments in green algae
   (d) Parasitic green algae
   (e) Difference between cell structure of green algae and those of higher plants
   (f) Origin of green algae
   (g) Asexual reproduction in green algae
   (h) Sexual reproduction in green algae
   (i) Structure of chloroplast in green algae
   (j) Alternation of generation in green algae
   (k) Economic importance of green algae
   (l) Shape of chloroplasts in chlorophyta
   (m) Zoospores of chloroplasts
10. In what respects does the protoplast of the cell of green algae show an advance over that of the blue green algae?
11. Describe the salient features of Chlorophyceae and name the various orders placed in the class.
12. Name an alga which is used as food.
13. Name the reserve food material in the cells of green algae.
14. Name an alga which has starch as reserve food.

15. Draw an ultrastructure diagram of T.S. of a typical eukaryotic flagellum.
16. Name a green alga which lacks vegetative reproduction.
17. Name a green alga which lacks flagellated cells.
18. Name an alga which shows false branching.
19. Name a chlorophycian alga with heterotrechous habit.
20. Name an alga which is rich in protein.
21. Name the alga where the thallus consists of an erect and postrate systems.
22. What is a coenobium?
23. Distinguish between
    (a) Chlorophyceae and Charophyceae
    (b) Chloroplast and Pyrenoid
24. Differentiate between isogamy, anisogamy and oogamy modes of reproductions.
25. Describe the structural details of neuromotor apparatus.

Objective Type

26. Fill in the blanks:
    (a) Fusion between gametes of unequal sizes is called ....
    (b) A motile flagellated asexual cell is known as ....
    (c) .... is fusion between gametes of equal sizes.
    (d) .... is the common name of *chara*.
    (e) Unicellular non motile spherical green alga is ....
    (f) An alga in which cup snails is called ....
    (g) An alga in which cup shaped chloroplast is present is known by the name ....
    (h) Chlorophyll a and b are present in the members of the class ....
    (i) The reserve food material of green algae is ....
    (j) coenobium is found in ....
    (k) Luxuriant growth of some algae in water often imparting colour to the water is called ....
    (l) The green algae may be broadly divided into classes namely .... and ....
    (m) Pyrenoids are meant for .... synthesis in various algae.
    (n) Zoospores spores which are motile and have two flagella are called ....

27. Select the correct answer:
    (i) Heterotrichy means having
        (a) Prostrate and erect branches
        (b) Rhizoidal and photosynthetic branches
        (c) Long and short branches
        (d) Branches differentiated into nodes and internodes.
    (ii) Reserve food is starch in
        (a) Chlorophyceae (b) Myxophyceae
        (c) Phaeophyceae   (d) Rhodophyceae.
    (iii) Asexual spores which are motile and have two flagella are called
        (a) Zoospores   (b) Synzoospores
        (c) Zygospores   (d) Chlamydospores.

(iv) Which of the following has reserve food as starch?
   (a) *Volvox*   (b) *Vaucheria*
   (c) *Ectocarpus*   (d) *Batrachospromum.*
(v) Chlorophyll a and b are together present in
   (a) Chlorophyceae   (b) Myxophyceae
   (c) Phaeophyceae   (d) Rhodophyceae.
(vi) Which of the following has heterotrichus habit?
   (a) *Ulothrix*   (b) *Coliochaete*
   (c) *Oedogonium*   (d) *Oscillatoria.*
(vii) Which of the following has heterotrichus habit?
   (a) *Volvox*   (b) *Vaucheria*
   (c) *Ectocarpus*   (d) *Ulothrix.*
(viii) Asexual zoospores are usually produced under
   (a) Unfavourable conditions   (b) Favourable conditions
   (c) Both of the above   (d) None of he above.
(ix) Zygospores are produced as a result of
   (a) Isogamous fusion   (b) Gametangial contact
   (c) Gametangial copulation   (d) Oogamy.
(x) Non motile thin walled spores of algae are called
   (a) Macrospore   (b) Microspore
   (c) Aplanospore   (d) Zygospore.
(xi) Fusion between gametes of unequal sizes is called
   (a) Isogamy   (b) Anisogamy
   (c) Planogametic fusion   (d) Oogamy.

QUESTIONS

Essay Type

1. Describe the structure of thallus, modes of nutrition and excretion in *Chlamydomonas*.
2. Describe in detail the life history of *volvox*.
3. Describe the structure and reproduction in *Chlamydomonas*.
4. Give an account of the thallus structure and reproduction in *Gonium*.
5. Describe the thallus structure and reproduction in *Pandorina*.
6. Describe the cell structure and reproduction in *Chlamydomonas*.
7. Give an account of thallus structure and reproduction in *Eudorina*.
8. Describe the various methods of reproduction in *Chlamydomonas*.
9. Describe the modes of reproduction in *Volvox*.
10. Describe sexual reproduction in *Chlamydomonas*. What light does it throw on the origin and evolution of sex in Algae?
11. With the help of suitable diagrams, describe the asexual and sexual reproduction in Volvox/ life cycle in *Volvox*.
12. Give an illustrated account of asexual/sexual reproduction in *Volvox*.
13. (a) Give an account of sexual reproduction in unicellular alga studied by you.
    (b) Discuss how Volvox is more advanced than *Chlamydomonas*.

14. Describe asexual reproduction in Volvox. Explain the term *coenobium*.
15. Describe the structure and method of reproduction in *Volvox*.
16. Describe the life history in *Chlamydomonas*.

Short Answer Type

17. Write short notes on
    (i) Coenobium in Volvox.
    (ii) Palmella stage
    (iii) Asexual reproduction in *Volvox*.
    (iv) Neuromotor apparatus
    (v) Asexual reproduction in *Chlamydomonas*
    (vi) Sexual reproduction in *Chlamydomonas*
    (vii) Daughter colony
    (ix) Structure an germination of zygote in *Chlamydomonas*
18. Draw neat and well labelled diagrams of the following:
    (a) Cell structure of *Chlamydomonas*
    (b) Monoecious coenobia of *Volvox*
    (c) Section through *Volvox colony*
    (d) Male coenobium in *Volvox*
    (e) Colony of *Volvox* showing antheridia and oogonia
19. Define coenobium. What is coenobium?
20. What is anisogamy?
21. Describe the structure of chloroplast in *Chlamydomonas*.
22. Give the name of the alga in which neuromotor apparatus in found.
23. Discuss the plakea stage in *Volvox*.
24. Describe Gonidia in *Volvox*.
25. What is palmella stage?
26. (a) Describe the structural details of neuromotor apparatus.
    (b) Differentiate between isogamy, anisogamy and oogamy modes of reproduction with examples.
    (c) Describe zygospore germination in *Chlamydomonas*.
27. Name the species of *Volvox* in which cytoplasmic strands are lacking.
28. Where does reduction division occur in *Volvox*?
29. (a) With the help of diagrams give the details of *Chlamydomonas* cell/thallus as seen under electron microscope.(Agra, 1995; Himachal Pradesh, 1996)
    (b) Give a brief account of sexual reproduction in *Volvox*.
30. Classify the following:
    (i) *Chlamydomonas*   (ii) *Volvox*
    (iii) *Gonium*   (iv) *Pandorina*
    (v) *Eudorina*.
31. Identify the following by giving one important character:
    (a) *Chlamydomonas*
    (b) *Volvox*

Objective Type

32. Fill in the blanks:
    (i) Thick walled resting spores formed after isogamy or anisogamy is called ....
    (ii) Coenobium is found in ....
    (iii) Spores produced by *Chlamydomonas nivalis* are coloured red by ....
    (iv) Number of chloroplasts in *Chlamydomonas* is ....
    (v) Neuromotor apparatus is present in ....
    (vi) Sexual reproduction ranges from isogamy to oogamy in ....
    (vii) Specialised cells meant for asexual reproduction in young coenobia are called ....
    (viii) The number of cells found in coenobium in *Volvox* ranges from .... to ....
    (ix) The species of *Chlamydomonas* which forms the red show is ....
    (x) *Chlamydomonas* belongs to the family ....
    (xi) *Volvox* belongs to the family ....
    (xii) The type genus of the family Volvocaceae is ....
    (xiii) The reduction division in *Chlamydomonas* occurs at .... stage ....
    (xiv) The number of contractile vacuoles in *Chlamydomonas* is ....
    (xv) Asexual reproduction in *Chlamydomonas* takes place by ....

33. Select the correct answer:
    (i) Coenobium is the characteristic feature of
       (a) *Chlamydomonas* (b) *Volvox*
       (c) *Spirogyra* (d) *Ectocarpus.*
    (ii) Unicellular motile thallus is present in
       (a) *Chamydomonas* (b) *Volvox*
       (c) *Spirogyra* (d) *Ectocarpus.*
    (iii) The chloroplast in Chlamydomonas is
       (a) Parietal (b) Spiral
       (c) Cup-shaped (d) Reticulate.
    (iv) Neuromotor apparatus is present in
       (a) *Chlamydomonas* (b) *Volvox*
       (c) *Spirogyra* (d) *Ectocarpus.*
    (v) The number of blepharoplasts in *Chlamydomonas* is
       (a) One (b) Two
       (c) Three (d) Four.
    (vi) In *Volvox* the number of cells in coenobium ranges between
       (a) 100–200 (b) 200–300
       (c) 300–500 (d) 500–1000.
    (vii) The type of genus of the family volvocaceae is
       (a) *Chlamydomonas* (b) *Volvox*
       (c) *Spirogyra* (d) *Ectocarpus.*
    (viii) In *Chlamydomonas* the reduction division occurs at
       (a) Zoospore stage (b) Thallus stage
       (c) Gamete stage (d) Zygospore stage

(ix) The asexual reproduction in Chlamydomonas takes place by
    (a) Zoospore   (b) Gamete
    (c) Zygospore   (d) Cell.

(x) The vegetative reproduction by daughter colonies takes place in
    (a) *Volvox*   (b) *Chlamydomonas*
    (c) *Spirogyra*   (d) *Ectocarpus*.

(xi) The gametes in Chlamydomonas are
    (a) motile and uniflagellate   (b) motile and biflagellate
    (c) non-motile and uniflagellate   (d) non-motile and beflagellate.

(xii) Palmella stage is characteristic of
    (a) *Chlamydomonas*   (b) *Volvox*
    (c) *Gonicium*   (d) *Ectocarpus*.

(xiii) The cytoplasmic strands are absent in
    (a) *Volvox auseus*   (b) *Volvox globator*
    (c) *V. mononae*   (d) *Volvox carteri*.

(xiv) Asexual reproduction in Volvox takes place by
    (a) Zoospores   (b) Aplanospores
    (c) Mitospores   (d) Daughter colonies.

(xv) Sexual reproduction ranging from isogamy to oogamy takes place in
    (a) *Chlamydomonas*   (b) *Volvox*
    (c) *Eudorina*   (d) *Pandorina*.

(xvi) In Volvox sexual reproduction is of advanced type known as
    (a) Isogamy   (b) Anisogamy
    (c) Herkogamy   (d) Oogamy.

(xvii) The reduction division takes place in Volvox
    (a) in daughter colonies
    (b) during formation of gametes
    (c) in zygote nucleus prior to germination
    (d) during formation of zoospores..

(xviii) The mode of nutrition in Chlamydomonas and Volvox is
    (a) Symbiotic   (b) Heterotrophic
    (c) Autotrophic   (d) Symbiotic

QUESTIONS

Essay Type

1. Write in detail on the phylogeny of chlorococcales.
2. Describe the structure and reproduction in *Hydrodictyon*.
3. Give an account of the occurrence, structure and reproduction in *Chlorella*. Briefly describe its economic importance.
4. Give a comprehensive account of structure, reproduction and life cycle in *Pediastrum*.
5. Describe the occurrence, structure and reproduction in *Scenedesmus*.
6. Describe mode of reproduction in *Hydrodictyon*.

Class: Xanthophyceae **185**

7. Describe the cell structure and reproduction in *Chlorella*.
8. Give a comprehensive account of reproduction/life history of *pediastrum/scenedesmus*.

Short Answer Type
9. Describe the characteristic features of the order Chlorococcales.
10. In what respects are the following different from each other:
    (i) *Volvox* and *Hydrodictyon*.   (ii) *Chlamydomonas* and *Chlorella*.
    (iii) Coenobium of *Sceneedesmus* and *Volvox*.   (iv) *Hydrodictyon* and *Vaucheria*.
    (v) *Chlorella* and *Scenedesmus*.
11. Draw well labelled diagram of the following:
    (i) Net of *Hydrodictyon*.
    (ii) Cell structure of *Chlorella*.
12. Name a green alga which lacks flagellated cells.
13. Name a nuicellular non-motile green alga.
14. Name an alga which is used as a source of food.
15. Name an alga popularly known as 'water net'.
16. Does vegetative reproduction occur in *Hydrodictyon*?
17. Write short notes on:
    (a) Sexual reproduction in *Hydrodictyon*.
    (b) Cell structure in *Chlorella*.
    (c) Economic importance of *Chlorella*.
    (d) Water net.
    (e) Reproduction in *Scenedesmus*.
    (f) Germination of zygote in *Hydrodictyon*.
    (g) Systematic position and evolution in Chlorococcales.

Objective Type
18. Fill in the blanks:
    (i) Unicellular, non-motile, spherical green alga is ....
    (ii) The alga which forms water net is ....
    (iii) In *Chlorella* the chloroplast is ....
    (iv) The sole method of reproduction in *Chlorella* is .... and ....
    (v) The alga which is used as a source of food is ....
    (vi) The chloroplast in *Hydrodictyon* is ....
    (vii) The zoospores in *Hydrodictyon* is ...., ... and ....
    (viii) Sexual and Asexual reproduction is present in water net also known as ....
    (ix) In *Pediastrum* the chloroplast is ....
    (x) In *Protosiphon* the thallus is ...., ... and ....
    (xi) The chloroplast in *Protosiphon* is reticulate or .... in position.
    (xii) Vegetative reproduction takes place by budding in ....
    (xiii) Chloroplast is laminate in ....
    (xiv) In *scenedesmus* vegetative reproduction takes place by ....
19. Select the correct answer:

(i) None- motile unicellular alga is
   (a) *Chlamydomonas* (b) *Chlorella*
   (c) *Pandorina* (d) *Volvox.*

(ii) The alga which may be used as a food in spaceship is
   (a) *Chlorella* (b) *Chlamydomonas*
   (c) *Chlorococcum* (d) *Haematococcus*

(iii) The alga which is popularly called 'water net' is
   (a) *Pediastrum* (b) *Hydrodictyon*
   (c) *Cladophora* (d) *Pithophora.*

(iv) Which of the following does not reproduce sexually?
   (a) *Chlamydomonas* (b) *Chlorella*
   (c) *Volvox* (d) *Hydrodictyon.*

(v) Reticulate or perforated pareital chloroplast is present in
   (a) *Chlorella* (b) *Hydrodictyon*
   (c) *Pediastrum* (d) *Protosiphon.*

(vi) Simple pareital band shaped chloroplast is present in
   (a) *Chlorella* (b) *Hydrodictyon*
   (c) *Pediastrum* (d) *Protosiphon.*

(vii) Cup or bell shaped chloroplast is present in
   (a) *Spirogyra* (b) *Nostoc*
   (c) *Chlorella* (d) *Hydrodictyon.*

(viii) Autospores are produced in
   (a) *Spirogyra* (b) *Nostoc*
   (c) *Chlorella* (d) *Hydrodictyon.*

(ix) The alga which is autophytic with its photosynthetic process similar to angiosperm is
   (a) *Spirogyra* (b) *Nostoc*
   (c) *Chlorella* (d) *Hydrodictyon.*

(x) In which of the following, the thallus is coenobium consisting of two, four or eight cells?
   (a) *Chlorella* (b) *Chlamydomonas*
   (c) *Scenedesmus* (d) *Volvox.*

QUESTIONS

Essay Type

1. Describe the habitat, mode of function and the structure of the cell in *Ulothrix*.
2. Describe the various methods of asexual reproduction in *Ulothrix*.
3. Describe the life history of *Ulothrix* commenting upon any point of morphological or physiological interest in it.
4. Give an account of sexual reproduction in *Ulothrix*.
5. Describe the structure and life history of *Ulothrix* and mention how its sexual method of reproduction throws light on the origin of the sex cells in the Algae.
6. Give a detailed account of the cell structure in *Ulothrix*.
7. With the help of diagrams only, describe the life cycle in *Ulothrix*.

Short Answer Type

8. Compare the structure of chloroplast in *Ulothrix* and *Chlamydomonas*.
9. Discuss the biological importance of *Ulothrix*.
10. Give a brief account of evolutionary trends in *Ulothrix*.
11. Discuss the evolutionary trends in Ulotrichales.
12. Describe the various types of zoospores in *Ulothrix*.
13. Compare the structure of zoospores and gametes in *Ulothrix*.
14. Write short notes on:
    - (a) Cell structure in *Ulothrix*
    - (b) Microzoospores
    - (c) Isogametes
    - (d) Asexual reproduction in *Ulothrix*
    - (e) Sexual reproduction in *Ulothrix*.

Objective Type

15. Select the correct answer:
    - (i) Girdle shaped chloroplast is found in
        - (a) *Chlamydomonas*
        - (b) *Volvox*
        - (c) *Oedogonium*
        - (d) *Ulothrix*.
    - (ii) Multicellular filamentous alga is
        - (a) *Ulothrix*
        - (b) *Vaucheria*
        - (c) *Chlamydomonas*
        - (d) *Pandorina*.
    - (iii) The life cycle of Ulothrix is
        - (a) haplontic
        - (b) diplontic
        - (c) haplobiontic
        - (d) diplobiontic.
    - (iv) Which of the following alga occurs as an attached form?
        - (a) *Volvox*
        - (b) *Zygnema*
        - (c) *Ulothrix*
        - (d) *Volvox*.
    - (v) The thallus of *Ulothrix* is filamentous and
        - (a) Branched
        - (b) Colonial
        - (c) Unbranched
        - (d) Solitary
    - (vi) *Ulothrix* produces
        - (a) Quadriflagellate microzoospores
        - (b) Biflagellate microzoospores
        - (c) Quadriflagellate macrozoospore
        - (d) All of the above.
    - (vii) In *Ulothrix*, the zygote is pear shaped and
        - (a) Uniflagellate
        - (b) Biflagellate
        - (c) Triflagellate
        - (d) Quadriflagellate
    - (viii) Meiosis in *Ulothrix* takes place at the time of
        - (a) germination of zygote
        - (b) formation of zoospores in the filament cells
        - (c) formation of gametes
        - (d) fragmentation.
    - (ix) The filaments of *Ulothrix* occur
        - (a) Only in marine water
        - (b) Only in fresh water
        - (c) Both in marine and fresh water
        - (d) On soil.

(x) Which of the following is produced in maximum numbers in *Ulothrix*?
   (a) Gametes
   (b) Quadriflagellate macrozoospore
   (c) Quadriflagellate microzoospore
   (d) Biflagellate microzoospore.

QUESTIONS

Essay Type

1. Describe the habit, structure and differentiation of cells in the filament of *Oedogonium*.
2. Give an account of the vegetative (thallus) structure and reproduction in *Oedogonium*.
3. Describe the thallus structure and reproduction in *Bulbochaete*.
4. Describe the habit, structure and methods of reproduction in *Oedogonium* and indicate the features of special interest in its life history.
5. Give an illustrated account of life-cycle of *Bulbochaete*.
6. Write in detail the life history/cycle of *Oedogonium*.
7. Describe the sexual reproduction in *Bulbochaete* with labelled diagrams.
8. Give an illustrated account of structure and reproduction in *Oedogonium/Polysiphonia*.
9. Describe the features of special interest in the structure and reproduction of *Oedogonium*. Explain the mode of formation of cap cells.
10. Describe mode of sexual reproduction in *Oedogonium*.
11. With the help of labelled diagrams describe the method of sexual reproduction in *Oedogonium*.
12. Describe the life cycle/sexual reproduction of nannandrous species of *Oedogonium*.
13. Give an account of the special features of structure and reproduction in *Oedogonium*.
14. Give diagrammatic representation of the life-cycle illustrating the relative strength of *Oedogonium*.
15. Give an account of structure and reproduction in Macrandrous species of *Oedogonium*.
16. How does the Macrandrous species of *Oedogonium* differ from nannandrous species in the method of sexual reproduction?

Short Answer Type

17. Write short notes on:
    (i) Dwarf males or Nannandria or Nannadrous species.
    (ii) Cell division in *Oedogonium*.
    (iii) Asexual reproduction in *Bulbochaete*.
    (iv) Zoospores of *Oedogonium* and *Vaucheria*.
    (v) Cell structure in *Oedogonium*.
    (vi) Reproduction in *Polysiphonia/Oedogonium*.
    (vii) Cap cells in *Oedogonium*.
    (viii) Cell division in *Oedogonium*.
    (ix) What is a Nannandrium?
18. Describe the following:
    (a) Asexual reproduction in *Oedogonium*.
    (b) Sexual reproduction in *Oedogonoin*.
    (c) Nannandrium.
    (d) Nannandrous and Macrandrous species.

(e) Dwarf male in *Oedogonium* with diagrams.
19. Identify the following giving one important characteristic
    (i) *Oedogonium* (b) *Bulbochaete*.
20. Differentiate between:
    (i) *Oedogonium* and *Bulbochaete*
    (ii) Antheridia of Macrandrous and Nannandrous species of *Oedogonium*.
    (iii) Macrandrous and Nannandrous species.
    (iv) Thallus structure of *Oscillatoria* and *Oedogonium*.
    (v) Sexual reproduction in *Bulbochaete* and *Oedogonium*.
22. Give one important character of the following:
    (a) *Oedogonium* (b) *Bulbochaete*
23. Elucidate the mode of cap formation in *Oedogonium*.
24. Explain the nannandrous forms in *Oedogonium*.
25. (a) Draw graphic life cycle of macrandrous species of *Oedogonium*.
    (b) Describe cell structure and asexual method of reproduction in *Oedogonium*.
26. Draw net and labelled diagrams of the following:
    (a) Reproductive organs of *Oedogonium*. (b) Reproductive organs of *Bulbochaete*.
    (c) Cell division in *Oedogonium* (d) Cell structure in *Bulbochaete*.
    (e) Cell structure in *Bulbochaete*.
27. Write a brief account of asexual reproduction in *Bulbochaete*.
28. Draw graphic life-cycle of *Bulbochaete*.
29. Describe briefly the sexual reproduction in *Bulbochaete*.
30. By means of labelled diagrams only describe the life cycle of *Oedogonium* or *Bulbochaete*.

Objective Type

31. Select the correct answer:
    (i) Chloroplast in *Oedogonium* is
        (a) Cup shaped (b) Reticulate
        (c) Spiral (d) Stellate.
    (ii) *Oedogonium* is a member of class
        (a) Chlorophyceae (b) Phaeophyceae
        (c) Myxophyceae (d) Rhodophyceae.
    (iii) *Oedogonium*, *Bulbochaete* and *Oedocladium* belong to the order
        (a) Volvocales (b) Conjugales
        (c) Oedogoinales (d) Siphonales.
    (iv) Cap cells are characteristic of the genus
        (a) *Spirogyra* (b) *Oedogonium*
        (c) *Vaucheria* (d) *Chara*.
    (v) *Bulbochaete* belongs to the order
        (a) Oedogoinales (b) Conjugales
        (c) Siphonales (d) Volvocales.
    (vi) The major difference between *Oedogonium* and *Spirogyra* is that in *Oedogonium*.
        (a) the cells are longer than broad
        (b) the chloroplast is spiral

(c) sexual reproduction is through conjugation
(c) cap cells are absent.
(vii) Stephanokentian type of zoospores are present in
  (a) *Oedogonium* (b) *Spirogyra*
  (c) *Vaucheria* (d) *Chara.*
(viii) While *Oedogonium* and *Bulbochaete* are exclusively fresh water forms, *oedocladium* is
  (a) also fresh water form (b) marine
  (c) terrestrial (d) halophytic
(ix) The attachment of *Oedogonium* to substratum is effected by
  (a) the lowest cell directly (b) holdfast to the attachment disc
  (c) both of the above (d) none of the above.
(x) The nannandrium develops from
  (a) Alanospore (b) Zoospore
  (c) Androspore (d) Hypnospore
(xi) The zoospores of *Oedogonium* are
  (a) nonciliate (b) uniciliate
  (c) biciliate (d) multicilliate.
(xii) Idiandrosporous forms of *Oedogonium* are species bearing
  (a) Antheridia
  (b) Oogonia
  (c) Androsporangia and oogonia on the same filaments
  (d) Androsporangia and oogonia on the separate filaments.
(xiii) The nannandrium is so called because
  (a) It is a small plant (b) it is a male plant
  (c) it produces small sperms (d) it is a small male plant.
(xiv) During cell division in *Oegonium* the cell that receives the cap is
  (a) the lower cell
  (b) the upper cell
  (c) both the lower and upper cells
  (d) sometimes the lower cell and sometimes the upper cell.

QUESTIONS

Essay Type

1. Describe the structure and reproduction of *Coleochaete*.
2. Describe the thallus structure and reproduction in *Fritschiella*.
3. Describe the methods of reproduction in *Fritschiella* and compare it with that of *Coleochaete*.
4. (a) Describe in detail the thallus organisation in *Coleochaete*.
   (b) Describe the occurrence and structure of thallus of *Coleochaete*.
5. Give a comparative account of the structure of thalli in five important genera of the Chaetophorales.
6. Describe the sexual reproduction in *Coleochaete* and comment on the post fertilization change.
7. (a) Give a detailed account of sexual reproduction in *Coleochaete*.
   (b) Describe the vegetative structure in *Coleochaete*.

8. Describe the various grades of heterotrichy met with in the order Chaetophorales.
9. Describe the thallus structure and reproduction in *Stigeoclonium*.
10. Write a short account of organisation of thallus in *Draparnaldia* and *Draparnaldiopsts*. Compare cell structure in both the genera.
11. Give a brief account of asexual and sexual reproduction in *Draparnaldia* and *Draparnaldiopsis*.
12. Write about the organisation of thallus in various species of *Coleochaete*.
13. Discuss the plenomenon of alternation of generation in *Coleochaete* and compare it with that of *Batrachospsermum*.
14. Describe briefly the life history of *Coleochaete*.
15. Discuss the affinities and systematic position of *Chaetophorales*.
16. Give an account of structure and reproduction in *Trentepohlia*.
17. Describe the structure of thallus and reproduction in *Pleurococcus*.
18. Describe the asexual reproduction or methods of pesennation in different person of chaetophorales.
19. Draw labelled diagrams of reproductive organs of *Coleochaete*.
20. Give the most important characteristic of the following:
    (i) *Coleochaete*
    (ii) *Stigeoclonium*
    (iii) *Draparnaldia*
    (iv) *Draparnaldiopsis*
    (v) *Fritschiella*
    (vi) *Trentepohlia*
    (vii) *Pleurococcus*
21. Write short notes on:
    (i) Thallus structure of *Coleochaete*
    (ii) Heterotrichous habit
    (iii) Sexual reproduction in *Coleochaete*
    (iv) Asexual reproduction ins *Stigeoclonium*
    (v) Methods of perennation *Fritschiella*
    (vi) Cell structure in *Draparnaldia*
    (vii) Sexual reproduction in *Draparnaldiopsis*
    (viii) Thallus structure of *Fritschiella*
    (ix) Characteristic features of *Trentepohlia* and *Pleurococcus*
    (x) Spermocarp.
22. Draw well labelled diagrams of the thallus and mention systematic position of the following.
    (i) *Coleochaete*
    (ii) *Fritschiella*
    (iii) *Stigeoclonium*
    (iv) *Draparnaldiopsis*
23. Draw labelled diagrams of the thallus of *coleochaete* with spemocarp.
24. Write arconcise account of reproduction of *Coleochaete*.
25. Discuss in brief the importance of alternation of generation in *Fritschiella*.
26. Describe the salient features of Chaetophorales.
27. Write an explanation note on the sexual reproduction in *Coleochaete*.
28. Name the fruit body of *Coleochaete*.
29. Explain the most fertilisation changes in *Coleochaete*.
30. Who is credited with the discovery of *Fritschiella*?
31. (a) Name the families of Chaetophorales.
    (b) Name the alga where the thallus consists of erect and prostrate systems.

Objective Type

32. Select the correct answer:
    (i) The genus *Frtschiella* was discovered by
        (a) V.J. Chapman
        (b) TV. Desikachary
        (c) BN Prasad
        (d) M.O.P. Iyengar.
    (ii) The heterotrichous habit means
        (a) Prostrate and erect branches
        (b) Long and short branches
        (c) Branches modified into leaves and air bladders
        (d) Rhizoidal and photosynthetic branches.
    (iii) Fruiting body in *Coleochaete* is called
        (a) Ascocarp
        (b) Basidiocarp
        (c) Spermocarp
        (d) Cremocarp
    (iv) The spermocarp of *Coleochaete* produces
        (a) Spermatozoids
        (b) Meiozoospores
        (c) Oospores
        (d) Mitospores.
    (v) Which of the following is produced due to post fertilisation change in *Coleochaete*?
        (a) Cystocarp
        (b) Mericarp
        (d) Pericarp
        (d) Spermocarp
    (vi) Which of the following is terrestrial?
        (a) *Fritschiella*
        (b) *Draparnaldiopsis*
        (c) *Draparnaldia*
        (d) *Stigeoclonium*.
    (vii) The chloroplast in *Stigeoclonium* is
        (a) Cup shaped
        (b) Girdle shaped
        (c) Reticulate
        (d) Ribbon shaped.
    (viii) In *Stigeoclonium* the asexual reproduction takes place by
        (a) Quadriflagellate macrozoospores
        (b) Quadriflagellate microzoospores
        (c) Both of the above
        (d) None of the above.
    (ix) The thallus in Chaetophorales is
        (a) Unicellular
        (b) Filamentous
        (c) Discoid
        (d) Heterotrichous.
    (x) In *Draparnaldia* the gametes are
        (a) Isogametes and quadriflagellates summers
        (b) Heterogamous and quadriflagellate swarmers
        (c) Isogametes and biflagellate swarmers
        (d) Heterogametes and quadriflagellate swarmers.
    (xi) In which of the following diploid thallus reproduce asexually by aplanospores and akinetes?
        (a) *Volvox*
        (b) *Draparnaldia*
        (c) *Draparnaldiopsis*
        (d) *Ectocarpus*.
    (xii) Macrozoospores and micromeiozoospores are produced in
        (a) *Stigeoclonium*
        (b) *Draparnaldia*
        (c) *Draparnaldiopsis*
        (d) *Volvox*.

(xiii) Curved plate like chloroplasts with 2-8 pyrenoid are present in the cells of
   (a) *Stigeoclonium*
   (b) *Draparnaldia*
   (c) *Draparnaldiopsis*
   (d) *Fritschiella.*

(xiv) The common method of perennation in *Fritschiella* is through
   (a) Daughter colonies
   (b) Protonema
   (c) Funnel shaped tuberous bodies
   (d) All of the above.

(xv) Quadriflagellate and biflagellate meiozoospores one produced in
   (a) *Stigeoclonium*
   (b) *Draparnaldia*
   (c) *Fritschiella*
   (d) *Draparnaldiopsis.*

(xvi) Cells which of the following has pectose caps and stratified cell walls in the cells?
   (a) *Stigeoclonium*
   (b) *Fritschiella*
   (c) *Trentepohlia*
   (d) *Draparnaldia.*

(xvii) Pedicellate, funnel shaped and sessile sporeangia are produced in
   (a) *Trentepohlia*
   (b) *Draparnaldia*
   (c) *Stigeoclonium*
   (d) *Fritschiella*

(xviii) In *Coleochaete* the chloroplasts
   (a) Cupshaped
   (b) Collar shaped
   (d) Parietal and laminate
   (d) Raniform

(xix) The sexual reproduction in *Coleochaete* is
   (a) Isogamous
   (b) Anisogamous
   (c) Oogamous
   (d) All of the above.

(xx) All the species of *Coleochaete* are
   (a) Marine
   (b) Fresh water
   (c) Terrestrial
   (d) Epiphytic.

QUESTIONS

Essay Types

1. Give an illustrated account of the life history of *Zygnema*.
2. Describe the vegetative structure of the thallus and mode of scalariform conjugation, in *Spirogyra*.
3. Give a brief account of lateral conjugation in *Spirogyra*. How will you distinguish a filament containing zygospores formed as a result of lateral conjugation from the one containing zygospores formed as a result of scalariform conjugation?
4. Enumerate the salient features in the life history of *Zygnema*, *Mougeotia* and *Spirogyra*.
5. Compare the thallus structure and life history of *Zygnema*, *Mougeotia* and *Spirogyra*. Point out the common characteristics in them.
6. Give an account of thallus structure, mode of reproduction and life history of *Closterium*.
7. Describe the thallus structure and reproduction in *Cosmarium*.
8. Compare the thallus structures of *Cosmarium* and *Spirogyra*.
9. Name any constricted desmid and describe its life history in detail.
10. Compare the life histories of constricted desmid with that of an unconstricted desmid.
11. Describe in detail the structure and life cycle of *Zygnema*.

12. Describe the cell structure of *Cosmarium* and explain the method of cell division with the help of suitable diagrams.
13. How does *Zygnema* reproduce sexually? Describe various methods.
14. Compare the cells structure in *Cosmarium*, *Spirogyra*, *Ulothrix* and *Oedogonium*.

Short Answer Type

15. Identify the following by giving one important character:
    (a) *Spirogyra*
    (b) *Zygnema*
    (c) *Mougeotia*
    (d) *Closterium*
    (e) *Cosmarium*
16. Draw neat and labelled diagrams of nay three of the following and mention some of their characters briefly:
    (a) *Spirogyra*
    (b) *Mougeotia* – cell structure
    (c) *Zygnema* thallus
    (d) Cell structure of *Closterium*
    (e) Cell structure of *Cosmarium*
17. (a) Describe the conjugation in *Zygnema*.
    (b) List the diagnostic features of the order Zygnematales or Conjugales.
18. The filaments of *Spirogyra* loose their colour after conjugation. Why?
19. Write short notes on:
    (a) Lateral conjugation
    (b) Differences between zygospores and oospores
    (c) Chloroplast in *Zygnema*
    (d) Differentiation of sex in *Spirogyra*
    (e) Conjugation
    (f) Cell structure in *Zygnema* and *Spirogyra*
20. Name an alga which shows lateral conjugation.
21. Name an alga which shows scalariform conjugation.
22. Name a constricted desmids.
23. Name an unconstricted desmid.

Objective Type

24. Fill in the blanks:
    (i) The botanical name of Pond Silk is ....
    (ii) The characteristic feature of the order conjugales is ....
    (iii) In .... the chloroplast is stellate.
    (iv) In .... the chloroplast is axile plate type.
    (v) Two parietal plate type chloroplasts are present in ....
    (vi) The common method of vegetative reproduction is through ....
    (vii) The desmids are .... organisms having two .... chloroplasts.
    (viii) Spiral chloroplast is the characteristic feature of ....
    (ix) Conjugation between two morphologically similar and physiologically dissimilar gametes results in a structure called ....
    (x) .... is the example of unconstricted desmid.

(xi) The example of a constricted desmid is ....
(xii) Azygospores are produced as a result of ....
(xiii) In *Cosmarium* the chloroplast is single, large and ....
(xiv) Generally numbers pyrenoids are present in the chloroplast of the members of ....
(xv) The median construction in *Cosmarium* is known as ....

25. Select the correct answer:
    (i) *Cosmarium* belongs to the order
      (a) Volvocales
      (b) Ulotrichales
      (c) Conjugales
      (d) Siphonales
    (ii) The shape of the chloroplast in *Zygnema* is
      (a) Ribbon shaped
      (b) Girdle shaped
      (c) Reticulate
      (d) Cup shaped.
    (iii) Scalariform conjugation takes place in
      (a) *Oedogonium*
      (b) *Zygnema*
      (c) *Ulothrix*
      (d) *Volvox*.
    (iv) The shape of the chloroplast in *Cosmarium* is
      (a) Axial with radiating plate
      (b) Spiral
      (c) Reticulate
      (d) Cup shaped
    (v) The zygospores of *Cosmarium* after germination gives rise to
      (a) One daughter cell
      (b) two daughter cells
      (c) Three daughter cells
      (d) four daughter cells.
    (vi) What is the diploid stage in *Spirogyra* called?
      (a) *Zygote*
      (b) Oospores
      (c) Aplanospores
      (d) Zygospores.
    (vii) Vegetative reproduction in *Zygnema* takes place by
      (a) Zygospores
      (b) Aplanospores
      (c) Fragmentation
      (d) Oospores.
    (viii) Pond silk or 'Water silk' is the common name of
      (a) *Spirogyra*
      (b) *Ulothrix*
      (c) *Mucor*
      (d) *Oscillatoria*.
    (ix) Sometimes a ladder like structure is exhibited in filaments of *Spirogyra*. This is due to
      (a) Direct conjugation
      (b) Scalariform conjugation
      (c) Lateral conjugation
      (d) Asexual reproduction.
    (x) The members of conjugales are
      (a) exclusively fresh water
      (b) exclusively marine
      (c) 50 percent fresh water
      (d) 90% fresh water.
    (xi) The name pond silk is given to *spirogyra* filaments because
      (a) The filaments secrete a silky substance
      (b) The cellulose layer secretes a mucilage like substance
      (c) The pectose layer of filaments become mucilaginous.
      (d) The filaments are not rough.
    (xii) Important feature of *Spirogyra* that distinguishes itself from *Oedogonium* is
      (a) the nucleus
      (b) its habitat
      (c) the size of vacuole
      (d) the spiral chloroplast.

(xiii) One of the important features of *Maugeotia* is
  (a) the spiral chloroplast
  (b) the plate like axial chloroplast
  (c) the reticulate chloroplast
  (d) the cup shaped chloroplast.

(xiv) When the zygospore in *Spirogyra* germinates
  (a) all the nuclei produced are functional
  (b) only one haploid nucleus is functional
  (c) only one nucleus is inactive and non functional
  (d) the diploid nucleus does not divide

(xv) On germination the zygospore nucleus in *zygnema* divides meiotically to produce
  (a) 4 haploid nuclei and all are functional
  (b) 4 haploid nuclei and only one nucleus is functional while others abort
  (c) 4 haploid nuclei, out of which two nuclei abort and two nuclei are functional
  (d) 4 haploid nuclei out of which only one nucleus aborts and three nuclei are functional.

(xvi) Nucleus in *Spirogyra* is suspended in the cell with the help of
  (a) Vacuole
  (b) cytoplasmic strands
  (c) chloroplast
  (d) all of the above

(xvii) Which of the following statements is correct about *Spirogyra*?
  (a) Asexual reproduction takes place by zoospores.
  (b) Filaments showing scalariform conjugation are homothallic.
  (c) Filaments showing lateral conjugation are homothallic.
  (d) Filaments showing lateral conjugation may be homothallic.

(xviii) Meiosis in *Spirogyra* takes place in
  (a) Zygospores at the time of their germination
  (b) Zygospores at the time of their formation
  (c) Zoospores when they are formed
  (d) Zoospores when they germinate.

QUESTIONS

Essay Types

1. Write an illustrated account of the sexual organs of *Chara* or *Nitella* and the germination of its oospores. Mention how it differs from other green algae.
2. Give an illustrated account of the method of sexual reproduction in *Chara* or *Nitella*.
3. Describe the structure and development of sex organs of *Chara* or *Nitella*.
4. Describe the structure and reproduction (life history) or *Chara* and discuss its position in alage.
5. Describe the structure and reproduction in *Chara* or *Chlamydomonas* or *Ectocarpus*.
6. Give an illustrated account of the sexual reproduction in *Chara*.
7. Describe the life cycle of *Chara* or *Ectocarpus* with the help of labelled diagrams only.
8. give an account of the life history of *Chara*.
9. Describe reproduction in *Chara* with the help of diagrams.
10. With the help of labelled diagrams, describe the sex organs of *Chara*.

11. Describe the sexual reproduction in *Nitella* with the help of labelled diagrams.
12. Describe the structure and reproduction in *Nitella*.
13. Describe the life cycle of *Nitella*. How does *Nitella* differ from *Chara*.
14. Describe the sex organs of *Nitella*.

Short Answer Type

15. Write short notes on:
    (a) Structure of sex organs in *Chara* and its systematic position.
    (b) Sex organs of *Chara* (Globule and Nucule)
    (c) Antheridicu or globule of *Chara*
    (d) Characteristic features of *Chara*.
    (e) Structure and morphology of antheridium in *Chara*.
    (f) Thallus structure of *Chara*.
    (g) Nucule of *Chara*.
    (h) Difference between Globule and Nucule.
    (i) Sexual reproduction in *Chara*.
    (j) Shrueture of male and female sex organs of *Chara*.
    (k) Vegetative reproduction in *Chara*.
    (l) Systematic position of Chara.
    (m) Female reproductive structure of *Chara*.
    (n) Cortication of *Chara*.
    (o) Thallus structure in *Nitella*.
    (p) Sexual reproduction in *Nitella*.
    (q) Post-fertilisation changes in *Nitella*.
    (r) Sex organs in *Nitella*.
    (s) Nucule
16. (a) Draw neat and labelled diagrams of *Chara*.
    (b) Draw the diagram of nucule of *Chara* and label the parts.
17. Draw neat and labelled diagrams of the following:
    (a) Structure of globule in *Chara*.
    (b) Sex organs of *Chara*.
    (c) L.S. Globule of *Chara*.
    (d) Thallus structure of *Chara*.
    (e) Branch of *Chara* with reproductive organs.
    (f) Globule and Nucule in *Chara*.
18. Why *Chara* is commonly called stonewort?
19. Describe post fertilisation changes in *Chara*.
20. Distinguish between nodal and internodal cells of *Chara*.
21. What are the functions of Crown cells in *Chara*?
22. Mention the characteristic features of *Nitella*.
23. Compare the thallus structure of *Nitella* with that of *Chara*.
24. How does *Chara* differ from *Nitella*?
25. Comment upon the features of special interest in *Chara* and *Nitella*.

Objective Type

26. Fill in the blanks:
    (i) Stonewort is the common name for _____ .
    (ii) _____ is the common name for *Chara*.
    (iii) In globule of *Chara* the centre of each shield cell extends towards node-like cylindrical cell known as _____ .
    (iv) There is no _____ formation in *Chara*.
    (v) Amylum stars are densely filled with _____ in *Chara*.
    (vi) Sex organs in *Chara* are produced at _____ .
    (vii) *Chara* is a _____ aquatic algae.
    (viii) The chloroplast in *Chara* lack _____ .
    (ix) The common mode of vegetative reproduction in *Chara* is by _____ .
    (x) In *Charales* the antheridium is known as _____ and the oogonium is known as _____ .

27. Select the correct answer:
    (i) The oogonium in *Chara* has five coronal cells while the number of coronal cells in *Nitella* is
       (a) 5   (b) 6
       (c) 8   (d) 10.
    (ii) The disc shaped chloroplast lack pyrenoids in
       (a) *Chara*   (b) *Chlamydomonas*
       (c) *Spirogyra*   (d) *Oedogonium*.
    (iii) The male sex organ in *Chara* is commonly known as
       (a) Antheridium   (b) Globule
       (c) Nucule   (d) Zygote.
    (iv) The female sex organs in *Chara* is known as
       (a) Antheridium   (b) Globule
       (c) Nucule   (d) Zygote.
    (v) Reproduction by Amylum stars, and secondary protonemata occurs in
       (a) *Chara*   (b) *Nitella*
       (c) *Oedogonium*   (d) *Spirogyra*.
    (vi) Manubrium is found in antheridium of
       (a) *Chara*   (b) *Chlamydomonas*
       (c) *Oedogonium*   (d) *Spirogyra*.
    (vii) The members of Charales are popularly called as stoneworts because
       (a) They grow on stones.
       (b) They have been recorded in Devonian and Silurian period.
       (c) The thallus bears an incrustation of lime.
       (d) There are cortical species.
    (viii) Discoid chloroplast occurs in
       (a) *Ulothrix*   (b) *Chara*
       (c) *Chlamydomonas*   (d) *Zygnema*.

(ix) Asexual reproduction by zoospores is completely absent in
- (a) *Oedogonium*
- (b) *Cladophora*
- (c) *Coleochaete*
- (d) *Chara*

5. Discuss/describe about the systematic position of *Vaucheria*.
6. Describe the structure and mode of reproduction in *Vaucheria*.
7. With the help of suitable sketches, describe the life history of *Vaucheria*. Assign it to its systematic position giving reasons.
8. Give an account of reproduction and systematic position of *Vaucheria*.
9. With the help of labelled diagrams only, illustrate the life cycle of *Vaucheria*.
10. Give a diagrammatic sketch of the life cycle of *Vaucheria*.
11. Describe the various modes of reproduction in *Vaucheria*.
12. Write an account of resemblances and differences in the life history of *Albugo* and *Vaucheria*.
13. Give an account of thallus structure and reproduction in *Vaucheria*.
14. Enumerate the structural and nutritional peculiarities of *Vaucheria* which distinguish it from green algae.
15. Describe structure and life history of *Vaucheria*. How does it differ from green algae in its nutritional peculiarities?
16. Give the distinguishing features of the class xanthophyceae. Also give its classification.
17. Give an account of habitat and structure of thallus of *Botrydium*.
18. Describe the modes of asexual reproduction in *Botrydium*.
19. Give a complete account of structure and life history of *Botrydium*.
20. (a) Give an illustrated account of thallus structure ad modes of reproduction in *Botrydium*.
    (b) Describe the life history of *Vaucheria*. Assign it to its systematic position giving reasons.

Short Answer Type

21. Identify the following and mention their important characters with the help of diagrams. Give their classification as well:
    (i) *Vaucheria*.
    (ii) *Botrydium*
22. Draw neat and labelled diagrams of the following:
    (i) Sex organs of *Vaucheria*.
    (ii) Antheridium and oogonium of *Vaucheria*.
    (iii) *Vaucheria* filaments with sex organs.
    (iv) Thallus structure of *Botrydium*.
    (v) Sex organs of *Botrydium*.
23. Differentiate between *Vaucheria* and *Botrydium*.
24. Describe the salient features of *Botrydium*.
25. Describe the thallus structure of *Botrydium*.
26. Write short notes on:
    (a) Synzoospore
    (b) Sexual reproduction in *Vaucheria*.
    (c) Fertilisation of *oogonium* in *Vaucheria*.
    (d) Mode of reproduction in *Vaucheria*.
    (e) Asexual reproduction in *Vaucheria*.

(f) Distinguishing features of xanthophyceae.
(g) Taxonomic position of *Vaucheria*.
(h) Sex organs of *Vaucheria*.
(i) *Vaucheria*
(j) Zoospore formation in *Vaucheria*.
(k) Filament of *Vaucheria*.
(l) Asexual reproduction in *Botrydium*.
(m) Sexual reproduction in *Botrydium*.
(n) Resting bodies.

27. Describe a siphonaceous thallus.
28. Write two characters to differentiate the thalli of *Vaucheria* and *Ectocarpus*.
29. Name an alga which produces synzoospores.
30. Compare the distinguishing feature of cyanophyceae, xanthophyceae and Phaeophyceae.
31. Name any syphonaceous alga and give its systematic position.
32. Give the structure of chloroplast/chromatophore of *Vaucheria*.
33. Give the botanical name of the alga in which main thallus is branched coenocytic and siphonaceous.
34. Differentiate between aplanospore and synzoospore.
35. Name the synzoospore producing alga.
36. Give the systematic position of *Vaucheria*.
37. Describe synzoospore of *Vaucheria*.
38. Describe sexual reproduction in *Vaucheria*.
39. Why is *Vaucheria* called 'Golden yellow Algae'? Discuss its systematic position.
40. Explain the structure of plant body in *Vaucheria*.
41. Define synzoospore.
42. Give an account of haploid and vegetative structure of *Vaucheria*.
43. Is the thallus of *Vaucheria* a gametophyte or sporophyte? Justify your statement.
44. Why are *Caulerpa* and *Vaucheria* put under different orders even though thallus of both are regarded as unicellular or siphonaceous?
45. *Vaucheria* is like a phycomycetous fungus but for chromatophores. Why?
46. Why is *Vaucheria* not placed under chlorophyceae although it contains chlorophyll?
47. Classify *Botrydium* and describe its salient features.

Objective Type

48. Fill in the blanks:
    (i) *Vaucheria* belongs to class ....
    (ii) Synzoospores are produced in ....
    (iii) *Botrydium* belongs to family ....
    (iv) The rhizoid branches bear cyst like tuberous structures in ....
    (v) The zoospores in *Botrydium* is ....
    (vi) In *Vaucheria* the most common method of asexual reproduction is through multiflagellate multinucleate structure known as ....
    (vii) In *Botrydium* sexual reproduction is ....
    (viii) In *Botrydium* sexual reproduction is ....

(ix) The food reserve in class xanthophyceae is ....
(x) The sexual reproduction in *Vaucheria* is ....

49. Select the correct answer.
    (i) Which of the following is coenocytic and multinucleate?
        (a) *Vaucheria*   (b) *Polysiphonia*
        (c) *Spirogyra*   (d) *Sargassum.*
    (ii) Reserve food material in *Vaucheria* is
        (a) Starch   (b) Protein
        (c) Oil   (d) All of the above.
    (iii) In *Vaucheria* cross walls are
        (a) never present
        (b) sometimes present
        (c) always present
        (d) Present only at the time of sexual reproduction.
    (iv) Oil droplets are found in
        (a) *Sargassum*   (b) *Cladophora*
        (c) *Polysiphonia*   (d) *Vaucheria.*
    (v) Asexual reproduction in *Vaucheria* takes place by zoospores which are
        (a) Uniflagellate   (b) Biflagellate
        (c) Quadriflagellate   (d) Multiflagellate.
    (vi) Which of the following has coenocytic thallus?
        (a) *Hydrodictyon*   (b) *Vaucheria*
        (c) *Ectocarpus*   (d) *Spirogyra.*
    (vii) *Botrydium* belongs to class
        (a) Chorophyceae   (b) Charophyceace
        (c) Xanthophyceae   (d) Bacillariophyceae.
    (viii) Synzoospores are present in
        (a) *Botrydium*   (b) *Oedogonium*
        (c) *Vaucheria*   (d) *Chlamydomonas.*
    (ix) Syphonaceous thallus is found in
        (a) *Vaucheria*   (b) *Zygnema*
        (c) *Spirogyra*   (d) *Ectocarpus.*
    (x) The alga in which thallus is branched, coenocytic and syphonaceous is
        (a) *Vaucheria*   (b) *Ectocarpus*
        (c) *Chlamydomonas*   (d) *Zygnema.*
    (xi) The multiflagellated zoospore in *Vaucheria* is generally interpreted as
        (a) Synzoospore   (b) sporangium
        (c) Zoosporangium   (d) All of the three.
    (xii) Resting bodies are present in
        (a) *Botrydium*   (b) *Vaucheria*
        (c) *Ectocarpus*   (d) *Spirogyra.*
    (xiii) In *Botrydium* the sexual reproduction is
        (a) Isogamous   (b) Anisogamous
        (c) Heterogamous   (d) Oogamous.

(xiv) In *Vaucheria* the sexual reproduction is
    (a) Isogamous   (b) Anisogamous
    (b) Heterogamous   (d) Oogamous

(xv) *Vaucheria* is usually terrestrial or
    (a) Fresh water forms   (b) Marine from
    (c) Saline forms   (d) Alkaline forms.

11 Bacillariophyceae

BACILLARIOPHYCEAE (DIATOMS)

Occurrence

Diatoms are exclusively found in salt waters as well as in freshwaters. They occur in all aquatic habitats forming an important part of the vegetation. Some of them are *planktonic* (free-floating) in fresh as well as marine waters. Others are submerged forms (benthic). They may be *epiphytic* on other aquatic vegetation. Epiphytic forms may produce brown fluffy growths and the plankton forms brown scums. They may also occur on damp soils. Many planktonic species are extremely sensitive to light and salinity of water. Diatoms are the chief source of food for marine fishes.

Diatoms may form fossil deposits due to the accumulation of siliceous cell walls. These deposits are known as diatomaceous earth. The deposits may orginate in freshwater or in oceans. The largest deposits on the surface of earth form beds of several miles length and 700 ft. in thickness.

Colonial Forms

Although diatoms are mostly unicellular, colonial species are also represented. Mucilage produced by the frustules help in forming colonies. The colonial forms show the following patterns:

1. Many cells may be enveloped in a common mucilage. In some genera, mucilage has a tubular structure. e.g., *Cymbella, Nitzchia,* etc. In *Navicula,* a large number of cells occur in tubular envelops which are abundantly branched forming a bushy colony.
2. Cells may produce mucilage locally to form stellate colonies as in *Asterionella,* filamentous structures as in *Melosira* or zig-zig colonies as in *Grammatophora.*
3. A colony may be formed by the cells joining together by special outgrowths like species as in *Chaetoceros.*

Structure

Most of the diatoms are unicellular varying extremely in shape (*Pinnularia, Biddulphia, Surirella,* etc).

Diatom Cell (Fig. 11.1)

Diatoms are most fascinating and offer us a sight of pleasure when viewed under a microscope, because of the very fine sculptured cell wall. The classification of the diatoms is based on the structure of the wall and ornamentation. The diatom cell is like a box having two halves, the upper half overlapping the bottom half. The whole diatom cell is also referred to as *frustule*. The upper half is the *epitheca* and it fits closely over the bottom half called the *hypotheca*. The diatom cell may be like an oblong box in the group *Pennales,* and like a circular box or petri dish in the second group called *Centrales*. Thus, a diatom cell has two valves and they correspond to the top and bottom parts of a

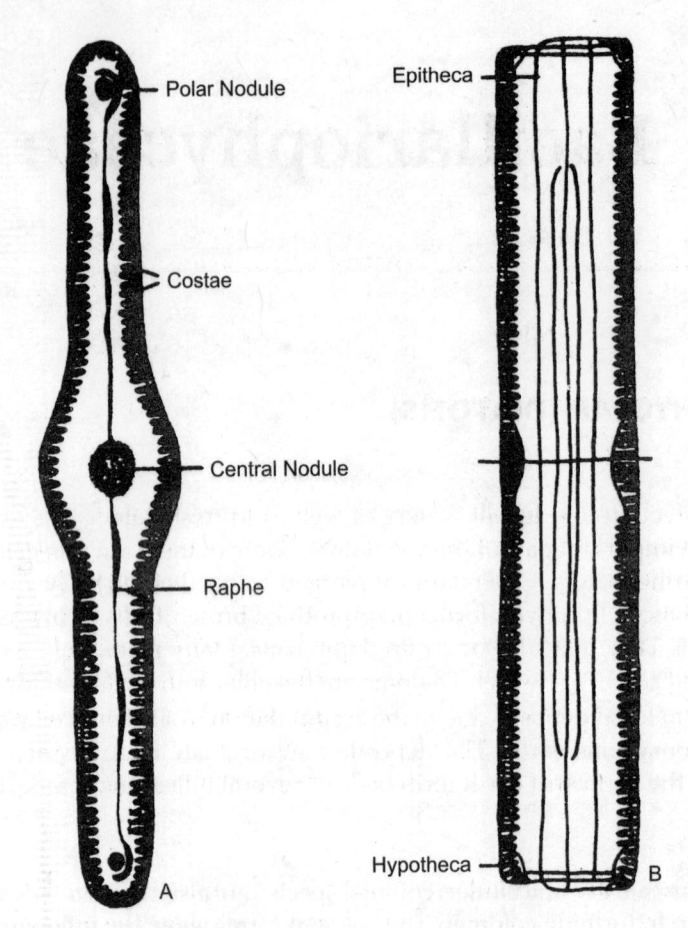

Fig. 11.1 *Diatoms*—Cell structure.

box. The edges of epitheca and hypotheca of diatom cell overlap forming a girdle. The edges are joined to connecting band or *cingulum*. Diatom cell can be seen from two views: *(a) valve view* from which the valve can be seen and *(b) girdle viewe* when the connecting band can be seen.

The valves are not much capable of enlargement, but the connecting band may grow, separating apart the epitheca and hypotheca more and more.

The centrales are more abundant in marine waters and much bigger than those inhabiting freshwater. They are radially symmetrical showing ornamentation around a central point. The pennales which are usually freshwater forms are bilaterally symmetrical along (1) the long axis and (2) in transverse plane at right angles to the previous one.

Structure of the Valves

Diatom frustules have silicified walls, Growth of diatoms is dependent on silicon concentration. The proportion of the silica in the walls varies in different species. The process of formation of a silicified wall is controlled by cytoplasm and requires energy. The organic component of the cell wall is described as pectin but there is no chemical evidence.

The cell wall shows characteristic ornamentations so that diatoms will be fascinating when observed under a microscope. The ornamentation is due to the markings on the valve surface. Electron microscope has revealed that a thickened wall shows four types of markings.

These markings are due to:
1. *Punctae* which are small perforations of the valve surface; they form fine lines called *striae*.
2. *Areolee* which are larger depressed structures.
3. *Costae* or *ribs* formed by heavy deposition of the silica, and
4. *Canaliculi* which are narrow tubular canals passing through the wall.

These types of markings may be radially arranged as in *centrales* or on either side of the axial field as in *pennales*. The *axial field* corresponds to the median longitudinal line of a valve. The narrow axial area is thickened at either ends, forming *polar nodules* and at the middle forming the *central nodule*. In some species, a narrow longitudinal slit is present in the axial field and it is called *raphe*. The raphe is asymmetrical in *Cymbella*. The raphe facilitates free exchange of gases between the protoplasm and the environment. In some species, a lighter region marks the axial field giving a superficial appearance of a raphe and it is called *pseudoraphe*. Both the valves of a diatom frustule may have a raphe *(Pinnularia)* or one valve may have a raphe and another a pseudoraphe *(Cocconeis)*.

Protoplast

Protoplasm forms a lining layer inner to the cell wall. There is a conspicuous central vacuole. In many *Centrales*, several strands of cytoplasm traverse across the vacuole; the nucleus lies in the cytoplasm inner to the cell wall, and there are several discoid chromatophores.

In *Pennales*, a broad band of cytoplasm cuts transversely across the vacuole and the nucleus is embedded in it; the chromatophores are 1 or 2 only and are like large lobed plates (they are several in Centrales).

Pyrenoids may be present embedded in the chromatophore or free from them. They are naked without starch sheath. Their function is not definite. Probably, they act like elaioplasts in the formation of oils. Mitochondria and Golgi bodies are also recorded.

Pigmentation

Typically, chromatophores are golden brown in colour but a few may be green. The green colour of chlorophyll *a* and *c* is masked by the xanthophylls. Fucoxanthin and dinoxanthin give a golden brown colour. A few diatoms are colourless with chromatophores and they lead a saprophytic life.

Assimilatory Products

Most commonly, the reserve food is in the form of droplets of fatty oils. These are formed as a result of photosynthetic activity. *Volutin* also may occur as globules. *Leucosin*, a type of carbohydrate may also be accumulated.

The Centrales and Pennales can be differentiated as follows:

Centrales	*Pennales*
1. Centrales are more widely distributed in sea than in freshwater,	Mostly freshwater forms but some are also represented in the sea.
2. They are radially symmetrical about a central point.	Pennate diatoms are bilaterally symmetrical or asymmetrical with reference to the axial strip.
3. Centrales have usually coarse markings.	Pennales have usually fine markings due to the punctae.
4. Chromatophores are usually many, discoid or irregular.	Chromatophores are usually one or two and lobed.

Contd.

Centrales	Pennales
5. The nucleus lies embedded in the cytoplasm next to the cell wall.	The nucleus is in the cytoplasmic strands connecting the two valves.
6. Centric diatoms never show movement.	Pennate diatoms with a raphe show jerky movement along the longitudinal axis.
7. Sexual reproduction is oogamous and spermatozoids are flagellate.	Sexual reproduction is isogamous and gametes are amoeboid.

Diatoms reproduce mainly by cell division. Sexual reproduction which leads to the formation of auxospores is sort of rejuvenation of the protoplast and it takes place after repeated cell divisions.

Cell Division (Fig. 11.2)

Cell division is a process by which diatoms multiply. Whenever cell division takes place, two daughter cells are formed in which one is smaller than the other. The nucleus divides first followed by chromatophores and pyrenoids. This is followed by the longitudinal division of the protoplast parallel to the valves. The two daughter protoplasts separate and so one of them gets the *epitheca* and the other the *hypotheca*. In both cases, the parent valve functions as the epitheca of the daughter cell and the new valve that is formed is always the hypotheca. Continuation of this process results in progressive diminution of size since the hypotheca of the parent becomes the epitheca of one of the daughter cells. This is known as Pfitzer's law. Decrease in size does not alter the nature of ornamentation.

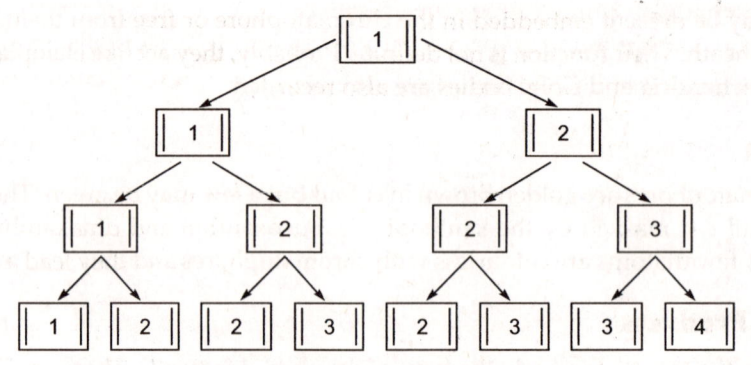

Fig. 11.2 *Diatoms*—Cell division.

Auxospores

When diatom cells become very small, they resort to sexual reproduction which leads to the formation of auxospores. The auxospores are without valves. They can enlarge very much and then form new valves so that the original size of the diatom is restored.

Auxospore Formation in Pennales

1. Two cells conjugate and form two auxospores (*Cymbella lanceolata*).
2. Two cells conjugate and produce a single auxospore (*Surirella splendida, Cocconeis placentular* var. *klinorachis*).
3. A single cell produces a single auxospore:
4. A diatom cell produces a single auxospore partenogentically (*Cocconeis placentula* var. *lineata*).

Auxospore Formation in Centrales

In Centrales, auxospores are not formed by conjugation. They are formed singly in each frustule. At one time, it was thought that auxospores are formed parthenogenetically. But in *Cyclotella meneghiniana*, meiosis was observed and two of the four haploid nucleus. Thus, auxospore formation is by *autogamy* comparable to that of *Amphora normani* of Pennales.

Microspores of Centrales *(Biddulphia mobilensis, Chaetoceros)*

In several members of centrales of planktons, microspores were recorded. Microspores are uniflagellate in some species and biflagellate in others. They are formed in numbers of 4, 8, 16, 32, 64 or 128 depending upon the species.

Resting Stages or Diatoms

Auxospores are not resting stages. They are the means by which the restriction in size is overcome. Diatom cells can withstand considerable desiccation. Some marine centrales (*Chaetoceros*) form endogenous cysts called *endospores* or *statospores*. They are formed by the contraction of the protoplast and secretion of a thick wall. The wall of the statospore may differ from the ornamentation of the parent cell.

Economic Importance

Diatoms inhabit fresh or marine waters. Diatoms form marine planktons useful as food to several fishes and other aquatic animals of oceans.

Diatomaceous earth is found in various parts of the world. These huge deposits were formed due to accumulation of siliceous cell walls at the bottom of water in which diatoms once lived in great numbers.

The diatomaceous earth is known as kieselguhr. It may originate in freshwaters or oceans. Deposits of marine species are formed due to geological changes. Sometimes these deposits extend for miles and may reach a height of 500 to 700 ft. Some of the subterranean marine deposits may be 2 to 3 thousand feet in depth.

Diatomaceous earth is commercially very useful. The following are the important uses:

1. It was used at one time for absorbing the nitroglycerine in making dynamite.
2. It is used in sugar refineries for filtration of liquids.
3. Used as an inert mineral filter in products like plastics.
4. Diatomaceous earth when added to paints, increases their visibility during nights.
5. It is used as an insulation against high temperatures in boilers and blast furnaces. It does not fail and will be resistant to shrinkage even at a temperature of 1000°F, and hence is preferred to asbestos or magnesia.
6. Used in preparing tooth pastes and silver polishes.

IMPORTANT QUESTIONS

Essay Type

1. Describe habitat, structure and reproduction in *Pinnularia*.
2. Give a general account of Diatoms.
3. Describe the modes of reproduction in Pennate diatoms.

4. Give an account of auxospore formation in Pennates.
5. Describe the distinguishing characters of Bacillariophyceae.
6. Describe the structure of pennate diatoms.
7. Describe cell structure of cell wall and reproduction in Diatoms.
8. Write a short essay on Economic importance of diatoms.

Short Answer Type

9. Write short notes on the following
    (i) Diatoms
    (ii) Auxospores
    (iii) Economic importance of Diatoms.
    (iv) Cell structure of Diatoms
    (v) Statospores
10. Write a note on phylogenetic relationships of Diatoms.
11. What are the distinguishing characters of Bacillariophyceae?
12. Why does the cell wall of dead *Pinnularia* not decompose?

Objective Type

13. Fill in the blanks:
    (i) The diatoms have been placed in the class _____.
    (ii) Of the diatoms, _____ forms are non motile.
    (iii) The example of diploid is _____.
    (iv) The saprophytic diatom is _____.
    (v) The cell wall in a diatom cell is _____.
    (vi) The diatom cell wall is often called as _____.
    (vii) In diatoms the cell division results in the formation of unequal _____.
    (viii) _____ are the characteristic spores of diatoms.
    (ix) Sexual reproduction is absent in _____.
    (x) The asexual method of reproduction in diatoms is _____.
    (xi) Statospores are usually formed in the genus _____.
14. Select the correct answer :
    (i) Auxospore formation is exhibited by
        (a) Desmids (b) Diatoms
        (c) Green algae (d) Red algae.
    (ii) The position of chromatophores in *Pinnularia* is
        (a) Axial (b) At the two ends
        (c) Opposed to the walls (d) Opposed to the girdle.
    (iii) Raphe in diatoms is visible in the
        (a) Girdle view (b) side view
        (c) Valve view (d) end view.
    (iv) The shape of the chromatophores in *Pinnularia* is
        (a) Discoid (b) Plate like
        (c) Star like (d) Rod like.

(v) The structure which is responsible for locomotion in diatoms is
   (a) Coste   (b) Raphe
   (c) Valve   (d) Girdle.

(vi) In which of the following Statospores are produced
   (a) *Melosira*   (b) *Cosmarium*
   (c) *Cladophora*   (d) *Spirogyra*.

(vii) In which of the following the daughter cells produced as a result of division are unequal?
   (a) Chlorophyceae   (b) Phaeophyceae
   (c) Xanthophyceae   (d) Bacillariophyceae.

(viii) Fresh water or marine planktons are largely made up of
   (a) Desmids   (b) Diatoms
   (c) Green algae   (d) Brown algae.

(ix) Oil chrysolaminarin and volutin are food reserves of
   (a) Desmids   (b) Diatoms
   (c) Green algae   (d) Brown algae.

(x) Diatomite which is economically important is obtained from the algae commonly known as
   (a) Green algae   (b) Blue green algae
   (c) Yellow algae   (d) Yellow brown algae.

12 Class: Phaeophyceae (Brown Algae): General Characters and Type Study

GENERAL CHARACTERS OF PHAEOPHYCEAE

The Phaeophyceae are distinguished by the following characteristic features:

1. The photosynthetic pigments are chlorophyll *a*, chlorophyll *c*, β-carotene, fucoxanthin, violaxanthin and other xanthophylls; generally, the carotenoids are more than the chlorophylls.
2. Food materials is stored in the form of laminarin and mannitol.
3. Fucosan vesicles which are whitish granules, are present.
4. The cell wall is formed of cellulose, fucinic acid and alginic acid.
5. The flagellated structures have two unequal lateral flagella of which the longer anterior one is pantonematic and the smaller which is posterior is acronematic.

Occurrence

The brown algae are marine forms with the exception of a few genera like *Pleurocladia, Bodonella*, etc. They are mostly cold water forms, but members of Dictyotales and *Sargassum* occur in warm waters. Members like *Laminaria, Dictyota, Cutleria* form thick vegetation in the sub-littoral zones. Several other members occur in the intertidal belt attached to the rocks. Some like *Sargassum* may be attached or free-floating.

Thallus Organization

Members of this order have heterotrichous filamentous body or pseudoparenchymatous or parenchymatous thallus. There are no unicellular or simple filamentous forms. The heterotrichous filamentous organisation is considered to be the simplest type e.g., *Ectocarpus*. Some members like *Macrocystis pyrifera* have a thallus which is nearly 100 metre long. *Postelsia* sp., resembles a palm. *Sargassum* has a 'stem' and 'leaves'. The thallus grows by means of an apical cell as in *Dictyota* or it may show characteristic trichothallic growth as in Ectocarpales. Intercalary growth is found in Laminariales. The inernal structure of some parenchymatous forms shows epidermis, cortex and medulla. A transverse section of stem or leaf of *Sargassum* shows an outer epidermis, cortex and central, medulla.

Cell Structure

The cell wall is formed of an outer mucilaginous layer containing alginic acid and fucinic acid and inner layer of cellulose. In the cytoplasm, small vacuoles and a few fucosan vesicles are present. The fucosan vesicles are refractile bodies and their nature is not well established. The cells contain parietal chromatophores. They are plate-like, discoid or ribbon-like, in members of Ectocarpales. The fine structure of the chromatophore shows that 3-4 thylakoids run in a parallel manner but they do not form grana. The cells are uninucleate with one or two nucleoli. Some members show centrosomes at the poles during the division of the nucleus. Mitochondria, Golgi bodies and endoplasmic reticulum are present.

The photosynthetic pigments are chlorophyll *a*, chlorophyll *c*, fucoxanthin and other xanthophylls. The carotenoids are in excess of the chlorophylls. The reserve foods are laminarin and mannitol.

Reproduction

The process of reproduction takes place by *vegetative, asexual* and *sexual* methods.

Vegetative reproduction. This is a very common method of reproduction. The thallus may break up into two or more parts by fragmentation and each part grows into an individual plant.

Asexual reproduction. In several members of Phaeophyceae, zoospores are formed which are motile biflagellate structures, as in *Ectocarpus*. In Dictyotales, non-motile tetraspores are formed and in Tilopteridales monospores are formed. The Fucales *(Sargassum)* lack asexual reproduction.

In Ectocarpales, zoospores are formed in *unilocular* sporangia and *plurilocular* sporangia. In the plurilocular sporangia formed on the sporophyte, zoospores are produced without meiosis. So, they are *diploid* and they give rise to the diploid plants. Thus, they help in the accessory reproduction of the same phase. The unilocular sporangia are also borne on the sporophyte. They produce zoospores after meiosis followed by a series of mitotic divisions. So, the zoospores are *haploid* and they give rise to the gametophytic plants. The zoospores formed in plurilocular as well as unilocular sporangia are similar in having two laterally inserted unequal flagella *e.g., Ectocarpus*. Some members of Phaeophyceae bear only unilocular sporangia.

In Dictyotales, the tetrasporic plant or sporophyte produces tetrasporangia. In each tetrasporangium, the diploid nucleus undergoes meiosis so that four uninucleate tetraspores are formed. These tetraspores germinate and produce gametophytic plants. In Tilopteridales, a single monospore is produced in each sporangium. The monospores give rise to the gametophytic plants.

Sexual reproduction. The process of sexual reproduction takes place through gametes. In Ectocarpales, the gametophytic plants have plurilocular sporangia which act as gametangia. These gametangia produce several biflagellate gametes that resemble the zoospores in external structure. The gametes are morphologically similar (isogamous) but in *Ectocarpus siliculosus*, they are physiologically anisogamous. The female gamete is sluggish while the male gametes are more active and motile. In *E. secundus*, the plurilocular gametangia are unequal in size. The megagametangia produce larger gametes and the microgametangia produce smaller gametes. The union between two such unequal gametes is known as anisogamy. The Ectocarpales are mostly isogamous but Cutleriales are distinctly anisogamaus. Several members of Phaeophyceae are oogamous. In Dictyotales, the antheridial sorus occurs on the male, and the oogonial sorus on the female gametophytes. But in Fucales, the antheridia and oogonia occur in cavities called conceptacles. The antheridia and oogonia may occur in the same conceptacle *(Fucus)* or in separate conceptacles as in *Sargassum*.

The antheridia of *Dictyota* are multilocular and each cell produces a single spermatozoid. Thus, the antheridium produces numerous spermatozoids. The spermatozoid is anomalous as it appears to have a single flagellum. But examination with Electron microscope revealed two basal granules but only the anterior flagellum is produced. The flagellum is not formed from the posterior basal granule. In Dictyotales, the egg is released from the oogonium and fertilization takes place externally. This union is oogamous. Dictyotales is unique in that the egg is not fertilized in the oogonium.

The Fucales are diploid plants and sex organs are produced in conceptacles. The antheridium is single-chambered and its nucleus first divides meiotically and then by mitotic divisions forming 64-128 biflagellate spermatozoids. The two flagella of the spermatozoid are lateral with the anterior one being of pantonematic type and the posterior one acronematic. The spermatozoids have a single nucleus and an eye-spot. The oogonia also occur in conceptacles and they are diploid. In *Fucus*, the oogonial nucleus divides meiotically and then by mitosis once, so that eight eggs are produced. In *Sargassum*, the oogonial nucleus divides like that of *Fucus*, but of the 8 haploid nuclei, three degenerate and the remaining nucleus along with the cytoplasm forms a single egg. Fertilization results in diploid zygote and germinates to give rise to the diploid plant.

Life Cycle

In Phaeophyceae, three types of life cycles are found. They are *isomorphic, heteromorphic* and *diploid types*. Ectocarpales are isogamous but Dictyotales are oogamous and both of them show isomorphic alternation of generations. The members of these orders have sporophytes and gametophytes which are morphologically similar. The Laminariales exhibit heteromorphic alternation of generations between a well developed macroscopic sporophyte and a microscopic filamentous gemetophyte. In the Fucales (*Fucus, Sargassum*), the plants are diploid and they lack a distinct alternation of generations. The sex organs are also diploid and the haploid stage in the life history is represented by the gametes. In Fucales, there is no gametophytic plant.

TYPE STUDY

Following type studies are discussed in detail:
1. *Ectocarpus*, 2. *Dictyota*, 3. *Sargassum*, 4. *Laminaria*, and 5. *Fucus*.

ECTOCARPUS

Class : Phaeophyceae

Order: Ectocarpales
Family: Ectocarpaceae

Occurrence

Ectocarpus is represented by several species. They are world-wide in distribution, but they are more common in colder seas. They appear as brown tufts attached to rocks, in littoral and sub-littoral regions.

Structure (Fig. 12.1)

Ectocarpus is considered to be the simplest *brown alga*. Typically, it exhibits *heterotrichy*, but sometimes such a character gets obscured by reduction of the erect system.

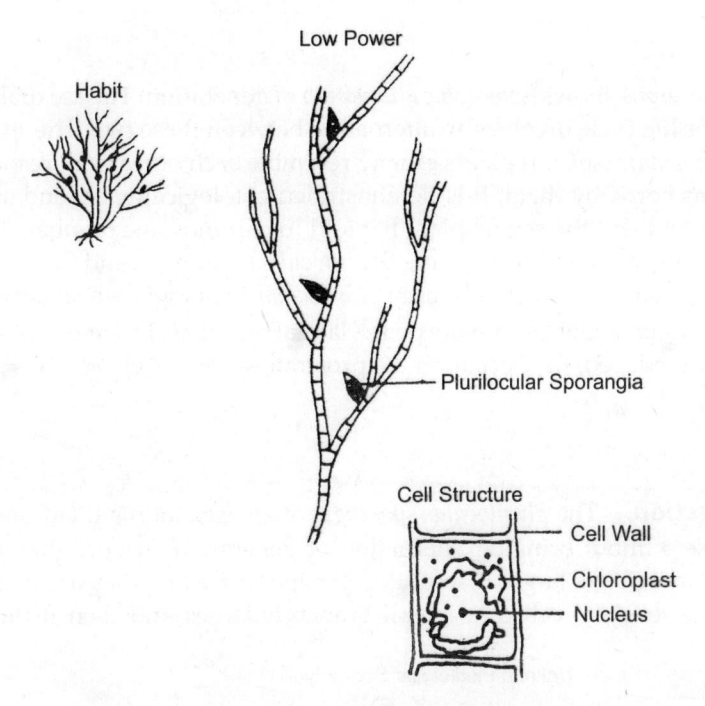

Fig. 12.1 *Ectocarpus*—Thallus and cell structure.

In *heterotrichous* forms, the prostrate system is represented by richly-branched filaments and the erect system by the copiously-branched upright threads, which form tufts. In some forms, the erect system is reduced to little-branched threads and the whole thallus looks as thought it is represented by the prostrate system only. The tips of the lateral branches usually end in long tapering colourless cells which form the hairs. In *E. siliculosus*, the lateral branches arise just below the septa. The thallus gets attached by basal rhizoids which, in some species, may even ensheath the older branches forming a sort of cortication.

The erect threads get intertwined together into a matted mass (*E. siliculosus*). The prostrate system wholly grows apically but the erect system, exhibits typically, *trichothallic* growth (intercalary growth) (Fig. 12.2). Growth is due to a series of meristematic cells present just below the hair. These ones cut off cells below and above; the cells formed below develop into typical vegetative cells with pigmentation and the cells cut off above, contribute to the growth of the hair in length.

The cells of *Ectocarpus* are *uninucleate* with many small disk-shaped or few band-like chromatophores (Fig. 12.1). The chromatophore contains a pyrenoid.

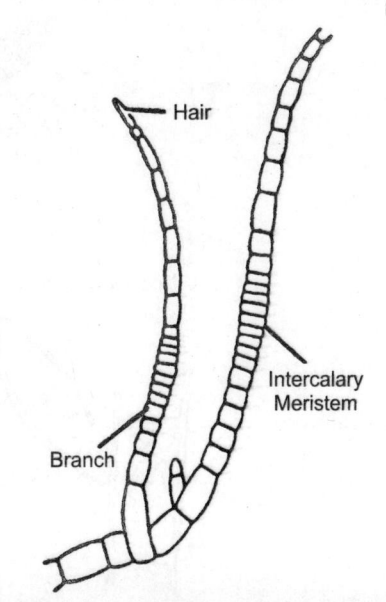

Fig. 12.2 *Ectocarpus*—Filaments showing intercalary meristem.

Life History

The life cycle of *Ectocarpus* shows *isomorphic alternation* of generation. The asexual and sexual plants are different and the life cycle involves an alternation between these two. The *sporophytic* (asexual) as well as the *gametophytic* (sexual) plants exactly resemble each other in all respects except for the reproductive organs borne by them. It is demonstrated cytologically, beyond any doubt that the asexual plant is diploid and the sexual plant haploid in chromosome number. The asexual plants produce two kinds of sporangia and they are: (i) *unilocular* sporangia, and (ii) *plurilocular* sporangia (Fig. 12.3). But the sexual plants produce only *plurilocular* sporangia which act as *gametangia*.

These sporangia occur singly, terminally on the lateral branches. In some species, where the erect system is very much reduced, they occur on the prostrate system itself which is pseudo-parenchymatous (*E. battersii*).

Reproduction

Asexual reproduction. The *plurilocular sporangia* of an asexual plant aid multiplication of the same asexual phase without bringing alternation of generations. Hence they are called *neutral sporangia*. These sporangia are larger than unilocular sporangia and they are multicellular. During its development the terminal cell of a lateral branch enlarges and then it undergoes repeated

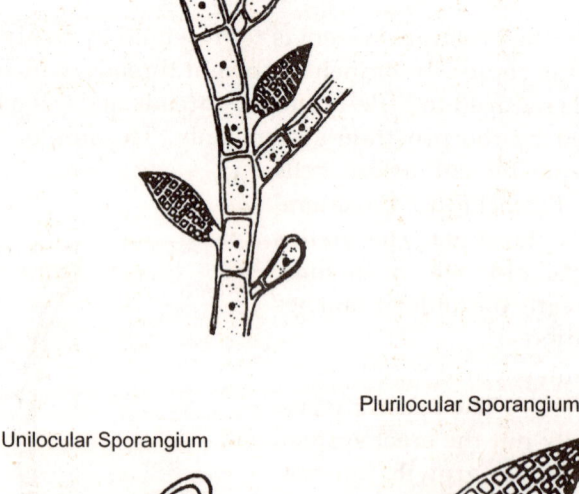

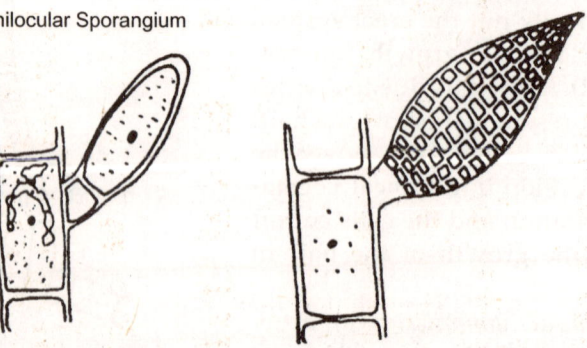

Fig. 12.3 *Ectocarpus*—Asexual reproduction—Unilocular and Plurilocular sporangia on filaments.

septation, in conjunction with division of nucleus (equational) and chromatophores. First the developing sporangium undergoes transverse septation into a row of 8 to 10 cells. Then, longitudinal divisions take place. The repeated transverse and longitudinal divisions bring about the formation of several hundreds of cells. These cells are arranged in transverse rows. The protoplast of each cell gets transformed into a *diploid biflagellate zoospore*. At the time of liberation, the septa disappear and the swarmers escape out through an apical pore gradually, in a slow and orderly procession.

The formation of zoospores in the neutral sporangia does not involve meiosis. So, the zoospores are diploid. They swim for some time, settle down by their anterior end and develop into diploid plants. Thus, the zoospores of plurilocular sporangia of asexual plants multiply the same phase.

The terminal cell of a lateral branchlet of a diploid plant gives rise to the *unilocular sporangium*. It enlarges into an ellipsoidal structure, which contains within it, several chromatophores. The cytoplasm increases and fills up the whole of the unilocular sporangium. The conspicuous diploid nucleus undergoes first a *meiotic division* which is later followed by equational divisions of the daughter nuclei. As a result of this, 32 or 64 haploid nuclei are formed. The cytoplasm cleaves into uninucleate bits which include 1 or 2 chromatophores. These 32-64 parts metamorphose into biflagellate *zoospores;* of the two laterally inserted flagella, the longer one is pointed forwards and the shorter one backwards. All these zoospores come out through an apical pore. They are extruded out *en masse*, in a mucilagenous mass. The zoospores, after a few seconds of inactive life, immediately separate out and swim away in different directions.

This haploid zoospore later directly settles down on a substratum, withdraws flagella, and gets rounded off to germinate into a filament. The new plant, thus produced from this haploid zoospores, belongs to the *sexual* or *gametophytic* phase.

Sexual reproduction. The haploid sexual plants (gametophytes) are exactly like the asexual plants or sporophytes. But they bear only plurilocular sporangia that produce gametes. So, these structures function as gametangia. These plurilocular gametangia arise in a manner similar to plurilocular sporangia of diploid plants. In *E. siliculosus*, all the cells of plurilocular gametangia are of equal size. Gametes are produced singly from each cell and they are all alike. The gamets resemble the zoospores. They are uninucleate and biflagellate and the flagella are laterally inserted with the longer one pointing forward, and the shorter one backward. The sexual union is *isogamous* but it may be physiologically anisogamous with the male gamete being more active than the sluggish female gamete.

In *E. secundus*, two kinds of gametangia are present. The megagametangia which have larger loculi (cells) produce large gametes. The microgametangia with smaller loculi produce smaller gametes. The sexual union between the macro- and micro-gametes is therefore anisogamous.

In *E. siliculosus*, a single passive female gamete will be surrounded by several male gametes which show vibrating movements (clump-formation). They fix themselves to the female gamete by the longer flagella which are pointed forwards. But only one of the males gametes will be brought into contact with the female gamete by the contraction of the flagellum. Later, it fuses with the female resulting in a zygote.

Thus, gametes fuse in pairs (isogamy) and form the zygotes. The zygote germinates directly into a new plant without undergoing meiosis. The resulting plant is therefore a diploid plant *i.e.*, sporophyte (2x).

The life cycle of *Ectocarpus* remained for a long time obscure, because of the close resemblance between accessory plurilocular sporangia of the sporophyte, and those which function as gametangia, present on the haploid gametophytic individuals. It is now established that in. *E. siliculosus*, the life cycle involves an alternation of *(i)* diploid individual bearing asexual unilocular sporangia and accessory plurilocular or *neutral* sporangia with *(ii)* a haploid individual bearing plurilocular sporangia which act as *gametangia*. In *Ectocarpus,* alternation of generations is typically *isomorphic*. The diploid plant bears plurilocular sporangia which are purely intended for the reduplication of the same phase. These are *neutral sporangia* and each cell of the plurilocular sporangium gives rise to a single zoospore which is biflagellate. Thus, several zoospores are formed. The zoospores escape out when the wall of the sporangium ruptures and germinate producing diploid plants. Thus, they aid in the accessory reproduction of the same phase and they are not concerned with alternation of generations.

But the *unilocular sporangia,* borne on the diploid plants (sporophytes) are mainly concerned with the alternation of generations. It produces 32-64 zoospores only after *meiosis,* and hence these are haploid. When they germinate, they produce haploid plants.

The haploid plants are just like the sporophytes, but they bear only plurilocular sporangia. These are just like the neutral sporangia borne on the diploid plants. But in function, they differ and they act as *gametangia* producing biflagellate gametes. Gametes which are liberated into the water, fuse in pairs to form the zygotes (sometimes, they may, parthenogenetically develop and produce the same haploid phase). The sexual union is isogamous. The zygote germinates and produces a diploid plant. Thus, in *Ectocarpus,* there is an *isomorphic alternation* of *generations,* but this has been made complicated by the accessory modes of reduplication of the same phase. The diploid plant can propagate indefinitely by zoospores produced in neutral sporangia (plurilocular sporangia). In haploid plants, reduplication of the same phase takes place by parthenogenetic development of the gametes.

DICTYOTA

Class : Phaeophyceae

Order: Ectocarpales
Family: Dictyotaceae.

Dictyota is a marine form found attached to the rocks in the intertidal zones. It is represented by about 35 species.

Structure

The thallus of *Dictyota* is ribbon-like and repeatedly branches in a dichotomous manner (Fig. 12.4). Sometimes, the typically dichotomous appearance may be lost due to unequal growth of the branches (*D. binghamiae*). The plant is attached by means of *rhizoids* arising from the basal cylindric part of the thallus or the rhizome which produces the erect thalli. The rhizoids are simple or branched and they arise in tufts. A number of colourless hairs or trichomes may arise from both surfaces. These hairs have a basal meristem and they are shed away at the time of reproduction. The lower cylindric part of the thallus may produce adventitious branches or *stolons*.

The thallus grows by means of single lenticular *apical cell* (Fig. 12.5). This cell cuts off a series of segments below, which broaden and undergo divisions twice, parallel to the surface forming three layers. The thallus forks when the apical cell divides vertically and the daughter cells begin to develop individually.

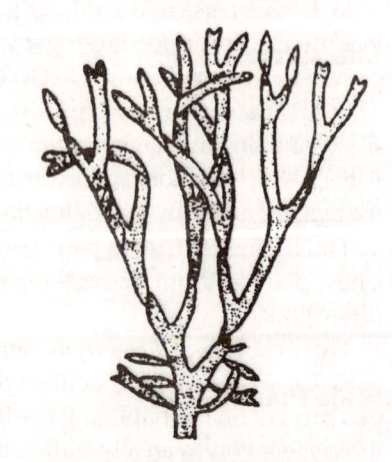

Fig. 12.4 *Dictyota*—Thallus.

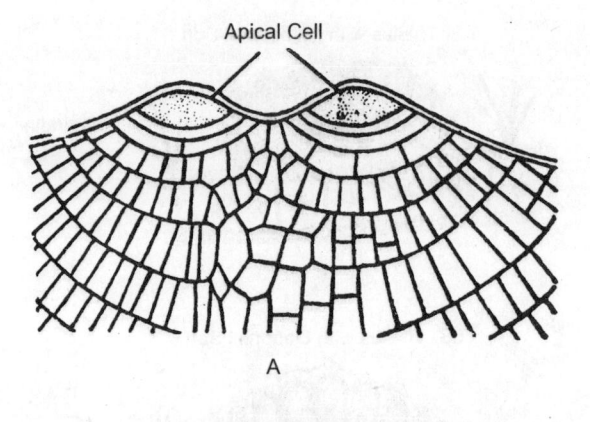

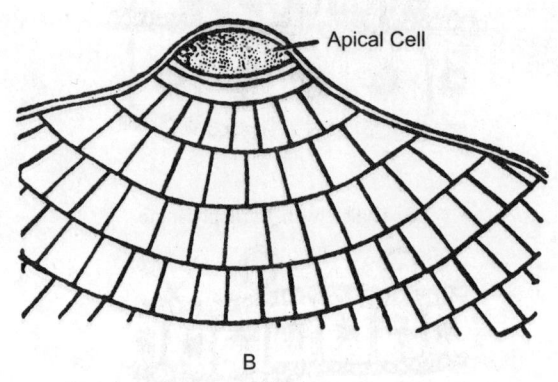

Fig. 12.5 *Dictyota*—Apical cell organization.

A transverse section of the thallus shows three layers. The middle layer consists of large cells which have very few chromatophores. These cells are uninucleate and contain refractive granules of stored food suspended by cytoplasmic strands. Apart from this, they contain *fucosam vesicles*. The middle layer of large cells is bounded by the upper and lower layers of small cells which are rich in chromatophores. Thus, the peripheral layers are useful in assimilation as the chromatophores contain chlorophyll and fuoxanthin. Laminarin, a polysaccharide, is stored as reserve food.

Life Cycle

There is regular alternation of generations between *haploid sexual plants* (gametophytes) and *diploid asexual plant* or the *tetrasporic plant* (sporophyte). The sexual plants are *dioecious*. The male plant produces antheridia and the female plant produces oogonia. The male, female and the tetrasporic plants look alike morphologically and they can be recognised only by their reproductive organs.

All these plants can reproduce by vegetative reproduction. They produce horizontally-growing *stolons* which develop adventitiously from the lower part of the thallus. Mostly, they arise from the injured parts and get separated after the death of the parent plant. This is a source of prolific mode of vegetative reproduction.

Male Plant (Fig. 12.6)

The sexual plants are dioecious. The male plant produces *antheridia* (100—150) on the surface of the thallus closely packed together; this *antheridial sorus* has towards the margins, elongate and undi-

Fig. 12.6 *Dictyota*—T.S. Thallus showing antheridial sori, oogonial sori, and tetrasporangia.

vided cells which retain their pigmentation of the chromatophores. These are considered to be sterile antheridia. The antheridia develop from the upper layer of small cell which are rich in chromatophores. But each cell, before developing into an antheridium, loses its pigmentation and elongates upwards. It cuts off a small basal cell which looks similar to the cells of surface layer. The upper cell first divides vertically and later both transversely and vertically forming a *plurilocular* structure. Each antheridium consists of numerous cells. A single cell produces one antherozoid or spermatozoid and hence numerous spermatozoids are formed from an antheridium. At the time of liberation, the spermatozoids look somewhat rounded. They lie within the antheridium in a mucilagenous mass when the cross-walls of the antheridia dissolve. The spermatozoid is *pear-shaped* and has an eye-spot present at the anterior end. The nucleus is large and it is posteriorly placed. The spermatozoid has a single flagellum but examination with electron microscope shows two basal granules one of which is without a flagellum.

Female Plant (Fig. 12.6)

The female plant produces *oogonia* in sori. In section, a *oogonial sorus* looks fan-like but when viewed from above, it is more or less elliptical. During development, some of the surface cells elongate upwards lifting up the cuticle which forms a covering. Each one cuts off a small basal cell and enlarges into an oogonium. The oogonial sorus contains 12-20 oogonia, which are all usually fertile. Occasionally, peripheral oogonia are sterile. Each oogonium produces a single egg which escapes out by gelatinisation of the apex. At the time of fertilization, several spermatozoids get attached to the egg by their flagella. One of the sperms fertilizes the egg forming a zygote. The zygote (diploid) does not undergo meiosis but develops into a *tetrasporic* plant by repeated equational cell divisions.

Asexual or Tetrasporic Plant

The tetrasporic plant is asexual and diploid and it produces *tetrasporangia*, the asexual reproducing bodies. They are spherical structures and may occur singly or in groups on both the surfaces. The small cell of the surface layer enlarges, cuts off a basal cell and develops into the tetrasporangium.

The nucleus of the sporangium enlarges in size and divides by meiosis forming four haploid nuclei. Later, the cytoplasm cleaves, giving rise to four naked tetraspores. The apex of the sporangium gelatinises and four spores escape out. Later, after becoming free, the naked spores secrete cell walls. Such spores germinate and produce the sexual plants. Of the four spores produced in a tetrasporangium, two develop into male plants and two into female plants. When tetraspore settles and germinates, it first undergoes transverse septation; the lower cell produces the rhizoid and the upper one acts as an apical cell of the thallus.

Alternation of Generations

In *Dictyota dichotoma*, there are *tetrasporic* plants which are *diploid* and *sexual* plants that are *haploid*. The sexual plants are dioecious. The male plants produce antheridial sori and the female oogonial sori. The spermatozoids are uniflagellate and they are liberated from the antheridium. It fertilizes the egg which is liberated from the oogonium. The zygote develops into the diploid tetrasporic plant. This plant produces *tetrasporangia* which produce *tetraspores* after meiosis (of the four spores formed from a sporangium, two develop into male plants and two others into female plants). Thus, it is quite clear that there is distinct *isomorphic alternation of generations*. In several instances, diploid tetrasporic plants are dominant, probably due to prolific vegetative reproduction. Sometimes, in a tetrasporangium, meiosis will not occur and the entire protoplast with the diploid nucleus escapes out and produces an asexual plant. Thus, it is an accessory method of multiplication of the diploid phase.

SARGASSUM

Class : Phaeophyceae

Order : Fucales
Family : Sargassaceae

Occurrence

Sargassum, popularly known as Gulfweed grows abundantly in tropical oceans. The chief centres of growth are Saragasso sea of South Central Atlantic and Gulf of Mexico. It occurs abundantly on the west coast of India.

Sargassum is found attached to rocks by means of the holdfast present at the base of the main axis. But detached portions of the plant occur as free-floating masses.

Structure

Sargassum is monopodially branched (Fig. 12.7). It has a holdfast, a short stipe and structures resembling stems and leaves of higher plants. The main axis is like a 'stem' and it bears 'leaves' at the nodes. Each 'leaf' is actually a vertically flattened and expanded lateral branch; it has a midrib and serrate or wavy margin. The leaves at the base of the plant have only rudimentary structures in their axis. But leaves of nodes have axillary structure consisting of an *air bladder* and a much branched *receptacle*. Thus, the leaf, the receptacle and the bladder together constitute one lateral branch system. The air bladder is useful for buoyancy. The leaves are the chief photosynthetic organs and the receptacles are the reproductive structures. Occasionally, in some species, the air bladder and the receptacles show essentially identical features (Fig. 12.8).

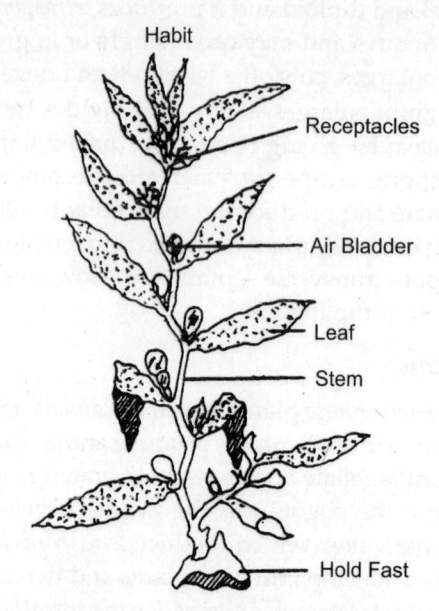

Fig. 12.7 *Sargassum*—Thallus.

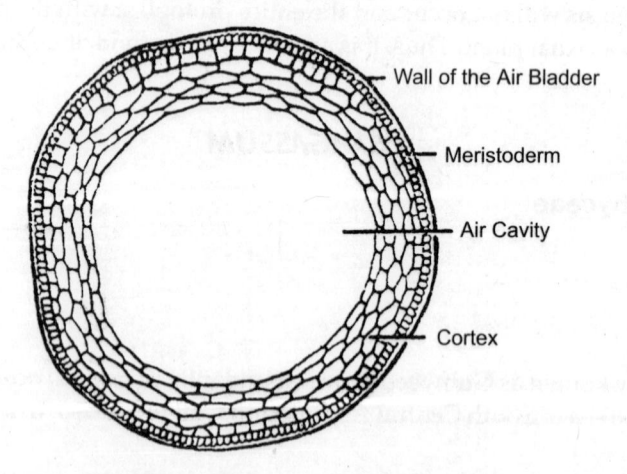

Fig. 12.8 *Sargassum*—T.S. air bladder.

The receptacles may contain either *male* or *female conceptacles* or both (Fig. 12.9 & 12.10). A fertile conceptacle is a cavity which contains the sex organs. The leaves also show dark dots which indicate the position of sterile cavities or conceptacles called *crytoblasts* or *cryptostomata*. These are aborted or sterile conceptacles and have a number of colourless hairs arising from the middle of its floor. They project in a tuft and grow in length by the activity of basal meristems. The occasional presence of antheridia in the cryptostomata of leaves shows that they are derived from the fertile conceptacles.

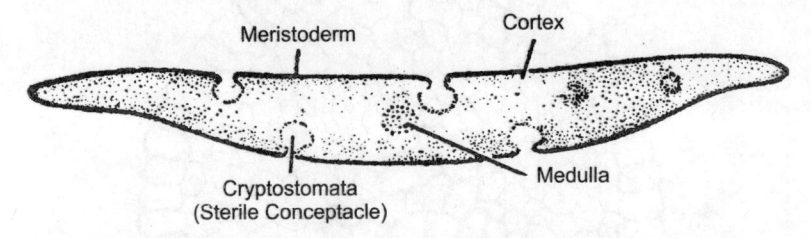

Fig. 12.9 *Sargassum*—T.S. leaf showing conceptacles.

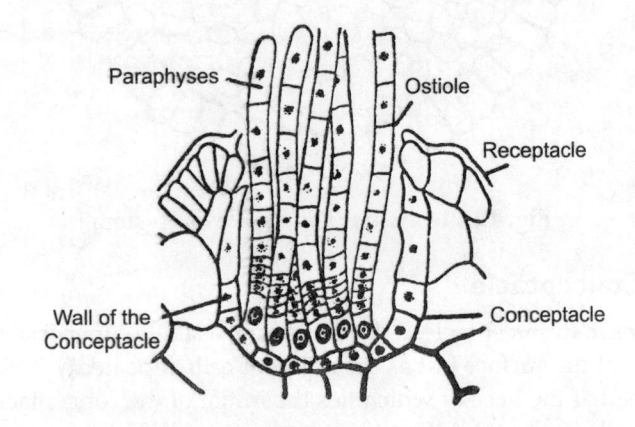

Fig. 12.10 *Sargassum*—L.S. Conceptacle.

Anatomy

The cross section of 'stem' i.e., axis, shows distinctions into three regions (Fig. 12.11). The *meristoderm* is the outer-most layer and it consists of a number of closely arranged cells which are in a single layer. These are richly pigmented and hence they are assimilatory in function. They are meristematic, and they always divide anticlinally. The *cortex* is the middle zone consisting of larger cells which store food material. The *medulla* is the central region consisting of narrow elongated cells which are useful in conduction. The leaf also shows a similar distinction, as it is also essentially a part of a lateral branch system. But is shows cryptoblasts. The air bladder also shows meristoderm and cortex but the medulla is replaced by a large air space.

Reproduction

Sargassum reproduces in a prolific manner vegetatively by fragmentation. *S. natans* and *S. hystrix* exclusively produce by this method.

Sexual reproduction. The sex organs are oogonia and antheridia. The oogonia occur in female conceptacles and the antheridia in the male ones. The female and male conceptacles may occur on the same receptacle (monoecious) or may occur on different plants (dioecious).

Sometimes, the conceptacles are abortive and they are known as *cryptoblasts*. These may occur in between fertile conceptacles in the *receptacle,* but they are mostly found on leaves.

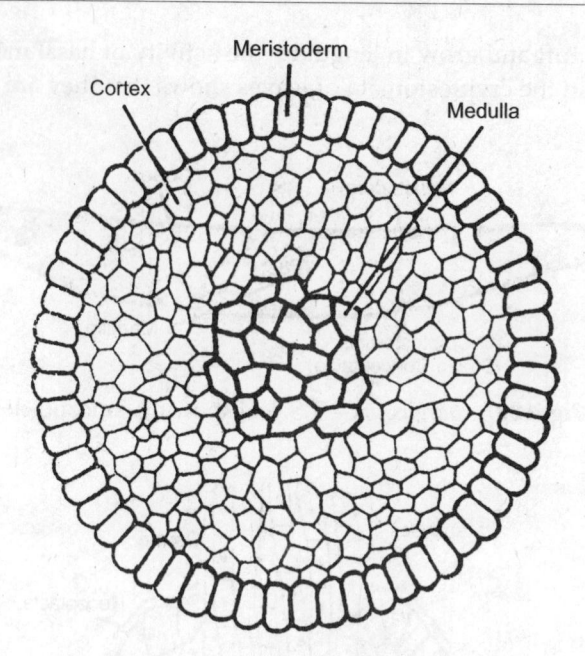

Fig. 12.11 *Sargassum*—T.S. axis or 'stem'.

Development of Conceptacle

A conceptacle develops in the receptacle. It develops from a single superficial cell. At first, the initial of the conceptacle lies at the surface but as the adjacent cells repeatedly divide and grow, a flask-shaped cavity is formed at the base of which lies the initial of the conceptacle. This initial divides transversely forming a lower basal cell and upper tongue cell. The tongue cell may disappear or develop into a hair. The basal cell, by a series of vertical divisions, forms a row of cells lining the floor of the cavity. These cells contain chromatophores. It becomes 2-3 layered by transverse divisions in which the superficial layer is fertile giving rise to either antheridia or oogonia. The conceptacle has a small opening called *ostiole*. Some of the cells near the ostiole may develop into hairs called *periphyses*.

Female conceptacle (Fig. 12.12).
Any superficial cell of the fertile layer lining the floor of the conceptacle can give rise to the oogonium. The superficial cell or oogonial mother cell divides into a lower stalk cell and an upper cell that develops into the oogonium. The oogonium increases in size and appears as if it is sessile. The nucleus of the oogonium is diploid. It undergoes meiosis and forms four haploid nuclei which divide again forming eight nuclei. But seven of the eight nuclei degenerate and only one nucleus remains. This nucleus along with the oogonial cytoplasm forms the egg. Usually 4-10 oogonia develop in the female conceptacle. Paraphyses are formed between the oogonia by some of the superficial cells of the floor of the conceptacle.

Male conceptacle (Fig. 12.13).
In the male conceptacle, several cells of the fertile layer lining the floor of the conceptacle function as antheridial mother cells. The antheridial mother cell grows into a papilla-like outgrowth and it is cut off by a cross wall at the base. The upper cell divides into a lower stalk cell and a terminal antheridium. The stalk cell may grow into the paraphyses pushing the antheridium to a side. Thus, the paraphysis of male conceptacle are different in origin from those of the female conceptacle.

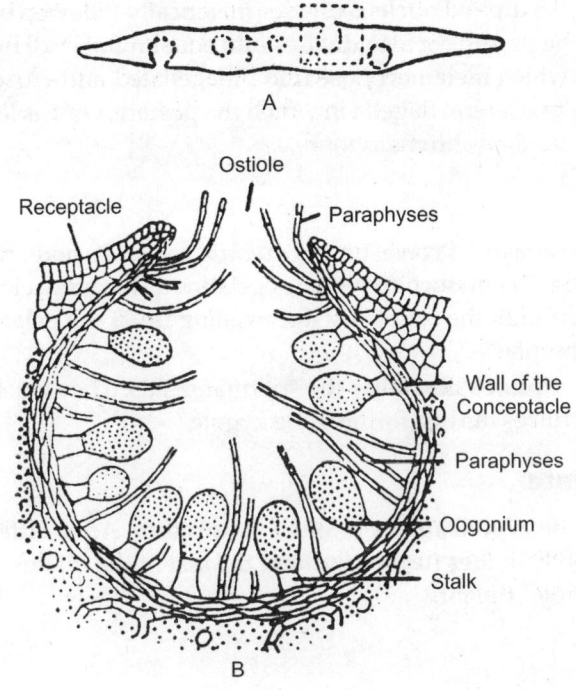

Fig. 12.12 *Sargassum*—(A) T.S. female receptacle, (B) T.S. female conceptacle enlarged.

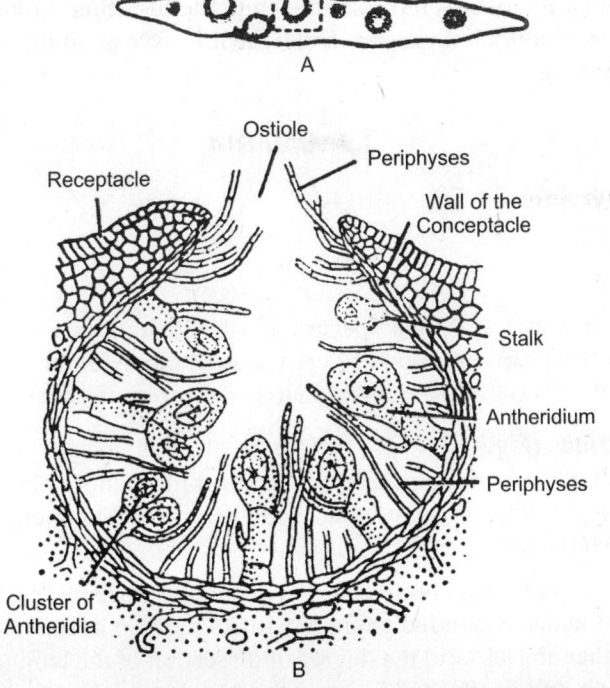

Fig. 12.13 *Sargassum*—(A) T.S. male receptacle, (B) T.S. male conceptacle enlarged.

In each antheridium, the diploid nucleus divides meiotically followed by mitotic divisions forming 64 haploid nuclei. The protoplast also divides into equal number of bits so that 64 uninucleate protoplasts are formed which metamorphose into biflagellated antherozoids. The spermatozoids are pear-shaped having two lateral flagella in which the posterior one is longer. It has a prominent nucleus, eye-spot and a vestigal chromatophore.

Fertilization

The oogonial wall is three-layered consisting of *exochite, mesochite* and *endochite*. The exochite imbibes water and ruptures. The mesochite becomes gelatinous forming a kind of stalk. The oogonia are slowly pushed up through the ostiole by the swelling up of the gelatinous material. They get attached outside the conceptacle.

The antheridial wall ruptures liberating the spermatozoids. They cluster round the oogonium but only one of them fertilizes the egg forming the zygote.

Germination of Zygote

The gelatinous wall of the zygote affixes it to the substratum. After differentiation of rhizoid towards the base, the zygote divides further forming the rest of the thallus. Thus, the zygote germinates and produces a diploid plant.

Life Cycle

Sargassum is diploid. Meiosis takes place in the oogonium during the formation of the egg and in the antheridium during the formation of spermatozoids. Fertilization is outside the conceptacle since both oogonia and spermatozoids come out through the ostiole. Sexual reproduction is oogamous and one of the spermatozoids fertilizes the egg. The resulting diploid zygote gives rise to a diploid plant by repeated mitotic divisions. Thus, the life cycle is diploid type and gametes alone represent the haploid stage.

LAMINARIA

Class : Phaeophyceae

Order : Laminaria
Family : Laminariaceae.

There are a number of common British species, all of which agree in general structure and life history, but as they differ from one another very markedly in external form it will be desirable to outline briefly these differences at the outset so far as the common species are concerned.

***Laminaria saccharina* (Fig. 12.14) (Tangles).** In this species the thallus consists of an undivided frond with a wavy margin, arising from a rather short, thin, round stalk, which is attached by a rhizoidal holdfast. It is perennial, grows up to 6 ft. long and occurs from low-water mark to a depth of 10 fathoms.

Laminaria digitata (Kelp). In this species there is a smooth, thick stalk which widens out gradually into a broad frond which is divided palmately into a number of separate fingers. It is a considerably larger species than the last and the degree of dissection of the lamina varies considerably. It may be upto 12 ft. long and is attached by an elaborate rhizoidal system. It occurs between low water and 15 fathoms.

Laminaria hyperborea (Tangles). This species resembles *L. digitata* in the shape of the thallus, but differs from it in that the stipe is not smooth and it expands abruptly into the palmate frond. It is a large plant, and though not as large as the last, grows up to 10 ft. long. It occurs between low-tide mark and 12 fathoms.

In all species the expanded lamina has no midrib and is borne on a basal stipe, which is attached to the rock surface by a holdfast of very variable form. This holdfast is made up of a number of separate branches of decreasing thickness which adhere very tightly, forming flattened discoid masses immediately in contact with the rock and attached by densely packed rhizoids. The lamina grows from a meristematic zone at its base, which annually forms a new frond, displacing that of the previous season, which then dies off.

Many of the species are used as food in Asia, especially by the Chines, Japanese and Russians. In Japan about ten species are eaten, and gathering between July and October is an important industry. These kelps contain considerable quantities of iodine, which is important for the functioning of the thyroid gland, and it is a noticeable fact that goitre is almost unknown among the Japanese. Apart from their use as food the kelps are perhaps the most important commercial source of iodine, which is extracted from the ash after the weeds have been burned.

Fig. 12.14 *Laminaria*—Thallus.

Structure of Thallus (Fig. 12.15)

The thallus of *Laminaria* shows an exceptionally complex structure, the most elaborate in any group of the Algae. Anatomically stipe and lamina are alike and both show a separation into three distinct zones although these are more clearly marked in the stipe owing to its greater thickness. The stipe may also show annual zones of growth. Near the apex the blade is only one cell thick, but it soon becomes two-layered, after which the primary tubes are formed, which constitute the medulla and separate the two external layers. By division of the cells of the outer layers parallel to the surface a zone of cortex is cut off. These cortical cells then elongate longitudinally and the common walls between them swell and so separate the cells from one another, except at certain points of union where they become drawn out into short secondary tubes. In this way three separate zones of tissue become differentiated. On the outside lies the external layer, the cell of which are primarily

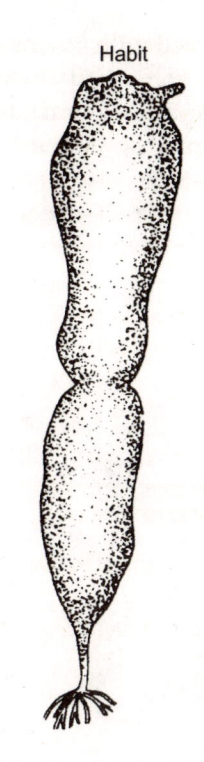

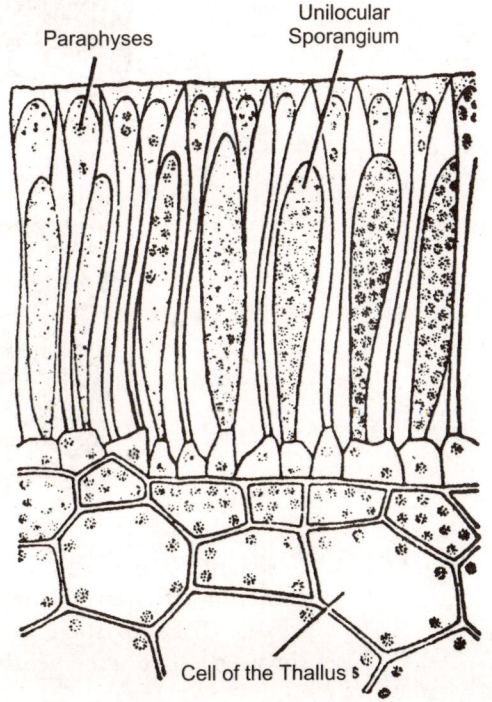

Fig. 12.15 *Laminaria*—T.S. Thallus.

concerned with assimilation. These divide only perpendicular to the surface. Inside this comes to cortex, composed of elongated cells separated from one another by mucilage, and finally in the middle is the central medulla composed of larger, longitudinally running filaments whose function appears to be primarily that of conducting materials in solution. In the stipe these zones can be clearly seen.

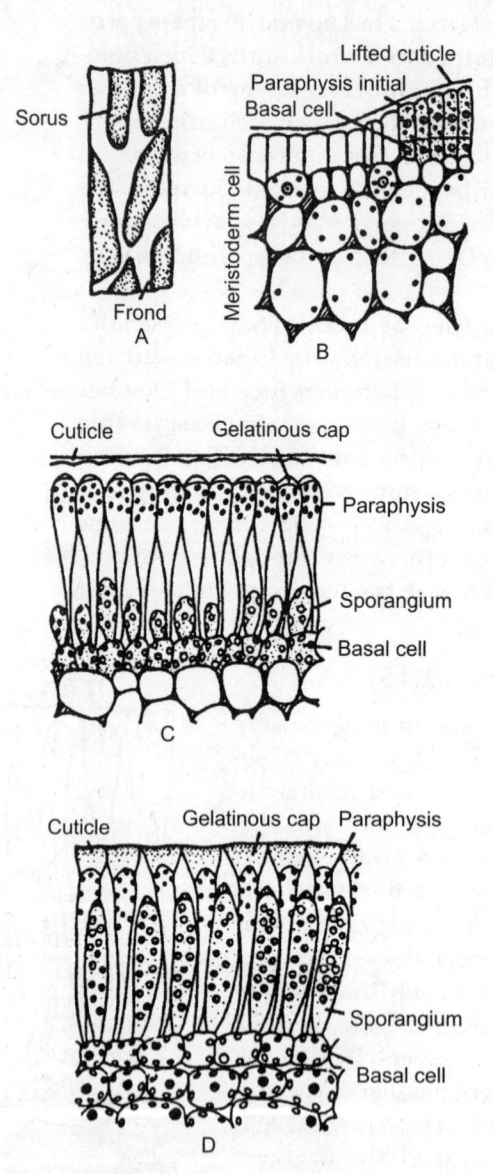

Fig. 12.15A *Laminaria.* Reproduction by meiospores and successive stages of development of sporangia. (A), part of a frond with sori; (B), Early development of sporangial sorus; (C) Latter stage in the development of sorus; (D), transverse section through a part of the nearly mature sorus.

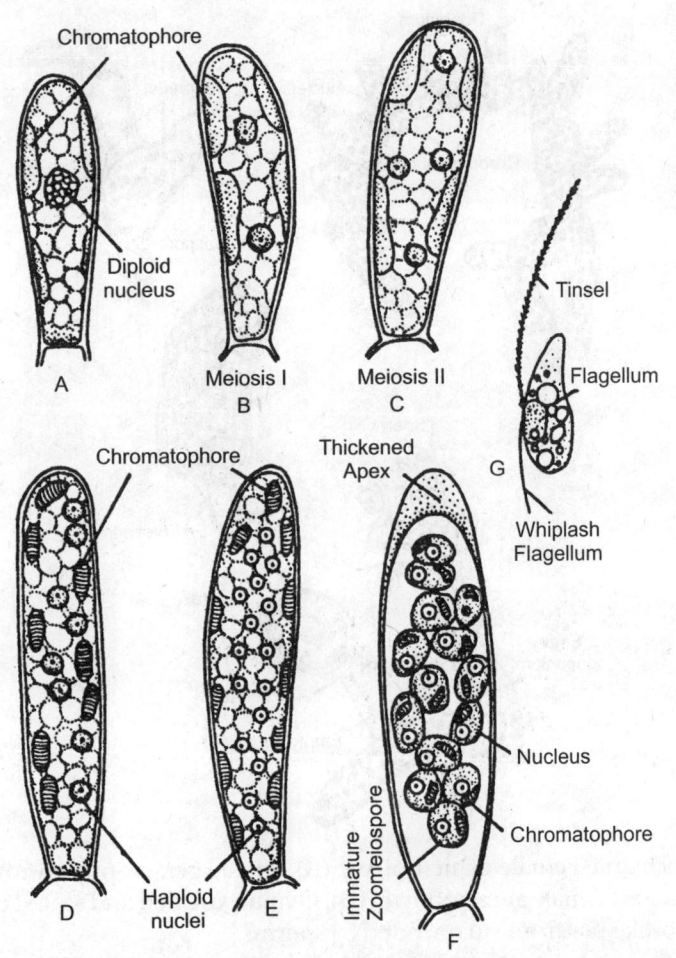

Fig. 12.15B *Laminariales.* Stages in the development of the unilocular sporangium and differentiation of meiozoospores; (A), young unilocular sporangium with a diploid nucleus; (B-C), meiotic division of diploid nucleus; (D), 8-nucleate stage; (E), 16-nucleate; (F), differentiation of meiozoospores; (G), liberated meiozoospore.

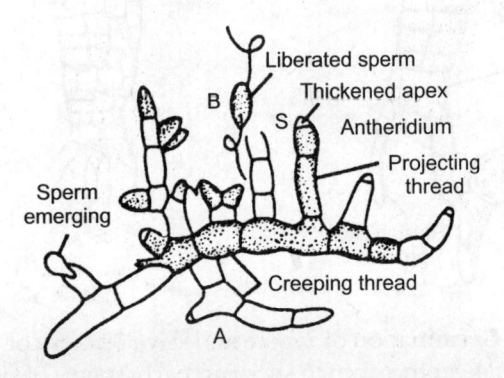

Fig. 12.15C *Laminaria.* Male gametophyte.

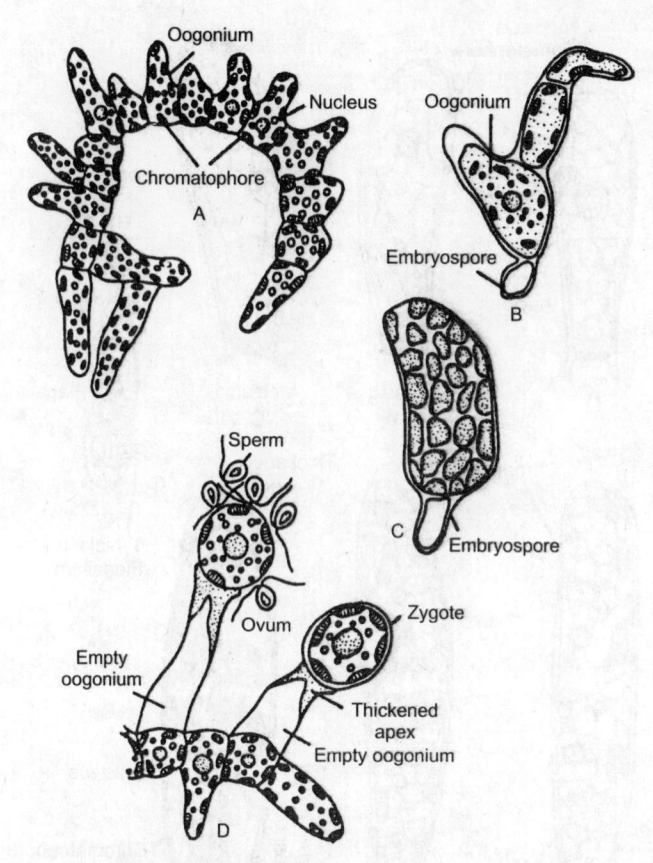

Fig. 12.15D *Laminaria.* Female gametophyte. (A), female gemetophyte with oogonia; (B-C), reduced female gemetophyte; (D), Ovum extruded and seated on the thickened, cup-like platfrom surrounded by sperms.

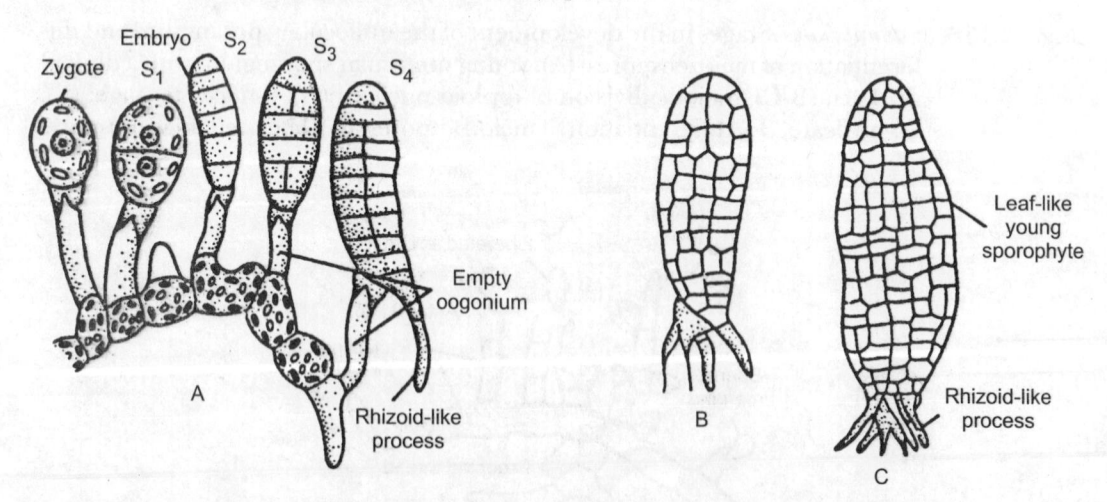

Fig. 12.15E *Laminaria.* Germination of zygote and development of young sporophyte. (A), Part of female gemetophyte showing early stages of development of embryo sporophyte (S_1, S_2, S_3 and S_4) still seated and attached to the cup-like apical platform of empty oogonium; (B-C), Later stages in the development of sporophyte.

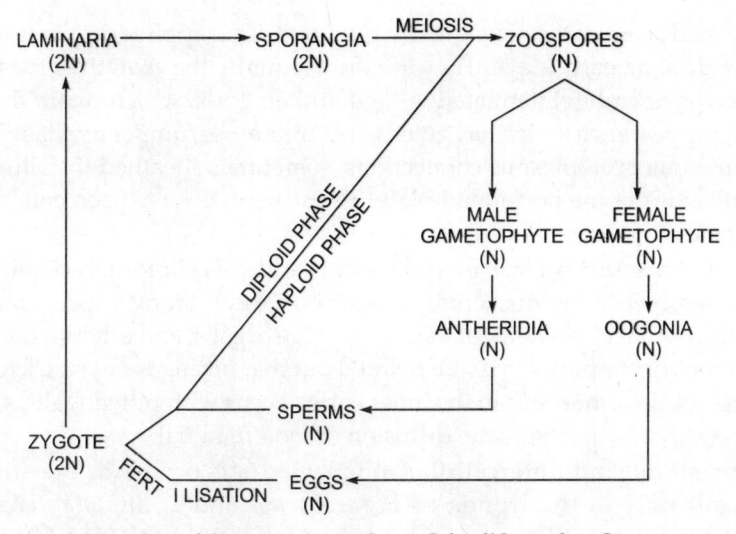

Fig. 12.15F Graphic representation of the life cycle of *Laminaria sp.*

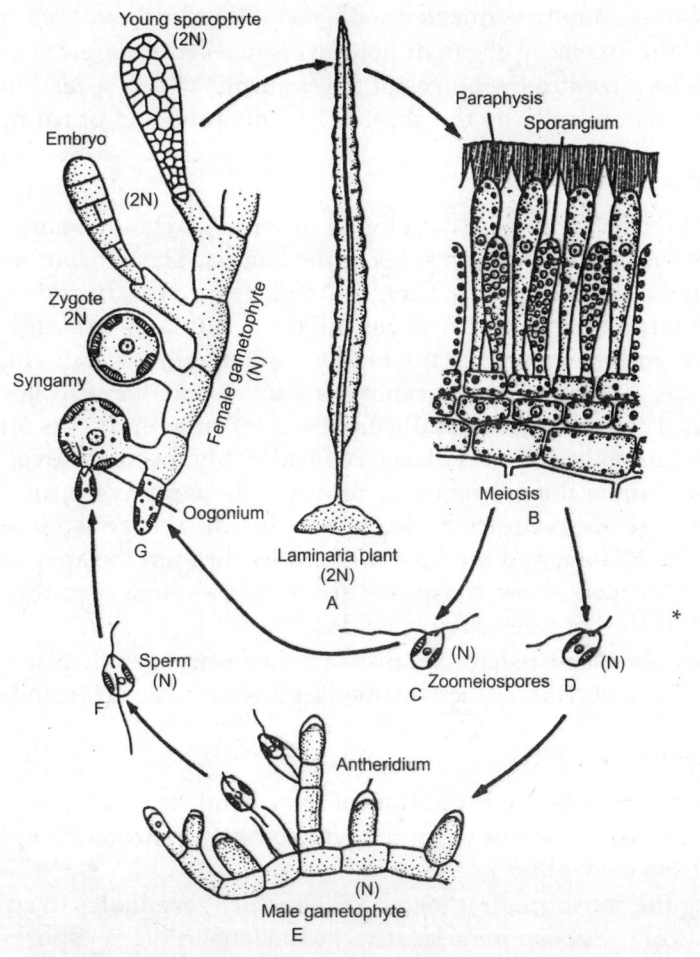

Fig. 12.15G *Laminaria sp.* Life cycle.

The cells of the medulla may become greatly modified. This applies particularly to certain cells which cease to divide at an early stage. They are drawn out by the growth of the tissues, into long straight filaments, considerably attenuated, except at their ends, which retain the original width, thus producing the appearance which has given them the name trumpet hyphae. The end walls are penetrated by numerous protoplasmic connections, sometimes sheathed in callus, which later extends to cover both sides of the perforated plate. The structures have been compared to the sieve tubes of higher plants.

Apart from this the trumpet hyphae possess spiral bands of cellulose thickening on their walls, a feature which again recalls the thickening of a wood vessel. Many suggest that they serve for conduction of fluids, while others think of them as storage organs, and others again prefer to regard them mainly as organs of support. It may be pointed out that pitting is not restricted to the trumpet hyphae, as in some species other cells in the inner cortex occur with pitted walls, similar to those in the medulla. These may also facilitate the diffusion of food material.

A system of anastomosing, intercellular mucilage ducts occurs in the stipe and frond of several species, but only in the fronds of *L. saccharina* and *L. digitata.* There are periodic openings from these ducts to the exterior and they are lined with isolated groups of secretory cells. They arise schizogenously between cells of the surface layer and deepen and extend with growth, becoming connected with each other into a continuous network.

The structure of the attaching organ or holdfast is markedly different from the rest of the thallus. Growth is localized in the apices of the branches which spread out in contact with the rock. It differs anatomically by the absence of a medulla and of trumpet hyphae.

Asexual Reproduction

The sexual reproductive organs are developed in widely extended sori, which may cover the greater part of the surface on both sides of the lamina. The zoosporangia arise from the superficial cells in the following way. Each cell divides into two, forming a basal cell and a terminal cell. The terminal cell is at first assimilatory, but later enlarges considerably, becomes club shaped and is invested at the top by a mucilaginous cap. This body becomes a paraphysis, and the caps of all the paraphyses adhere and serve to keep them together. Meanwhile the basal cell enlarges laterally and from its outer ends cuts off two cells, one on each side of the terminal cell. Each enlarges considerably, becomes oval and forms a sporangium. These sporangia thus come to lie between the paraphyses, and the whole is covered by the mucilage derived from the latter. Inside the zoosporangium thirty-two zoospores are differentiated, and are finally liberated through the apex of the sporangium. According to most workers these zoospores are all of the same size, though in one or two cases zoospores of different sizes have been described.

The zoospores are almost pear-shaped bodies with two long, laterally placed flagella, and each may possess a tiny eye spot. They are activity motile but soon settle down and germinate.

The Gametophytes

The result of germination is the formation of very small male and female gametophytic plants, both of which are filamentous and differ completely from the sporophytic thallus. They also differ from each other in shape and size.

On germinating the zoospore first forms a tube which terminates in an enlargement into which the contents of the zoospore migrate. The nucleus in the zoospores divides, and one daughter nucleus passes into the enlargement while the other degenerates. The cell formed by the enlargement divides several times, and the mature gametophyte consists of a short filament of cells with shorter branches. It may even be reduced to two or three cells. The male gametophyte is composed of smaller cells than the female gametophyte.

The sex organs arise from the lateral branches. The antheridium is a small, more or less spherical, cell which gives rise to antherozoids. The antherozoid is a small, oval cell provided with a pair of fairly long, unequal, laterally placed flagella. The oogonium is also formed from a cell of a lateral branch, inside which is a single oosphere. The oogonium is surrounded by a thick mucilaginous investment which is prolonged at the apex into a cup comparable with an egg cup. The oosphere emerges from the oogonium and lies in the cup with the greater part of the oosphere exposed.

Fertilization

Fertilization is effected in the usual way by the migration of the antherozoid to the oosphere and the union of the male and female nuclei, after which a wall is formed around the oosphere resulting in the formation of an oospore. Male and female gametophytes are found in equal numbers, and it has been shown experimentally that from any zoosporangium an average of sixteen male and sixteen female gametophytes will be produced. It follows that normally each oosphere should become fertilized, and this takes place quite rapidly. Should the oosphere fail to receive a male gamete it may be capable of independent development producing a parthenogenetic sporophyte.

The development of the oospore begins with its division into two cells, from which is produced a filament with an apical cell. At first the young plant remains attached to the top of the oogonium, but it later becomes detached and one or more of the basal cells of the filament to the substratum. Next the upper part widens into a monostromatic blade on a filamentous stipe. The blade then becomes distromatic and the stipe becomes polysiphonous. A new meristematic region appears between stipe and blade; apical growth gradually ceases and the apex of the frond is eroded. The first appearance of the cortex is as a single layer of cells between the two layers of the blade. The cells are large and parenchymatous and are increased by the addition of new cells in the meristematic zone. The medullary hyphae appear as the cortex becomes double-layered, and the expansion of the trumpet hyphae takes place in the intercellular spaces.

Alternation of Generations (Fig. 12.15G)

Laminaria thus exhibits an alternation of generations comparable with that of *Dictyota*, but whereas in *Dictyota* the two generations are morphologically alike, in *Laminaria* the gametophyte is reduced to a minimum, producing little more than the essential sex organs, while, on the other hand, the sporophytic generation shows the greatest tissue elaboration found in the Phaeophyceae. It must not be thought, however, that there is a definite progression within the Phaeophyceae towards a reduction of the gametophyte, for in certain other orders the gametophyte is large and fairly elaborately developed, while the sporophyte is reduced to a filamentous structure bearing the zoosporangia. It might appear therefore that in the Phaeophyceae there have been two contrasting tendencies operating, the one to reduce the sporophyte and the other to reduce the gametophyte.

Laminaria exhibits an advanced type of alternation of generations. The main plant is the sporophyte (A). It is differentiated into the holdfast, cylindrical stipe and the broad lamina. It represents the diploid phase in the life cycle. The diploid individual or the sporophyte bears clusters of sporangia called the sori (B). They are present on both sides of the blade or the lamina. The diploid nucleus of each sporangium undergoes meiosis. The four haploid nuclei divide mitotically to produce 32 nuclei. Due to cleavage of the cytoplasm 32 haploid meiospores which may be called the meiozoospores are formed. Out of these 16 germinate to produce the male gametophytes (F) and the other 16 to form the female gametophytes (G). It is important to note that meiospores, on germination, produce the gametophytes and not the sporophyte plants. The gametopytes are heterotrichous in habit. They are much smaller than the sporophyte. There is a sharp morphological differentiation between the two. The male and the female gametophytes produce antheridia and oogonia respectively. These in turn produce the sperms (E) and the eggs. The gametophyte phase, therefore is represented by two distinct, male (E) and the female (G) filamentous individuals which differ from the sporophytes in morphological as well as cytological details—a feature in sharp contrast to the other classes of Algae.

The male and the female gametes fuse (G, *a*) to form the diploid zygote (G, *b*). The zygote on germination produces the diploid sporophyte and not the gametophyte. Two crucial points in the life cycle of *Laminaria* are the meiotic division of the diploid nucleus of the sporangium and the fusion of the gametes to form the zygote. At these points one generation switches on to the other generation. The large sporophytes regularly alternate with microscopically male and female gametophytes. This is called alternation of generations. Such an alternation of generations in which the two alternating individuals is the life cycle differ morphologically, physiologically and in genetic constitution in designated and heteromorphic. The life cycle of *Laminaria* which is characterised by distinct alternation of generations with sporogenic meiosis is called diplohaplontic.

Economic Importance

The sporophytes of some species are eaten as food. Kelps are a valuable source of iodine and potassium fertiliser. Iodine is still manufactured from *Laminaria*, *Ecklonia* and *Eisenia* in Japan. They are as well the source of algin and alginate which are of so great importance in the production of plastics and artificial fibers.

Salient Features

1. The diploid thallus is large in size reaching upto 6 to 8 feet in length and differentiated into a holdfast, stipe and a blade.
2. The growing region lies in the transitional zone between the strip and the blade thus growth is intercalary.
3. The lamina is renewed at the base each year. The old blade is pushed away and dies.
4. The thallus sporophyte shows a high degree of morphological and anatomical differentiation.
5. The trumpet hyphae with their perforated cross walls recall the sieve elements of the vascular plants. The presence of cross-connections between the cells of the inner cortex and medulla is another anatomical feature unique to the algae.
6. The sporangia are arranged in distinct groups called sori. The mode of development of sporangia is unique.
7. Each sorus consists of paraphyses intermingled with unilocular sporangia. The diploid protoplast of each sporangium undergoes meiosis to produce 32 meiozoospores.
8. Out of these, 16 germinate to produce the male and the other 16 to produce the female gametophytes.
9. The gametophytes are heterotrichous in habit and are very much reduced as compared with the large sporophytes.
10. The male gametophytes are much branched and have smaller cells with scanty or no chromatophores.

11. The antheridia occur in terminal clusters on the erect filaments. Each antheridium produces a single biflagellate antherozoid. It is liberated through an apical pore formed by the thickening or gelatinization of the wall at the apex of the antheridium.
12. The female gametophytes are scarcely branched. They possess large cells filled with chromatophores. Any cell can function as an oogonium. The oogonial cell enlarges and becomes tubular or pyriform. Its contents round off to form a single ovum or egg.
13. The apical portion of the oogonial wall gets thickened and splits to allow the single ovum to escape. As the ovum is coming out the split wall again meets and forms a platform for the ovum to rest on.
14. The zygote starts germinating while still attached to the apex of the oogonium. The nucleus divides by simple mitosis. There is no meiosis.
15. The zygote does not undergo any period of rest.
16. There is heteromorphic alternation of generations in the life cycle which is diplohaplontic.

Taxonomic Position

Division	:	**Phaeophyta** (Phaeophycophyta)
Class	:	**Heterogeneratae**
Order	:	**Laminariales**
Family	:	**Laminariaceae**
Genus	:	*Laminaria*
Species	:	*laminariodes*

FUCUS

Class : Phaeophyceae

Order : Fucales
Family : Fucaceae

Fucus is a very common Brown Alga which occurs all round in the British coast attached to rocks, and lives between high and low tide marks, the various species showing distinct zonation. It is attached to the rocks by means of a specialized basal disc called the holdfast. The thallus is dark brown in colour of almost black when dried, and is dichotomously branched, the lower part being narrow and almost round, while the upper parts are flatter and broader but with a thick midrib (Fig. 12.16). The plants very considerably in size according to the conditions under which they are growing. If the plants are living on exposed and rather dry situations near high-tide mark they are rarely more than 6 in. in length, but if living under conditions of more complete immersion they may be anything up to 3 ft. in length, according to their age.

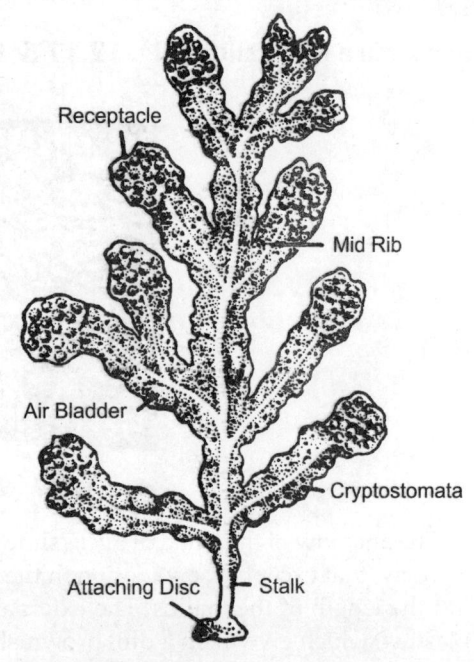

Fig. 12.16 *Fucus*—Thallus.

Apical Growth

When the plants are in a reproductive condition the ends of the branches become enlarged and covered with tiny wart-like projections, each of which has a minute pore at the centre. These bodies mark the position of cavities in the tissue of the thallus, which are called *conceptacles*, and in which the reproductive organs are borne. At the tip of each branch there is a small groove, at the bottom of which lies the single four-sided apical cell by means of which the thallus grows. The apical cell cuts off broad segments on two sides and thinner segments on the sides at right angles to the first, thus producing a flattened thallus. From time to time it also divides equally into two halves, thus starting a dichotomy of the shoot. Cells are also cut off from the base of the apical cell and these become the medullary filaments. In *Fucus* we have an example of an Alga which grows only at the tips of the fronds. The cells cut off from the apical cell form at first a close tissue, but lower down the central ones elongate into filaments, while the superficial layers remain parenchymatous.

Because it possesses an apical cell, the thallus structure of *Fucus* is usually placed in contrast with the filamentous and trichothallic growth of lower types like *Ectocarpus* and *Cutleria*, but in reality it is based upon a filamentous plan. The young plant has a group of apical hairs, formed at the growing point, each of which has a basal region of active cell division. At a later stage one or more of these hairs, usually that in the centre, dies off to its base, and the basal cell gives rise to the apical cell of the future thallus.

When the thallus is injured the inner tissue near the wound grows out into a tuft of hairs and the same process is repeated, producing an apical cell from which new tissues are regenerated.

Scattered along the length of the thallus in close association with the thickened midrib are *air bladders*, which serve very much the same function as the air spaces which we shall meet with in certain aquatic flowering plants, namely, that of assisting the plant to float.

Small tufts of hairs grow from cells in the apical groove and also from small cavities, called *cryptostomata*, scattered over the surface of the thallus. These hairs may act as absorbing organs.

Structure of Thallus (Fig. 12.17 & 12.18)

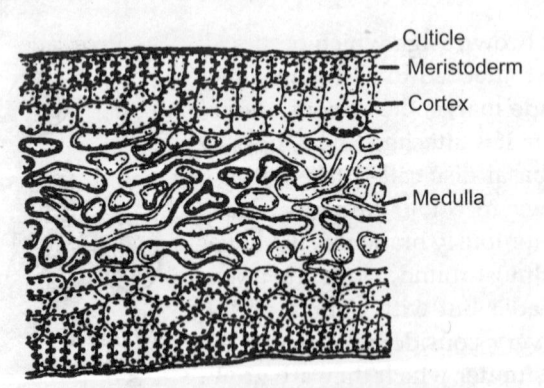

Fig. 12.17   *Fucus*—T.S. wing (a portion).

The anatomy of the frond of *Fucus* shows little difference in structure wherever we may cut it, that is to say, that except for a variation in the quantity of the tissues the structure is the same throughout the length of the thallus. The external layer consists of small rectangular cells with abundant plastids, which give them a dull brownish colour, and it is in this layer that the bulk of the assimilation takes place. The cells of this layer remain capable of active division all over the plant. Within this external layer comes a varying thickness of *cortex* composed of thin-walled parenchymatous cells. These also continue to divide actively, so that the cortex increase in thickness towards the base of the plant. The inner cortical cells also grow out into thick-walled *hyphae*, which grow in among the original filaments of the central zone or *medulla*, filling this space with a mass of closely interwoven cells, among which the original filaments can be picked out by their thin walls. Towards the base, in the region of the stipe, the medulla contains nothing but these thick hyphae, and the same is true of the flattened edges of the thallus.

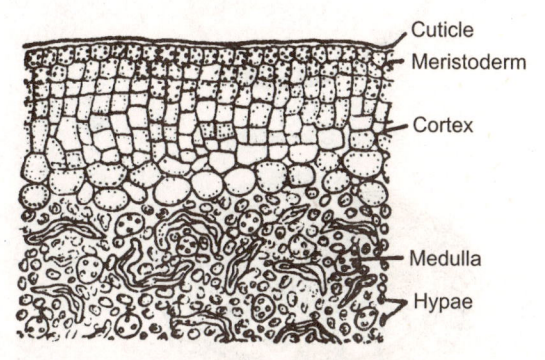

Fig. 12.18 *Fucus*—T.S. midrib (a portion).

The original medullary filaments are to be seen in the midrib of the thallus, and they run more or less straight vertically, recalling the similar elements in *Laminaria*. They are said to act as food-conducting elements and their cross-walls are probably perforated, like sieve-plates in the higher plants. Thus we see that *Fucus* contains tissues which are specialized for absorption, photosynthesis, conduction and attachment respectively, which implies a remarkably high grade of organization in Thallophyta.

Reproduction

In *Fucus* there is no asexual method, but reproduction through vegetative propagation may take place by the regeneration of the missing parts from bits of the thallus which are broken off from the parent plant.

Sexual reproduction. The reproductive organs develop within the *conceptacles*. In some species only one kind of reproductive organ develops in any one conceptacle, and in such instances, moreover, the sexes are separated on different plants. This varies in different species; in some, *e.g.*, *F. serratus*, the two kinds of sex organs are found on different plants; in others, *e.g.*, *F. spiralis*, both types of sex organs may occur in the same conceptacle. The gamets are of two quite distinct types, that is to say, the plants show an advanced condition of oogamy.

The conceptacles arise from superficial cells very near the growing point. A cell becomes depressed below the surface and then divides transversely. The upper cell elongates like a tongue towards the opening of the depression and degenerates early, while the lower cell divides vertically to form a group of cells which become the lining layer of the mature conceptacle.

The sex organs consist of *oogonia* and *antheridia* (Figs. 12.19, 12.20, 12.21 & 12.22). The oogonium arises from a superficial cell of the conceptacle, which divides to give a basal cell, but divides no more, but the oogonium enlarges and its nucleus divides into eight, with reduction of the chromosome number to one half. The eight nuclei separate and the cytoplasm splits into eight portions, each with one nucleus. Each of these eight cells is termed as an *oosphere*. When they are mature the outer wall *(exochiton)* of the oogonium breaks and the eight cells are liberated as a package

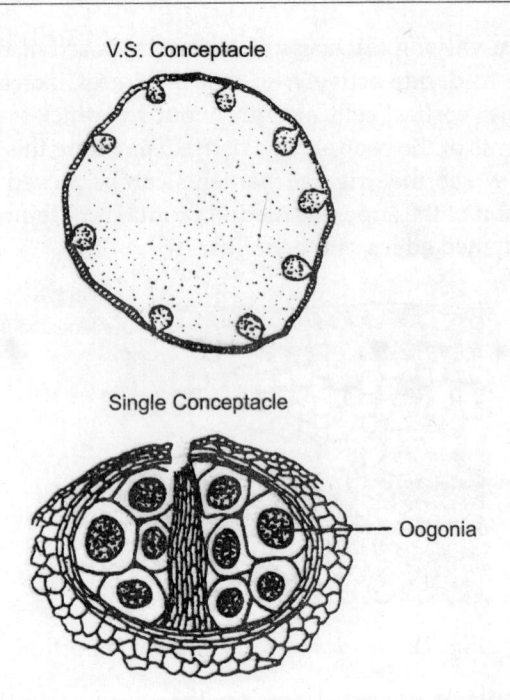

Fig. 12.19 *Fucus*—V.S. conceptacle showing oogonia.

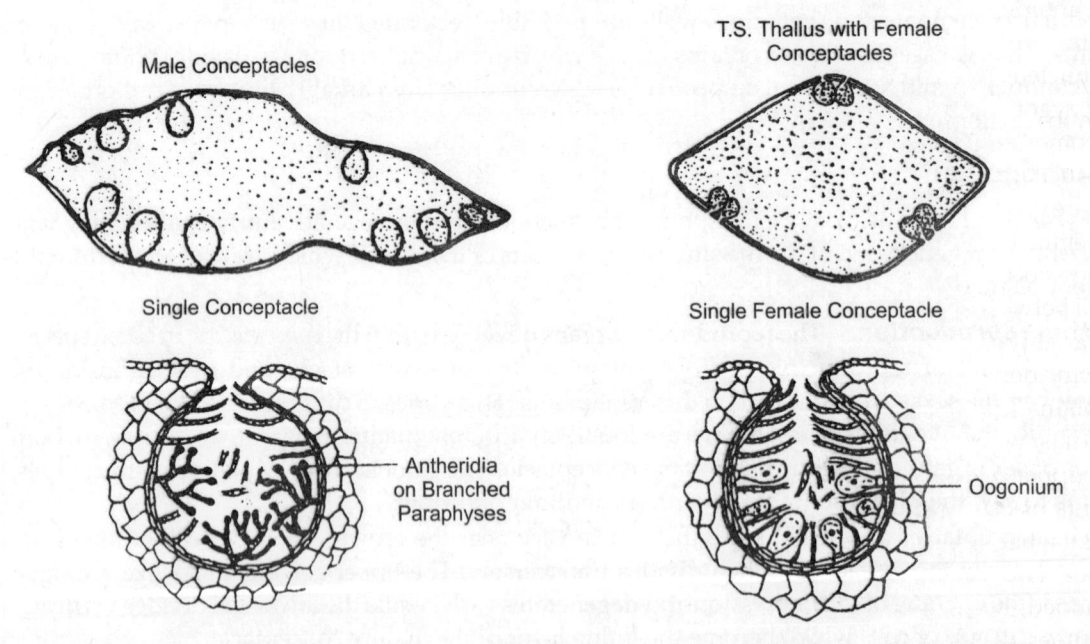

Fig. 12.20 *Fucus*—V.S. male conceptacle.   **Fig. 12.21** *Fucus*—T.S. Thallus with female conceptacles.

contained in an inner wall *(mesochiton)*. In the sea water this inner wall also opens, and inner-most coat of mucilage *(endochiton)* disperses and the eight oospheres escape as passive spherical bodies. Each oosphere contains many plastids, giving it an olive-green colour. Interspersed with the oogonia in the conceptacles are long sterile hairs, or *paraphyses*, which arise all over the surface of the walls of the conceptacle, while their free ends project out of, or at any rate towards, the pore or *ostiole* of the conceptacle.

The male reproductive organs or antheridia arise in much the same way as the oogonia from the surface of the male conceptacles. A superficial cell grows into a short branched structure, with both sterile and fertile branches. The latter are the antheridia. The conceptacle is almost filled by these densely branched structures. At first the antheridium contains a single nucleus, but this is rapidly increased by division until sixty-four are formed. Around each nucleus a membrane is formed, enclosing a small quantity of protoplasm and a single orange chromoplast; thus sixty-four antherozoids are formed. Each antherozoid is pear-shaped and has two lateral flagella, one forward and one backward. They are unequal, with the shorter flagellum forwards, contrary to the general rule among the lower Phaeophyceae. The orange-coloured chromoplasts give a bright colour to the mass of antherozoids which escape through the ostiole.

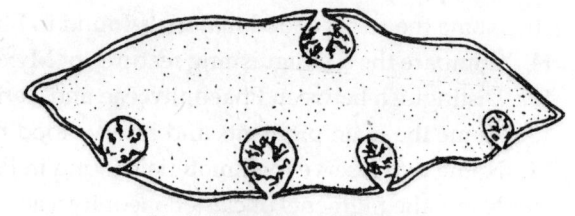

The antheridium, like the oogonium, has a multiple wall, though its layers are not so clearly defined. The outer wall opens apically and releases an oval inner sac, containing the antherozoids, in which they emerge through the ostiole of the conceptacle. The inner sac then swells and opens at its ends, and the antherozoids swim out into the sea water.

The oospheres and antherozoids are liberated with a quantity of mucilage at ebb-tide and are picked up by the advancing edge of the flood-tide, and it is here that fertilization takes place. The membrane around the antherozoids bursts when they have been liberated from the conceptacle, and they swim off by means of their flagella. The antherozoids soon seek out an oosphere, and immediately attach themselves to its surface in large numbers. They attach themselves by the tips of the anterior flagella and the movements of the posterior flagella cause the oosphere to rotate in the water, with its attached swarm of males. Eventually one antherozoid penetrates the oosphere and its nucleus travels inwards through the cytoplasm and fuses with that of the oosphere. The resulting zygote immediately secretes a wall which prevents penetration by any other antherozoid and becomes an oospore. The wall of the oospore is mucilaginous and fixes it to the substratum before growth begins. After twenty-four hours it begins to divide, the nucleus dividing first, followed by the laying down of the first-cell wall between the daughter nuclei. This wall is invariably at right angles to the direction of the incident light. The lower of the two cells contains few plastids and develops into the holdfast while the rest of the plant grows from the upper cell. As growth continues a few filaments of cells develop at the lower end and form what are called rhizoids, while at the other end, the thallus, which is at first rounded, becomes flattened. The plant soon becomes permanently anchored to a rock or similar support by the aid of the rhizoids from which later the holdfast is developed.

In the life-history of *Fucus*, therefore, we see a highly developed sexual method of reproduction, which is not associated with any kind of asexual one.

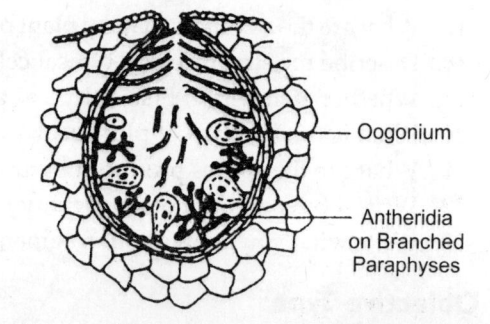

Fig. 12.22 *Fucus*—T.S. Thallus with bisexual conceptacles (male and female).

IMPORTANT QUESTIONS

Essay Type

1. Mention the general characteristics of Phaeophyceae and discuss the various types of life cycles in the group.
2. Describe alternation of generations in Phaeophyceae.
3. Describe various types of life cycles in phyaeophyceae.
4. Discuss the phenomenon of alternation of generations in the members of Phaeophyceae studied by you. Discuss their inter relationships and evolutionary tendencies.
5. Write an account of the range of thallus organisations in brown algae and discuss their evolutionary tendencies.
6. Describe methods of sexual reproduction and structure of sex organs in Phaeophyceae.
7. Give an account of cell structure and modes of reproduction in Phaeophyta.
8. Write what you know about range of vegetative structure in brown algae.

Short Answer Type

9. Write short notes on :
    (a) Distinguishing features of Phaeophyceae.
    (b) General characters of Phaeophyceae.
    (c) Economic importance of brown algae.
    (d) Modes of sexual reproduction in Phaeophyceae.
    (e) Unilocular sporangium.
    (f) Zoospore.
    (g) Hetertrichous condition.
    (h) Cryptoblasts.
10. Name the reserve food materials found in Phaeophyceae.
11. Compare the distinguishing features of Myxophyceae, Xanthophyceae and Phaeophyceae.
12. Distinguish between Phaeophyceae and Xanthophyceae.
13. Name the main pigments and reserve food material in the class Phaeophyceae.
14. Name two reserve food materials found in Phaeophyceae.
15. Name the pigments of class Phaeophyceae.
16. What are unilocular and plurilocular reproductive bodies? Where do they occur? What is their importance?
17. What are the advances that the plant body in Phaeophyceae shows over that of Chlorophyceae?
18. Describe the structure of a typical cell in Phaeophyceae.
19. Whether members of Phaeophyceae are fresh water or marine?
20. What is the function of plurilocular sporangia?
21. What are the spores produced by unilocular sporangia?
22. Write a brief account of anisogamy in Phaeophyceae.
23. Write what you know about fragmentation.

Objective Type

24. Fill in the blanks
    (i) In Phaeophyceae reserve food material is _____.

(ii) The general term for large brown algae is _____.
(iii) The Carbohydrate food reserve of brown algae is _____.
(iv) The principal pigment importing brown or olive brown colour to the thallus of Phaeophyceae is _____.
(v) Alginic acid is present in the cell walls of class _____.
(vi) In Phaeophyceae _____ vesicles are present.
(vii) Algin is obtained from _____.
(viii) All the brown algae except _____ show distinct alternation of generation.
(ix) Phaeophyceae are usually _____.
(x) Sea palm is the common name given to _____.

25. Select the correct answer :
   (i) The principal pigment importing distinctive brown or olive brown colouration to the thallus of Phaeophyceae is
      (a) Siphonoxanthin
      (b) Fucoxanthin
      (c) Necoxanthin
      (d) Flavoxanthin.
   (ii) In class Phaeophyceae the reserve food is stored in the form of
      (a) Laminarin
      (b) Glucose
      (c) Fructose
      (d) Glycogen.
   (iii) Which are the two pigments seen in Phaeophyceae?
      (a) Chlorophyll and Xanthophyll
      (b) Chlorophyll and Fucoxanthin
      (c) Phycocyanin and Phycoerythrin
      (d) Phycocyanin and Xanthophyll.
   (iv) A pigment that is absent in Phaeophyceae is
      (a) Chlorophyll *a*
      (b) Chlorophyll *b*
      (c) β-Carotene
      (d) Chlorophyll *c*.
   (v) Hetrogeneratae is so called because
      (a) It has two different types of sporangia
      (b) It has heteromorphic alternation of generations
      (c) It has two different types of sex organs
      (d) It has two types of zoospores.
   (vi) Unilocular sporangia is so called because it has only
      (a) One zoospore
      (b) Haploid zoospores
      (c) One meiotic division in it
      (d) One chamber.
   (vii) Plurilocular sporangia is so called because they
      (a) Produce numerous zoospores
      (b) Have numerous compartments
      (c) Have many divisions of their contents during zoospore formation.
      (d) Have numerous vacuoles.
   (viii) Which of the following is commonly called sea palm?
      (a) *Ectocarpus*
      (b) *Postelsiapalmaeformis*
      (c) *Fucus*
      (d) *Dictyota*.
   (ix) Algin is produced by the numbers of class
      (a) Chlorophyceae
      (b) Myxophyceae
      (c) Phaeophyceae
      (d) Rhodophyceae.

(x) The reserve food is stored in the form of manitol and laminarin in
  (a) Chlorophyceae
  (b) Myxophyceae
  (c) Rhodophyceae
  (d) Phaeophyceae.
(xi) Plurilocular sporangia are produced on
  (a) Haploid plants
  (b) Diploid plants
  (c) Triploid plants
  (d) Polyploid plants.
(xii) The zoospores are pyriform and
  (a) Uniflagellate
  (b) Laterally biflagellate
  (c) Apically biflagellate
  (d) Aflagellate.
(xiii) Unilocular sporangia produce
  (a) Mitozoospores
  (b) Meiozoospores
  (c) Both of the above
  (d) None of the above.
(xiv) The members of class Phaeophyceae are
  (a) Freshwater forms
  (b) Exclusively marine
  (c) Both freshwater and marine
  (d) Terrestrial.

QUESTIONS

Essay Type

1. Describe the structure and mode of reproduction in *Vaucheria* or *Ectocarpus*.
2. Give a brief account of the life history of Ectocarpus and explain the alternation of generation this alga.
3. With the help of labelled diagrams only, describe the life cycle of *Ectocarpus*.
4. Discuss alternation of generations in *Ectocarpus* and mention where reduction division takes place.
5. What is mean by alternation of generations? Describe in detail with reference to an alga (*Ectocarpus*) of class Phaeophyceae. Give its systematic position upto order.
6. Describe habitat, vegetative structure and reproduction of *Ectocarpus*.
7. Give an account of salient features of Phaeophyceae and describe in brief the structure and reproduction in *Ectocarpus*.
8. Give an account of environmental effect on the life of *Ectocarpus*.
9. Give an account of structure and reproduction in *Ectocarpus*.
10. Write about habitat, thallus structure and different types of life cycles in *Ectocarpus*.
11. Establish that *Ectocarpus* belongs to Phaeophyceae.
12. Bring out the role played by sporangia and gametangia in the life-cycle of *Ectocarpus*.
13. Give a brief illustrated account of structure and reproduction of any brown alga which you have studied.
14. Describe the life history of *Ectocarpus*.
15. What are unilocular and plurilocular sporangia? Is there any difference between the zoospores produced by the two? Give reasons.
16. Define Alternation of Generations. Discuss the phenomenon with reference to the life-cycle of *Ectocarpus*.

Short Answer Type

17. Write one important character for the pair which distinguishes its component genera
    (a) *Ectocarpus* and *Polysiphonia*
    (b) *Plurilocular* and *unilocular sporangia*
    (c) *Ectocarpus* and *Batrachospermum*
18. Give only brief answer :
    (i) Meiosis in *Ectocarpus*
    (ii) Write the systematic position and important morphological features of *Ectocarpus*.
19. Write short notes on :
    (a) Plurilocular sporangia of *Ectrocarpus*
    (b) Asexual reproduction in *Ectrocarpus*
    (c) Sexual reproduction in *Ectocarpus*
    (d) Unilocular and Plurilocular sporangia
    (e) Structure of cell in *Ectocarpus*
    (f) Unilocular sporangium
20. Draw labelled diagrams of the thallus and mention systematic position of *Ectocarpus*.
21. Why the alternation of generation in *Ectocarpus* is called Isomorphic?
22. (a) Give the structure of chromatophore in *Ectocarpus*
    (b) Give the classification of *Ectocarpus*
23. Draw well labelled diagrams of the following :
    (a) Reproductive structures in *Ectocarpus*
    (b) Structure of the thallus in *Ectocarpus*
    (c) Cell structure in *Ectocarpus*.
24. What are unilocular and plurilocular reproductive bodies and where do they occur? What are they concerned with?

Objective Type

25. Select the correct answer :
    (i) In *Ectocarpus* during fertilisation the male and female gametes attach themselves to the female gamete by the anterior flagellum which is
        (a) Long and whiplash   (b) Long and tinsel
        (c) Short and whiplash   (d) Short and tinsel.
    (ii) Thallus construction in *Ectocarpus* is
        (a) Filamentous unbranched   (b) Heterotrichous
        (c) Heterothallic   (d) Trichothallic.
    (iii) In the motile reproductive bodies of *Ectocarpus* the anterior flagellum compared to the posterior one is
        (a) Tinsel type and longer   (b) Whilplash type and shorter
        (c) Tinsel type and shorter   (d) Whilplash type and longer.
    (iv) Isomorphic alternation of generation is observed in
        (a) *Hydrodictyon*   (b) *Batrachospermum*
        (c) *Pinnularia*   (d) *Ectocarpus*.

(v) Plurilocular sporangia are characteristic of
   (a) *Sargassum*   (b) *Dictyota*
   (c) *Ectocarpus*   (d) *Ulothrix*.

(vi) In the life cycle of *Ectocarpus* meiosis occurs during
   (a) Zoospore formation in plurilocular sporangia
   (b) Zoospore formation in unilocular sporangia
   (c) Gamete formation in the gametangium
   (d) Germination of the zygote.

26. Of the four tetraspores from each sporangium two give rise to the male plants and the other two to the female plants.
27. The diploid and the haploid plants are morphogically alike.
28. There is thus **isomorphic alternation of generations** in the life cycle which is **diplohoplantic**.

Taxonomic Position

Division	:	**Phaeophyta** (Phaeophycophyta)
Class	:	**Isogeneratae**
Order	:	**Dictyotales**
Family	:	**Dictyotaceae**
Genus	:	*Dictyota*
Species	:	*dichotoma*

QUESTIONS

Essay Type

1. Describe the habitat and structure of the thallus of *Dictyota*.
2. Describe sexual reproduction in *Dictyota*.
3. Explain the phenomenon of alternation of generations in *Dictyota*.
4. Give a diagrammatic representation of the life cycle of *Dictyota* illustrating the relative lengths of the haploid and diploid generations.
5. Give significant features in the structure and methods of reproduction in *Dictyota*.
6. What are tetrasporangia? Where are they produced and to what structures do they give rise to? What are the functions of the plants produced by the tetraspores?
7. Describe the morphology of thallus of *Dictyota* (both external and internal). Discuss the occurrence, structure, development and functions of tetrasporangia.

Short Answer Type

8. Diagrammatically give the life cycle of *Dictyota*.
9. Draw well labelled diagrams of T.S. of *Dictyota* passing through male and female sori.
10. Describe the internal structure of *Dictyota* thallus.
11. Write short notes on :
   (a) Oogonial sorus of *Dictyota*
   (b) Fertilisation in *Dictyota*
   (c) Formation of thallus of *Dictyota*
   (d) Sexual reproduction in *Dictyota*
   (e) Asexual reproduction in *Dictyota*
   (f) Internal structure of thallus of *Dictyota*.

12. The gametophytic thalli of *Dictyota* indica are often found missing on our shores. Explain its absence.
13. Discuss the structure of Tetrasporangia in *Dictyota*.
14. Define Tetraspore.
15. Compare life cycles of *Dictyota* and *Fucus*.

Objective Type

16. Select the correct answer :
    (i) The antheridium in *Dictyota* is
        (a) Unicellular
        (b) Bicellular
        (c) Multicellular
        (d) Acallular.
    (ii) Which of the following is exclusively marine?
        (a) *Ulothrix*
        (b) *Spirogyra*
        (c) *Nostoc*
        (d) *Dictyota*.
    (iii) The sperm in *Dictyota* is
        (a) Monoflagellate with tinsel type of flagella
        (b) Monoflagellate with whiplash type of flagella
        (c) Biflagellate with tinsel and whiplash flagella
        (d) Aflagellate.
    (iv) In which of the following the thallus is differentiated into prostrate irregularly shaped disc shaped hold fast and upright fond?
        (a) *Ectocarpus*
        (b) *Sargassum*
        (c) *Dictyota*
        (d) *Coleochaete*
    (v) The process of sexual reproduction in Dictyota is
        (a) Isogamous
        (b) Anisogamous
        (c) Oogamous
        (d) None of the above.
    (vi) In which of the following alga, the thallus is dioecious?
        (a) *Spirogyra*
        (b) *Ulothrix*
        (c) *Dictyota*
        (d) *Ectocarpus*.
    (vii) Where does meiosis take place in *Dictyota*?
        (a) During the formation of gametes
        (b) During the formation of tetraspores
        (c) During the production of sperms
        (d) During the germination of zygote.
    (viii) Tetraspores are characteristic of
        (a) *Ectocarpus*
        (b) *Oedogonium*
        (c) *Dictyota*
        (d) *Spirogyra*.
    (ix) In sporophytic thalli of *Dictyota* asexual reproduction takes place by
        (a) Mitospores
        (b) Zoospores
        (c) Akinetes
        (d) Heterocyst.
    (x) Fertilisation in *Dictyota* is
        (a) Internal
        (b) External
        (c) Both
        (d) None.
17. Each sorus consists of paraphyses intermingled with unilocular sporangia. The diploid protoplast of each sporangium undergoes meiosis to produce 32 meiozoospores.
18. Out of these, 16 germinate to produce the male and the other 16 to produce the **female gametophytes**.

19. The gametophytes are heterotrichous in habit and are very much reduced as compared with the large sporophytes.
20. The male gametophytes are much branched and have smaller cells with scanty or no chromatophores.
21. The antheridia occur in terminal clusters on the erect filaments. Each antheridium produces a single biflagellate antherozoid. It is liberated through an apical pore formed by the thickening or gelatinization of the wall at the apex of the antheridium.
22. The female gametophytes are scarcely branched. They posses large cell filled with chromatophores. Any cell can function as an oogonium. The oogonial cell enlarges and becomes tubular or pyriform. Its contents round off to form a single **ovum** or **egg**.
23. The apical portion of the oogonial wall gets thickened and splits to allow the single ovum to escape. As the ovum is coming out the spilt wall again meets and forms a platform for the ovum to rest on.
24. The zygote starts germinating while still attached to the apex of the oogonium. The nucleus divides by simple mitosis. There is no meiosis.
25. The zygote does not undergo any period of rest.
26. There is heteromorphic alternation of generations in the life cycle which is diplohaplontic.

Taxonomic Position

Division	:	**Phaeophyta** (Phaeophycophyta)
Class	:	**Heterogeneratae**
Order	:	**Laminariales**
Family	:	**Laminariaceae**
Genus	:	*Laminaria*
Species	:	*laminariodes*

QUESTIONS

Essay Type

1. Give an illustrated account of the structure of the thallus and reproduction in *Laminaria*.
2. By means of labelled sketches only describe the life-cycle of *Laminaria*.
3. Give an illustrated account of the gametophyte phase in *Laminaria*.
4. What are the cross connections? How are they formed? Describe in detail.
5. Give an account of the structure of thallus in *Laminaria*. How are new blades formed?
6. Write what you know about the formation and structure of the male organ in *Laminaria*.
7. Describe the formation of the female gametophyte and the structure of the female sex organ in *Laminaria*.

Short Answer Type

8. How is fertilisation affected in *Laminaria*? Describe briefly the germination of zygote of *Laminaria*.
9. Describe briefly the anatomy of stipe in *Laminaria*.
10. What are the trumpet hyphae? How are they comparable to phloem of higher plants? Describe their probable functions.
11. Describe the germination of meiozoospores to form male gametophyte.

Class: Phaeophyceae (Brown Algae): General Characters and Type Study

12. Write short notes on :
    (a) Sporophyte of *Laminaria*
    (b) Alternation of generations in *Laminaria*
    (c) Distinguishing characters of the order Laminariales
    (d) Fertilisation in *Laminaria*
    (e) Male gametophyte of *Laminaria*
    (f) Female gametophyte of *Laminaria*
    (g) L.S. and T.S. of stipe
    (h) Meiozoospores.
13. Give a brief account of sexual reproduction in *Laminaria*.

Objective Type

14. Select the correct answer :
    (i) *Laminaria* is a
        (a) Xerophyte
        (b) Hydrophyte
        (c) Epiphyte
        (d) Lithophyte.
    (ii) The members of the order Laminariales are commonly known as
        (a) Pond silk
        (b) Net algae
        (c) Kelps
        (d) Sea anemone.
    (iii) In which of the following the adult plant is a sporophyte
        (a) *Spirogyra*
        (b) *Oedogonium*
        (c) *Laminaria*
        (d) *Coleochaete*.
    (iv) Where does meiosis take place in Laminaria?
        (a) In the sporangium
        (b) During formation of mitospores or zygote
        (c) During the formation of gametes
        (d) During the formation of meiospores.
    (v) Which of the following statements are true regarding the behaviour of 32 meiospores of *Laminaria saccharina*?
        (a) All the 32 meiospores give rise to a monoecious gametophyte
        (b) While 24 meiospores form the male gametophyte, only 8 meiospores form the female gametophyte.
        (c) 16 meiospores form male gametophyte and 16 meiospores form the female gametophyte.
        (d) All the 32 meiospores form the male gametophyte.
    (vi) Algin or Alginate are produced from
        (a) *Ectocarpus*
        (b) *Fucus*
        (c) *Laminaria*
        (d) *Sargassum*.
    (vii) Kelps are valuable source of
        (a) Nitrogen
        (b) Iodine
        (c) Sodium
        (d) Copper.
    (viii) The trumpet hyphae are characteristic of
        (a) *Polysiphonia*
        (b) *Sargassum*
        (c) *Ectocarpus*
        (d) *Laminaria*.

(ix) The alternation of germination in *Laminaria* is
   (a) Isomorphic and diplohaplontic   (b) Heteromophic and diplohaplontic
   (c) Isomorphic and haplontic   (d) Heteromorphic and haplontic.
(x) Subclass heterogeneratae is so called because
   (a) It has two different types of sporangia
   (b) It has heteromorphic alternation of generations
   (c) It has two different types of sex organs
   (d) It has two types of zoospores.

QUESTIONS

Essay Type

1. Give an account of structure and reproduction of *Sargassum*.
2. Enumerate distinguishing features of phaeophyceae. Describe male and female conceptacles of *Sargassum* with suitable diagrams.
3. With the help of suitable diagrams, describe the life cycle of *Sargassum*.
4. Give an account of habitat and sexual reproduction in *Sargassum*.
5. Describe the sexual reproduction of *Sargassum*.
6. Give a brief account of structure and distribution of *Sargassum*.
7. Give an account of the structure of the thallus and mode of reproduction in *Sargassum*.
8. Describe the post fertilisation changes in the life history of *Sargassum*.
9. Describe in detail the nuclear changes and alternation of generations in *Sargassum*.
10. Give diagrammatic representation of the life cycle of Sargassum illustrating the relative lengths of the haploid and diploid generations.
11. State the salient features of the lie cycle of *Sargassum*.
12. Describe sexual reproduction with the help of self-explanatory sketches, in *Sargassum*. Point out its differences from *Fucus*.
13. *Sargassum* exemplifies a reduced life-cycle. Explain.
14. Describe the habitat, thallus structure and reproductive methods in *Sargassum*. Compare habitat with that of *Gracilaria*.
15. Give a self-explanatory diagram of *Sargassum* plant and its graphic life cycle.
16. Discuss the reproduction in *Sargassum*.

FUCUS

17. Describe the structure and reproduction in *Fucus*.
18. Give an account of the occurrence, thallus structure and reproduction in *Fucus*.
19. Give an account of the habit and habitat of Fucus and explain how its form and structure is adapted to the environmental conditions.
20. Give an account of the life history of *Fucus* and discuss the question of alternation of generations in the genus.
21. Give a diagrammatic account of the life cycle of *Fucus* illustrating the relative length of the haploid and diploid generations.
22. Write an illustrated account of the sexual reproductive in *Fucus*. Describe sexual reproduction in Fucus.

23. Give an account of the habit, structure of thallus and mode of reproduction in *Fucus*.
24. Write an account of the life history of *Fucus*.
25. Give a brief account of the life history of *Fucus*.
26. Give an account of vegetative structures of *Fucus* and *Sargassum*.
27. Give illustrated account of the thallus structure of *Fucus*.

Short answer type

28. Write short notes on
    (a) Male conceptacle of *Sargassum*.
    (b) Sexual reproduction in *Sargassum*.
    (c) Thallus structure of *Sargassum*.
    (d) Conceptacles of Sargassum.
    (e) Structure of conceptacle.
    (f) Sex organs of *Sargassum*.
    (g) Male and female conceptacles in *Sargassum*.
    (h) Cryptoblamata.
    (i) Structure and development of sex organs in *Fucus*.
    (j) Habit and habitat of *Fucus*.
    (k) Conceptacles in *Fucus* and *Sargassum*.
    (l) Conceptacles and Receptacles.
    (m) Vegetative thallus of *Fucus*.
29. Where does reduction division occur in *Sargassum*, *Volvox* and *Ectocarpus*?
30. What are the functions of air bladders in *Sargassum*?
31. Draw labelled diagrams of the thallus and give the most important characteristic and classification of *Sargassum*.
32. With the help of labelled diagrams, show the structure of *Sargassum*.
33. Draw well labelled diagrams of the following :
    (a) V.S. male conceptacle of *Sargassum*.
    (b) V.S. female conceptacle of *Sargassum*/mature female conceptacle.
    (c) Male and female conceptacles of *Sargassum*.
    (d) Conceptacles of *Sargassum*.
    (e) Male conceptacle of *Sargassum*.
    (f) External morphology of *Sargassum* thallus.
    (g) Diagrammatic life cycle of *Sargassum*/*Fucus*.
34. Distinguish between :
    (a) *Sargassum* and *Chara*.
    (b) *Sargassum* and *Fucus*.
    (c) *Ectocarpus* and *Sargassum*.

Objective Type

35. Select the correct answer :
    (i) The receptacles are found in
        (a) *Ectocarpus*   (b) *Sargassum*
        (c) *Polysiphonia*   (d) *Nemalion*.

(ii) The life cycle of Sargassum is
   (a) Haplontic
   (b) Diplontic
   (c) Haplo diplobiontic
   (d) None of the above.
(iii) Which of the following is typically a marine alga?
   (a) *Ulothrix*
   (b) *Spirogyra*
   (c) *Chlamydomonas*
   (d) *Sargassum*.
(iv) The alga commonly called Rockweed is
   (a) *Ectocarpus*
   (b) *Fucus*
   (c) *Sargassum*
   (d) *Laminaria*.
(v) *Fucus* plant is
   (a) Haploid
   (b) Diploid
   (c) Triploid
   (d) Tetraploid.
(vi) Reduction division in *Fucus* takes place during
   (a) germination of zygote
   (b) growth of thallus
   (c) formation of sex organs
   (d) gametogenesis.
(vii) The spermatozoid in Fucus is a
   (a) minute, slightly elongated pear shaped structure
   (b) large terminally uniflagellate structure
   (c) minute ovoid biflagellate cells
   (d) large pear-shaped terminally biflagellate cell.
(viii) The mature oogonium in *Fucus* contains
   (a) one egg
   (b) two eggs
   (c) four eggs
   (d) eight eggs.
(ix) Which of the following is called the Gulf weed?
   (a) *Ectocarpus*
   (b) *Sargassum*
   (c) *Fucus*
   (d) *Nemalion*.
(x) In *Sargassum* growth of the thallus is by the activity of a
   (a) a three sided apical cell
   (b) a four sided apical cell
   (c) an intercalary meristem
   (d) all cells of the thallus.
(xi) In *Sargassum* the part of the oogonium that forms the pseudostalk is called
   (a) Exochite
   (b) Mesochite
   (c) Endochite
   (d) Stalk cell.
(xii) In which of the following alga the *oogonia* are raised on gelatinous stalks out of the conceptales?
   (a) *Ectocarpus*
   (b) *Sargassum*
   (c) *Fucus*
   (d) *Polysiphonia*.
(xiii) Sterile flask-shaped cavities present in the leaves of Sargassum are called
   (a) Conceptacles
   (b) Cryptoblasts
   (c) Cryptoblamata
   (d) Cryptostomata.
(xiv) The photosynthetic region in the thallus of Sargassum is
   (a) epidermis
   (b) Cortex
   (c) Medulla
   (d) Cuticle.

(xv) The product of photosynthesis is Sargassum is
   (a) Starch
   (b) Protein
   (c) Manitol
   (d) Glycerol.
(xvi) *Sargassum* conceptacles occur
   (a) In the reciptacles
   (b) In the holdfast region
   (c) On the stem
   (d) On the leaf.
(xvii) In the eight-celled nucleated oogonium of *Sargassum*, the number of diploid nuclei that develop into eggs are
   (a) only one
   (b) only four
   (c) less than four
   (d) all the eight nuclei.
(xviii) In *Sargassum* the fertilisation of the female gametes takes place when the egg is
   (a) in the oogonium
   (b) in water
   (c) outside the conceptacle still attached to the plant
   (d) inside the conceptacle still attached to the plant.

13 Class: Rhodophyceae (Red Algae): General Characters and Type Study

SALIENT FEATURES OF RHODOPHYCEAE

The Rhodophyceae or the red algae are characterized by the following distinguishing features:
1. There are no motile or flagellated stages.
2. The photosynthetic pigments include chlorophyll *a*, chlorophyll *d*, biliproteins like *r*-phycoerythrin and *r*-phycocyanin, and the xanthophyll called taraxanthin.
3. The reserve foods are floridean starch and galactoside floridoside.
4. The cell wall contains polysulphate esters in addition to cellulose and pectin.
5. They exhibit a characteristic process of sexual reproduction in which the male gamete (spermatium) is passively transported to the trichogyne of the female sex organ, the carpogonium.

Occurrence

Members of Rhodophyceae are exclusively marine forms except a few freshwater genera like, *Batrachospermum* and *Lemanea*. The marine forms grow in the intertidal and sub-littoral zones. They grow in all marine waters, but they are more abundant in warmer seas. They grow as lithophytes attached to the rocks and some of them even grow at a depth of 100 metres. *Corallina* is a calcareous alga that contributes to the formation of coral reefs. A few like *Choreonema* are parasitic on other marine algae. The red algae can grow at different depths because they can show chromatic adaptation complementary to the colour of the light so that maximum absorption of the incident light takes place.

Thallus Organization

The thallus of Rhodophyceae shows several variations. They may be unicellular (*Porphyridium*), filamentous and simple (*Erythrotrichia*) of filamentous and branched (*Goniotrichum*) or heterotrichous (*Batrachospermum*). In some, they are parenchmatous as in *Porphyra*. In general, the thallus of Rhodophyceae is not as large as that of brown algae and is more delicate in texture.

The filamentous thalli may be uniaxial (*Batrachospermum*) or multiaxial (*Polysiphonia*). The uniaxial filaments may show cortication. In multiaxial forms, the filaments may appear polysiphonous e.g., *Polysiphonia*. In some members, a pseudoparenchymatous thallus may be formed by the coalescence of several filaments.

The growth of the thallus may be diffuse (*Bingoideae*) or it may be apical as in (*Polysiphonia*). Anatomically some foliose forms may be differentiated into an outer photosynthetic zone, a middle cortical zone storing food material, and a central medulla.

Cell Structure

The cell wall is formed of an outer layer of pectin and an inner layer of cellulose. It may have an outer coating of mucilage. In Bangioideae, each cell has a single, axile stellate chloroplast with a central naked pyrenoid. But in Florideae, several discoid parietal chromatophores occur and they lack pyrenoids. The fine structure of chromatophore shows widely separated single thylakoids. The floridean starch occurs as scattered grains in the cytoplasm. The cells are uninucleate and rarely multinucleate as in *Griffithsia*. The photosynthetic pigments are chlorophyll *a*, chlorophyll *d*, *r*-phycoerythrin, *r*-phycocyanin and some xanthophylls. The reserve food is floridean starch.

Reproduction

Vegetative reproduction takes place in unicellular forms. In multicellular forms, asexual and sexual reproduction occur. These are complementary to each other for the completion of the life cycle.

The male and female sex organs are *spermatangia* (antheridia) and *carpogonia* which are unicellular. The spermatangia are produced on short branches of limited growth (*Batrachospermum*) or a special trichoblasts (*Polysiphonia*). In each spermatangium, a single non-motile spermatium is formed. The carpogonium is sessile in Bangioideae (*Porphyra*) but in Floridae, it is terminally produced on a carpogonial filament called procarp (*Polysiphonia*). The carpogonium has a neck like trichogyne. During fertilization, spermatia released from the spermatangia come into contact with the trichogyne. The spermatial nucleus passes down the trichogyne and unites with the nucleus of the egg.

The process of sexual reproduction upto fertilization is more or less uniform in Rhodophyceae. But the post-fertilization changes are complex and they form the basis for the classification of Florideae. In Bangioideae, a carposporophyte is not formed and the zygote divides directly forming carpospores. The carpospores germinate to form *conchocelis-like* plants. In *Porphyra*, the conchocelis stage is regarded as the prostrate system and the thallus of *Porphyra* as the erect system of an originally heterotrichous alga.

In the more advanced Rhodophyceae, like the Florideae, elaborate post-fertilization changes take place forming a *carposporophyte* which is parasitic on the gametophyte. In Nemalionales (*Batrachospermum*), the diploid zygotic nucleus divides by meiosis. The haploid nuclei migrate into the gonimoblast initials that arise from the base of the carpogonium. These initials give rise to the gonimoblast filaments on which carposporangia are terminally produced. The gonimoblast filaments become enveloped by sterile filaments arising from the surrounding vegetative cells. Thus, a haploid *corposporophyte* or *cystocarp* is produced. Later, the haploid carpospores which are formed singly in the carposporangia are released into the water. In Gelidiales, the zygotic nucleus divides mitotically and hence the gonimoblast filaments are *diploid*. Therefore, the carposporophyte also is diploid.

In Florideae, the diploid zygote nucleus migrates into the auxiliary cell through a tubular connection e.g., *Polysiphonia*. In *Polysiphonia*, the supporting cell cuts off an auxiliary cell. But in *Gracilaria*, the supporting cell itself functions as an auxiliary cell. In the auxiliary cell, the zygote nucleus always divides by mitosis. So, the gonimoblast filaments produced from the auxiliary cell are diploid. The resulting carposporophyte is also diploid and, therefore, the carpospores released from the carposporangia are diploid.

In *Batrachospermum*, the carpospores germinate to give rise to juvenile stages called *Chantransia* stages. These stages can multiply through monospores produced in monosporangia. Thus, the monospores serve as an accessory means of multiplication of chantransia stage. Some lateral branches of these stages develop into *Batrachospermum* plants.

In Florideae *(Polysiphonia, Gracilaria)*, the diploid carpospores germinate to give rise to *tetrasporophytes*. The tetrasporophytes are asexual and they produce *tetrasporangia*. In the tetrasporangium, the nucleus divides meiotically so that four haploid *tetraspores* are formed. These ones give rise to the gametophytic plants.

Life Cycle

In the life cycle of *Porphyra,* it is not definite when meiosis takes place. Probably, it takes place during the division of the zygote.

In *Batrachospermum,* the thallus is gametophytic and the cystocarp or carposporophyte is produced after meiosis of the zygote nucleus. The carposporopbyte, carpospores as well as the chantransia stages are haploid and the life cycle is described as *haplobiontic.* But in Florideae *(Polysiphonia),* the carposporophyte is diploid. The carpospores are diploid and they produce tetrasporophytes on which tetrasporangia are produced. Meiosis takes place in tetrasporangia so that haploid tetraspores are produced. These ones give rise to the gametophytes. Thus, there are three generations—haploid gametophyte, a diploid carposporophyte and a diploid tetrasporophyte. This type of life cycle is described as triphasic or diplobiontic.

TYPE STUDY

Following type studies in *Rhodophyceae* are discussed in detail:
 1. *Batrachospermum,* 2. *Polysiphonia,* 3. *Ceramium.*

BATRACHOSPERMUM

Class : Rhodophyceae

Order : Nemalionales
Family : Batrachospermaceae

Members of Rhodophyceae (Red Algae) are mostly marine. But *Batrachospermum* is a filamentous form inhabiting cool freshwater streams which are relatively well aerated and clean. It is found attached to stones. It appears like delicate beaded threads.

Structure (Figs. 13.1)

Batrachospermum is 10-15 cm. long and it is bluish-green violet or olive-green and soft to touch. It appears like branched beaded chains. The thallus is *heterotrichous* having a prostrate and erect system. The prostrate system is formed by the older branches which help in attachment. The erect system is formed of 5-20 cm. long unisexual filaments which are frequently branched in a *monopidial* manner. The filaments have beaded appearance due to *glomerules* or whorls of short laterals at the nodes (Fig. 13.1). The short laterals have limited growth and they arise just below the septa of the main axis. They are formed of *moniliform* cells and sometimes they end in hairs. At some nodes, branches of unlimited growth are produced which repeat the structure of the main axis. The long axis grows by means of an apical cell which cuts off cells below by a series of transverse divisions. These cells grow in length and are broader at the septa. They are uninucleate and contain several parietal chromatophores each with a naked pyrenoid. An axial cell below the apex cuts off 4-6 peripheral cells below the septum. They will be the initials that grow into the short laterals. The basal cells of short laterals may produce downwardly growing, corticating threads which may envelop the main axis completely. The corticating threads arising from the basal nodes may even reach the substatum and serve in fixation. The corticating cells may also produce whorls of shorts laterals so that the beaded appearance of the axis may be lost.

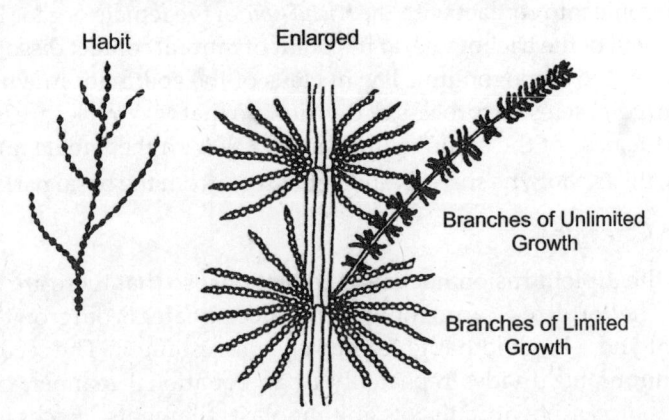

Fig. 13.1   *Batrachospermum*—Habit and a portion enlarged.

Life History

Among *Florideae*, *Batrachospermum* is considered to be simple, since it shows the *uniaxial* type of construction and sexual reproduction of the simplest type.

The sex organs are *carpogonia* (female) (Fig. 13.2) and *spermatangia* or *antheridia* (male). They are produced on the same plant (monoecious) or on different plants (dioecious).

Female Sex Organ (Fig. 13.3)

The *carpogonia* are produced terminally on small branches. These are formed from the basal cells of laterals of limited growth, and very rarely they are even formed on the cortical threads.

The female sex organ, the carpogonium, is one-celled, but it terminally prolonged into a club-shaped structure called *trichogyne*.

Between the trichogyne and the basal part of the carpogonium, there is a slight constriction. The cytoplasm occupies the entire inner part of the carpogonium, but the nucleus is situated in the basal part only. The trichogyne has no nucleus.

Male Sex Organ

The antheridia or spermatangia are borne on branches of limited growth on the terminal or subterminal cells which function as spermatangial initials. The spermatangia or the antheridia can be easily distinguished as they are generally colourless.

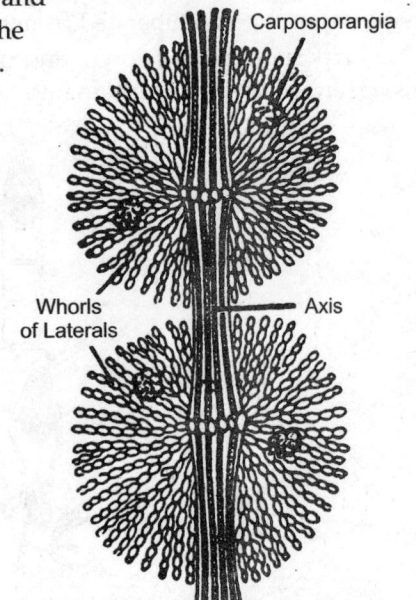

Fig. 13.2   *Batrachospermum*—Thallus showing carposporangia.

The antheridium or spermatangium is unicellular and its cytoplasm shows a conspicuous nucleus and few chromatophores. The entire protoplast of the antheridium escapes out as a single spermatium through an opening of the split wall. The spermatia are naked and have no flagella.

Fertilization

The *spermatia* are produced in large numbers to ensure fertilization. They float in the waters, passively and by chance come into contact with the *trichogyne* of the female organ. The membrane of the spermatium and the wall of the trichogyne, at the point of mutual contact dissolve, forming an open passage into the base of the carpogonium. The nucleus of the spermatium which is in ptophase at the time of liberation, completes the process of division, so that two nuclei are formed. But only one of them migrates to the base of the carpogonium and fuses with the female nucleus.

After fertilization, the trichogyne shrivels and gets cut off from the basal part of the carpogonium.

Post-fertilization Changes

The first division of the diploid fusion nucleus is reductional so that there are two daughter nuclei which are haploid. The fertilized carpogonium produces a lateral outgrowth, the first initial of gonimoblast filament and a haploid daughter nucleus passes into it. The second nucleus remains within the carpogonium and divides repeatedly in an equational manner, contributing haploid daughter nuclei to successive initials of gonimoblast filaments. Each initial gives rise to gonimoblast filaments. The terminal cells of the gonimoblast filaments later differentiate into *carposporangia*. Along with this development, a number of sterile filaments grow from some of the vegetative cells surrounding the basal portion of the carpogonium. They completely ensheath the gonimoblast filaments and carposporangia, forming a characteristic fructification called *cystocarp*. The cystocarp or carposporophyte is haploid and is said to be parasitic on the female gametophyte. A carposporangium formed terminally on the gonimoblast, gives rise to a single *carpospore*. Thus, several carpospores are liberated from the carposporangia.

The carpospore settles down, and on germination, produces an elongate process which gets transversely divided by a septum into two cells. This develops into the prostrate system which later

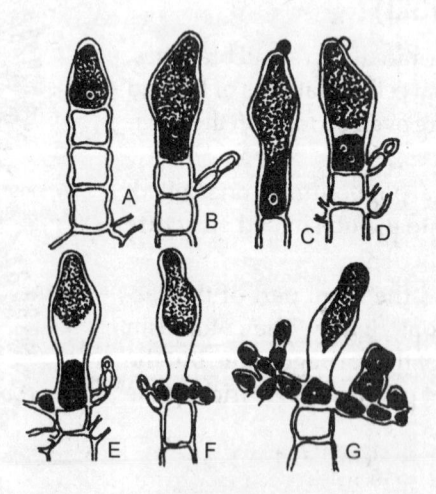

Fig. 13.3 *Batrachospermum*—(A and B) carpogonium, (C) fertilization, (D) male and female nuclei lying at the base of the carpogonium, (E) meiotic division of the zygote nucleus and formation of first gonimoblast, (F and G) gonimoblast filaments forming carpospores.

produces the erect filaments. These juvenile (young) stages were previously believed to be an independent alga called *Chantransia*. But as they are found to be a stage in the life history of *Batrachospermum*, they are called chantransia stages.

The chantransia stages may get repeatedly propagated by means of *monospores* produced in monosporangia. The structure and germination of monospores closely resembles that of carpospores. The monospores are not directly concerned with the alternation of generations and they are only useful in the accessory reproduction of the juvenile stages *(chantransia* stage).

From the chantransia stage, *adult shoots* arise either from the prostrate or erect system. The adult shoots clearly show the growth of the main axis by means of an *apical cell*. It shows at the septa, whorls of short laterals of limited growth. This develops into the *Batrachospermum* plant.

Life Cycle

Batrachospermum plant is haploid. It produces carpogonia and spermatagnia. During sexual reproduction, the male nucleus of spermatium unites with the female nucleus forming a diploid zygote nucleus. This nucleus immediately undergoes meiosis contributing haploid nuclei to the gonimoblast initials. The gonimoblast filaments produce carposporangia that produce carpospores. A sterile sheath develops around the gonimoblast filaments and carposporangia forming the cystocarp. The *cystocarp* is haploid and it is parasitic on the gametophyte. The carpospores germinate and produce young stages called *chantransia* stages which are also haploid. These stages can multiply through monospores produced in monosporangia. *Batrachospermum* plant arises as a special shoot of the Chantransia stage.

Thus, the life history of *Batrachospermum* is represented mostly by the haploid phase consisting of the adult plant, cystocarp and chantransia stages and diploid stage is represented only by the fusion nucleus. So, the life cycle is described as *haplobiontic* by Svedelius.

POLYSIPHONIA

Class: Rhodophyceae

Order: Ceramiales
Family: Rhodomelaceae

Polysiphonia is a common alga of *Rhodophyceae* having a wide distribution. *P. urceolata* is an epiphyte on *Laminaria*. *P. fastigata* grows on *Ascophyllum* semiparasitically. *P. elongata* is a lithophyte. Some species grow in polluted waters on the roots of mangroves (*P. variegata*). In India, *P. variegata* and *P. platycarpa* occur on the west coast.

Structure

The name *Polysiphonia* is given to the genus as the thallus is multiaxial or polysiphonous, consisting of an axial row of central cells surrounded, by a layer of 4-20 pericentral cells (Figs. 13.4, 13.5 & 13.6). The thallus is heterotrichous having a filamentous prostrate system and an erect system of branches. The prostrate system is attached to the substratum with the help of elongated unicellular rhizoids arising from the pericentral cells facing the substratum. The tips of the rhizoids are flattened into haptera or lobed discs. In some, the whole plant may be erect getting attached by rhizoids arising at the base (Fig. 13.7).

Habit

Fig. 13.4 *Polysiphonia*—Thallus.

The thallus has two kinds of lateral branches. They are:

(1) *Polysiphonous* branches and (2) *trichoblasts*. The polysiphonous branches of unlimited growth are like the parent axis in having central and pericentral siphons which are pigmented due to the uniseriate, colourless branches of limited growth. They are dichotomously branched and their tapering ends appear like hairs. As new trichoblasts are formed, the older ones fall away. Hence, the thallus shows more polysiphonous branches and few trichoblasts.

The main axis as well as the polysiphonous branches of unlimited growth are uniaxial towards the tip where pericentral cells are not yet formed. They grow by means of an apical cell. The apical cell forms a row of axial cells by transverse divisions. Later, the axial cells undergo periclinal divisions forming a definite number of pericentral cells surrounding the axial row of central cells.

The number of pericentral cells is constant for a species. In *P. elongate* 4 pericentral cells are formed whereas in *P. nigrecens,* they are 20 in number. The central cell and pericentral cells of a tier are derived from the same axial cell and they elongate equally so that they are of the same size. In surface view, the polysiphonous branches appear to consist of a number of tiers in an orderly manner. There are pit connections between and among the central and pericentral siphons. In older parts, the pericentral cells cut off cortical cells to the outer side which corticate the filaments (*P. violaceae*).

The polysiphonous branches or the uniseriate trichoblasts arise as lateral initials. A lateral initial is cut off by an axial cell which is 3-5 cells below the apical cell. The axial cell forms pericentral cells only after giving rise to an initial that may grow into a trichoblast or polysiphonous branch. Therefore, these branches appear as if they arise from the pericentral cells. The initial may give rise to the colourless uniseriate branch (trichoblast) or may function like an apical cell and form a row of axial cells from which pericentral cells are later formed so that a polysiphonous branch is formed.

In *P. nigrescens*, polysiphonous branches may arise from the basal cells of trichoblast. This shows that polysiphonous laterals and trichoblasts are homologous. Some trichoblasts are fertile and they bear sex organs. In the male plant, the fertile trichoblast is conspicuos, but in the female plant it is much reduced.

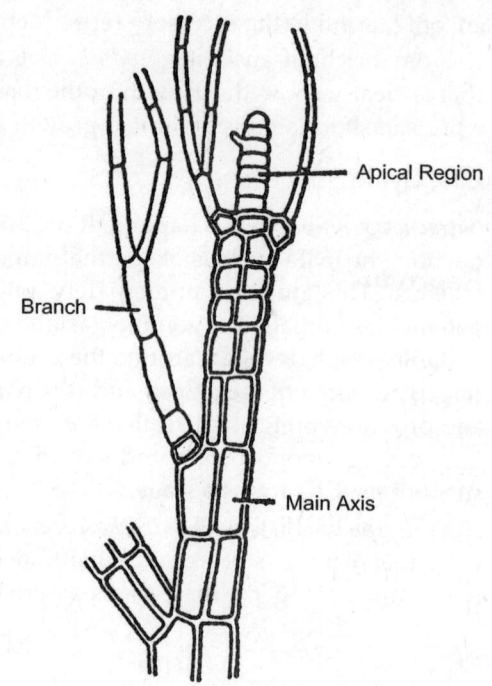

Fig. 13.5 *Polysiphonia*—Thallus portion enlarged.

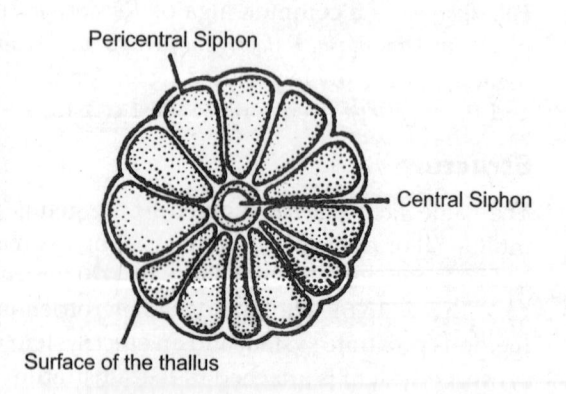

Fig. 13.6 *Polysiphonia*—Thallus basal portion.

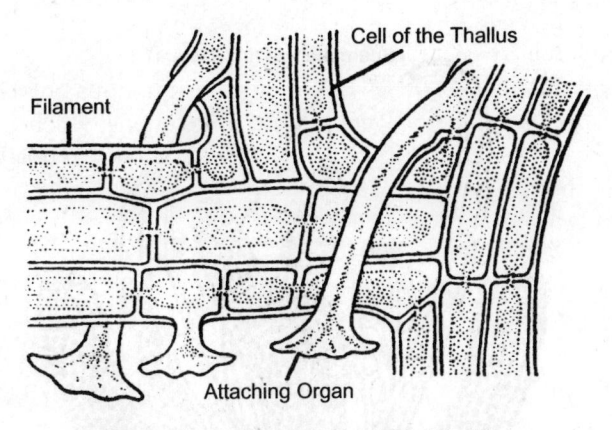

Fig. 13.7 *Polysiphonia*—T.S. Thallus.

Reproduction

Polysiphonia has three kinds of plants which are all similar in external appearance. They are: 1. male plants, 2. female plants and 3. tetrasporophytes. The male plant produces *antheridia* and the female plant produces *carpogonia*. These sexual plants alternate with diploid asexual *tetrasporophytes*.

Male Sex Organs (Fig. 13.8 A & 13.8B).

The spermatangia or antheridia are the male sex organs and they are produced on fertile trichoblasts borne near the apex. A fertile trichoblast branches dichotomously two or three cells from the base. Usually, one of the arms develops into a fertile axis and the other develops into a sterile one which is dichotomously branched. The fertile axis is unbranched and many-celled in length. Except the two lowest cells, all the cells of fertile axis cut off central and pericentral cells. The fertile axis is thus actually polysiphonous. Each pericentral cell cuts off spermatangial or antheridial mother cells on the outer side, each of which produces 3 or 4 antheridia or spermatangia. The wall of the spermatangium ruptures and the contents escape as a spermatium.

Female Sex Organs (Fig. 13.9)

The fertile trichoblast of the female plant is short as compared to the male one. It is 5-7 celled long and it is formed from an axial cell which is 3-4 cells below the apical cell before it cuts off pericentral cells. The trichoblast is uniseriate but the three basal cells cut off pericentral cells functions as the supporting cell. This supporting cell produces a four-celled carpogonial filament called *procarp*. The terminal cell of the procarp develops into the *carpogonium* which has a swollen base with uninucleate egg and an erect long trichogyne. The supporting cell, besides forming the carpogonial filament, also cuts off sterile initials, one towards the base and another towards a side. The basal sterile cell or initial does not divide for a time but the lateral initial commences to divide almost immediately.

Fertilization

Fertilization occurs at this stage. The spermatium attaches itself to the trichogyne. The wall at the point of contact dissolves and the male nucleus passes through the trichogyne and fuses with the female nucleus.

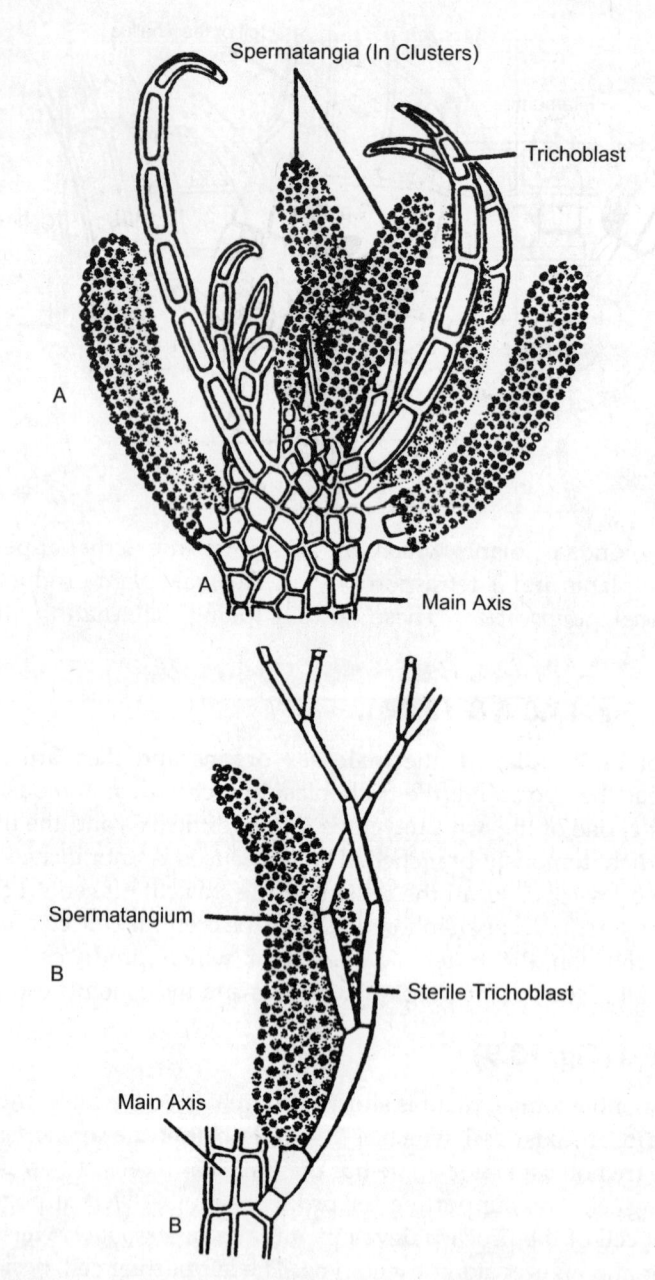

Fig. 13.8 *Polysiphonia*—Male sex organs—(A) Spermatangia. (B) Single spermatangium.

Post-Fertilization Changes

Soon after fertilization, the lateral initial gives rise to a 5-10 celled sterile filament; the basal initial similarly develops into a two-celled filament. This cell forms a tubular connection with the base of the carpogonium. The diploid fusion nucleus migrates into the auxiliary cell. The trichogyne degenerates and the pericentral cells adjacent to the supporting cell begin to form a sterile sheath.

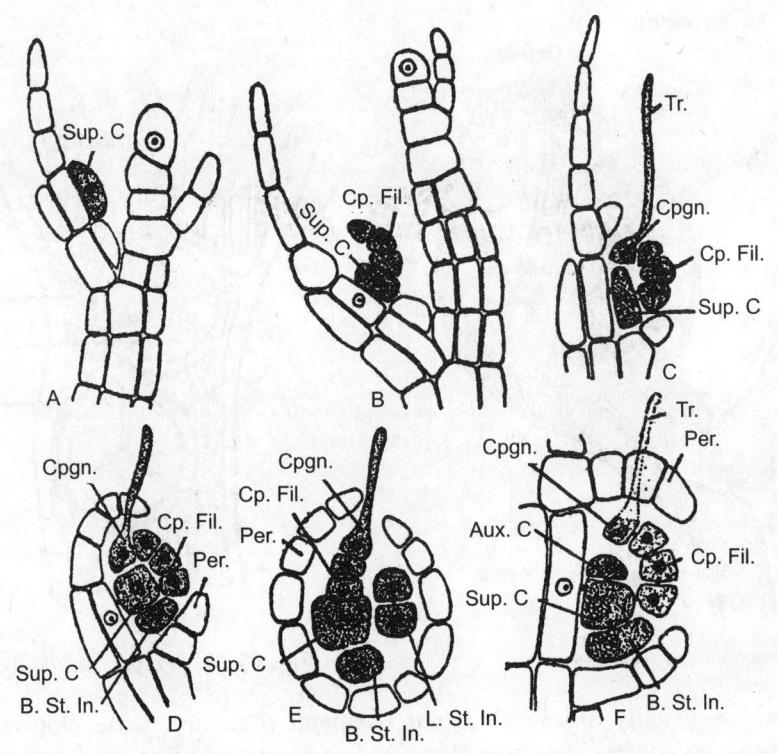

Fig. 13.9 *Polysiphonia*—Procarp formation (Carpogonium)—(A) before development of carpogonial filament. (B and C) developing and mature carpogonial filaments, (D) after cutting off of sterile filament initials, (E) tangential view. (F) after fertilization and formation of auxiliary cell. (Aux. C—auxiliary cell; B. St. In—basal sterile initial; Cpgn-carpogonium; Cp. Fil—Carpogonial filament; L. St. Fil—lateral sterile filament; Per—Pericarp; Supp. C—Supporting cell; Tr—Trichogyne).

From the upper side of the auxiliary cell, a gonimoblast initial is formed, and this will be followed by others. The fusion nucleus of the auxiliary cell divides mitotically several times, forming several diploid nuclei. These nuclei get distributed into the gonimoblast initials which give rise to the gonimoblast filaments. As a result of this, each cell of gonimoblast filament contains one diploid nucleus. The terminal cell of the gonimoblast filament functions as a *carposporangium*.

Fusion takes place between the supporting cell, auxiliary cell and sterile filaments, forming a large irregular *placental cell*. The carpogonial filament simply withers away and disorganises. The resultant structure, after all these changes, is known as *carposporophyte*. This carposporophyte gets surrounded by a sterile sheath called *pericarp*, which is formed by pericentral cells of the fertile trichoblast present adjacent to the supporting cell.

The mature carposporophyte is known as *cystocarp* (Fig. 13.10). The cystocarps are urn-shaped and they have a small apical aperture, the *ostiole*. The carposporangia form *carpospores*, germinate and give rise to diploid *sporophytes* which produce tetrasporangia. This asexual plant is known as tetrasporophytes.

Tetrasporophyte (Fig. 13.11)

The tetrasporophyte bears sporangia known as *tetrasporangia*.

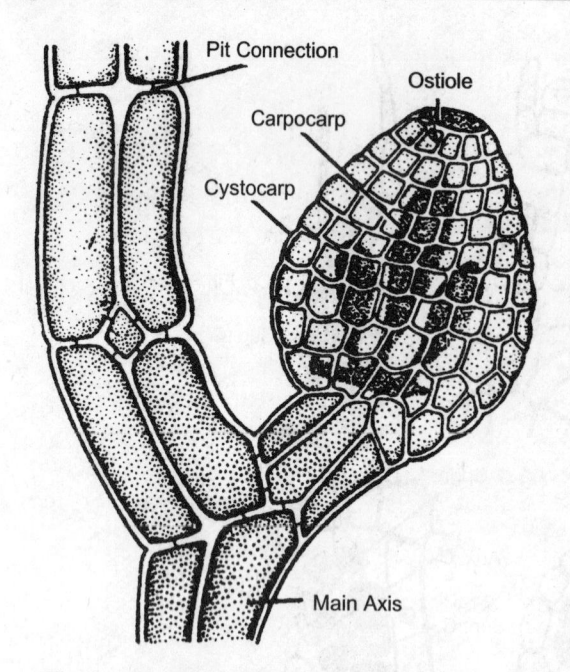

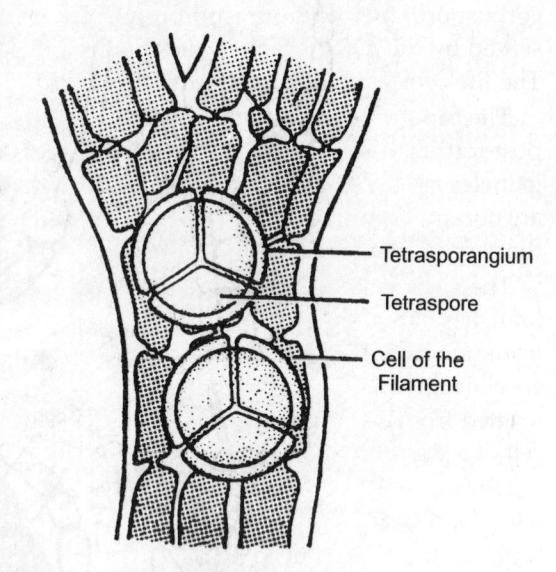

Fig. 13.10 *Polysiphonia*—Cystocarp.

Fig. 13.11 *Polysiphonia*—Tetrasporophyte.

The *tetrasporangia* usually follow a regular basipetal sequence in development. The fertile branches bearing sporangia are known as *stichidia*. The sporangia often cause swellings within the stichidia Tetrasporangia are formed in successive transverse tiers. But always, only one pericentral cell of each transverse tier can form a tetrasporangium. The fertile pericentral cell forms a tetrasporangium with a short stalk, cutting off one or two cover cells.

The tetrasporangium is the seat of meiosis. The diploid nucleus of the tetrasporangium enlarges considerably and divides meitotically forming four haploid nuclei. Then the cytoplasm undergoes cleavage so that four uninucleate protoplasts are formed. Thus, four tetrahedrally-arranged tetraspores are formed. They get liberated by the rupture of the wall of sporangium. The tetraspores, on germination, produce gametophytes (sexual plants) half of which are male and the other half female.

Life Cycle

In *Polysiphonia*, the male and female gametophytes which are haploid are morphologically similar to the diploid tetrasporophytes. The male plant produces spermatangia on fertile trichoblast and the female plant produces a carpogonium at the end of procarp. When fertilization occurs, diploid zygote nucleus is formed. It migrates into the auxiliary cell which cuts off initials of gonimoblast filaments. The diploid nucleus divides mitotically contributing daughter nuclei to these initials which give rise to the gonimoblast filaments. The carposporangia formed at the tips of gonimoblast filaments produce carpospores. In the meanwhile, a sterile sheath envelops the gonimoblast filaments and carposporangia forming the *cystocarp* or *carposporophyte* which is diploid as it is formed after fertilization without meiosis. The carposporophyte is considered to be a special generation, parasitic on the female gametophyte. The carpospores are diploid and they germinate forming tetrasporophytes. These diploid plants produce tetrasporophytes which produce tetrasporangia after meiosis. Four haploid tetraspores are formed which give rise to male and female gametophytes.

Thus, in *Polysiphonia*, the life cycle involves three generations, namely gametophyte, corposporophyte and tetrasporophyte. So, it is described as *triphasic*. The haploid phase is represented by the gametophytes and the diploid phase by the carposporophyte and tetrasporophyte. The life cycle is described as *diplobiontic*.

The process of reproduction in *Batrachospermum* and *Polysiphonia* is elaborate involving complex post-fertilization changes. In their life cycles, motile flagellate cells are never formed. The male gametes are the spermata and they are formed singly in spermatangia or antheridia. The spermatia are non-motile and they are passively carried to the female sex organ, the carpogonium. The base of the carpogonium has a uninucleate egg and the neck-like trichogyne has cytoplasm.

The spermatangia of *Batrachospermum* are produced from the terminal or subterminal cells of branches of limited growth. In *Polysiphonia*, a large number of spermatangia are formed in a cluster from one of the branches of the fertile trichoblast. The carpogonium is produced terminally on a special branch called carpogonial filament or procarp. In *Batrachospermum*, the carpogonial filament formed from one of the short laterals. But in *Polysiphonia*, the procarp is formed by the supporting cell of a short trichoblast.

During fertilization, the spermatia which are released from the spermatangia come into contact with the trichogyne. The male nucleus passes down the trichogyne and fuses with the egg nucleus the base of the carpogonium.

The process of reproduction is relatively same upto this point be the variations are complex in the post-fertilization changes, leading to the formation of a new generation, the *carposporophyte* which supposed to be parasitic on the gametophyte.

Comparison Between *Batrachospermum* and *Polysiphonia*

After fertilization, in *Batrachospermum*, the zygote nucleus divide meiotically in the carpogonium itself, but in *Polysiphonia* it migrate into the auxiliary cell through a tubular connection and divides in mitosis. In *Batrachospermum*, the gonimoblast initials which are on off from the base of the carpogonium receive haploid nuclei form after meiosis of the zygote nucleus. But in *Polysiphonia*, gonim blast initials are formed from the auxiliary cell and they receive nuclei formed by mitotic divisions of the zygote nucleus. Therefore the gonimoblast filaments, the carposporangia formed are haploid in *Batrachospermum*, whereas they are diploid in *Polysiphonia*. In *Batrachospermum*, some of the vegetative cells surrounding the carpogonium form sterile filaments ensheathing the gonimoblast filaments and carposporangia, so that a haploid cystocarp carposporophyte is formed. In *Polysiphonia*, the sterile sheath pericarp is formed from the pericentral cells adjacent to the supporting cell. The auxiliary cell, supporting cell and the cells of basal an lateral sterile filaments fuse to form a large placental cell. The fruiting body or the *cystocarp* is diploid as the zygote nucleus has not undergone meiosis during its formation. The carpospores are haploid or diploid depending upon the nature of the carposporophyte. *Batrachospermum*, they are haploid and they germinate producing juvenile stages called chantransia stages. *Batrachospermum* plant develops as a special branch of the chantransia stage. The chantransia stages can multiply through monospores produced in the monosporangia borne at the tips of the branches.

In *Polysiphonia*, the diploid carpospores grow into tetrasporophytes which are morphologically similar to the male and female gametophytes. The tetrasporophytes produce tetrasporangia in which meiosis takes place forming haploid tetraspores, which give rise to the male and female plants.

In *Batrachospermum*, division of the zygote nucleus is mitotic and so post-fertilization changes lead to the formation of haploid carposporophyte. The carpospores and chantransia stages are haploid. The gametophyte and cystocarp (carposporophyte) are haploid generations and the zygote nucleus alone is diploid. This life cycle is known as *diphasic* or *haplobiontic*.

In *Polysiphonia*, the life cycle is *triphasic* involving male and female gametophytes, carposporophyte and tetrasporophyte. The carposporophyte and tetrasporophyte are diploid and the life cycle is *diplobiontic*. Meiosis occurs during the formation of tetraspores from tetrasporangia.

PORPHYRA

Distribution   It is common marine red alga found in the intertidal zone on the rocky seashores. It occurs on both the coasts, Atlantic and Pacific of North America. Some species grow attached to rocks (lithophytes) and others are epiphytic.

Thallus   The plant body is a plate or sheet-like parenchymatous blade apparently resembling that of a green marine alga *Ulva*. The margin of the blade ranges from, smooth wavy to greatly convoluted. The thallus is often unbranched and attached to the substratum by a small, basal disc or cushion-like holdfast. Depending on the species the blade-like thallus is one to two cells thick. Frequently it is one cell layer thick.

Cell Structure (Fig. 13.11A)   The cells are elongated. They are cubical or ellipsoidal in form and lie embedded perpendicular to the thallus surface in the tough gelatinous matrix derived from the cell walls. The outer walls of the cells are strongly thickened and covered with cuticle. The cells are uninucleate. Each cell has a single, large, axile stellate chromatophore with a centrally located pyrenoid. In certain species there are two chromatophores in each cell. The nucleus lies adjacent to the chromatophores but between the adjacent cells. The growth is intercalary. It takes place by the division of cells at intervals irrespective of their position.

Asexual Reproduction (Fig. 13.11A)   It takes place by means of asexual spores formed by direct transformation of vegetative cells. These are known as the neutral spores. The process of conversion mostly commences at the thallus apex. The spores are formed by the anticlinal division of vegetative cells. Since the division is in a plane perpendicular to the surface, the thallus remains single layered and neutral spores lie in a monostromatic layer. After liberation the neutral spores exhibit amoeboid changes.

Sexual Reproduction   It is oogamous. Most of the species are dioecious but some are monoecious. The male sex organ is called spermatangium and the female carpogonium. Both are formed from single vegetative cells.

(a) Spermatangia.   They are formed by repeated division of a vegetative cell (A-C). The process starts at the apex and spread towards the base of the thallus. A vegetative cell of the male thallus divides in three planes. The first wall is parallel to the surface (A). The daughter cells undergo two vertical divisions. (B-C). Further transverse divisions result in a mass of 32-128 (D) small colourless cells called spermatangia. They are arranged in several superimposed tiers of 4,8 or 16. When the plants are reflooded by the incoming tide the walls of spermatangia become gelatinised. The male elements are extruded at the thallus margin (D). The liberated naked uninucleate protoplast of each spermatangium is called a spermatangium. It is a nonmotile male gamete. On liberation the nonmotile spermatia are carried passively by the water currents in all directions. Some of them may be drifted to and lodged against carpogonia. Several spermatia may be found adhering to the female thallus (Fig. 13.11A).

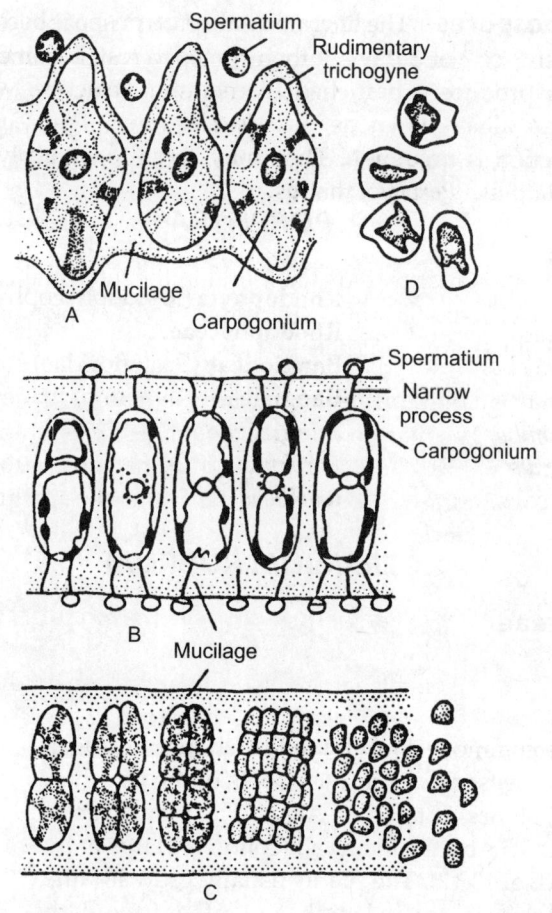

Fig. 13.11A *Porphyra*. (A), V.S. thallus of *P.* tenera with carpogonia; (B), V.S. Female thallus of *P. umbilicus* with spermatia adhering to it; (C), stages in the development of carpospores; (D), carpospores in the amoeboid condition.

(b) Carpogonia. The carpogonium is formed by slight modification of an ordinary vegetative cell of the female thallus (A). It increases in size. The swollen cell undergoes no division. Its protoplast functions directly as an egg. In certain species (*P. perforata*) the ellipsoid carpogonium extends to the thallus surface at one or both the ends by giving out a small papillate outgrowth which is considered a rudimentary trichogyne (A). In species which lack the protuberance or trichogyne (*P. umbilicus*), the spermatium itself puts forth a narrow process containing a thin stream of cytoplasm (B). It establishes a connection between the spermatium and the carpogonium. The spermatial nucleus migrates through it into the carpogonium to fuse with the egg nucleus.

(c) Formation of carpospores. Immediately after fertilisation the zygote nucleus undergoes meiosis. The four haploid nuclei divide mitotically to form 8, 16 or 32 nuclei. Cleavage of the protoplast at each nuclear division leads to the formation of a group of 8-32 uninucleate haploid meiospores known as the carpospores. The mature naked, nonflagellate carpospores are released by the breakdown of the surrounding cell walls when the female thallus is submerged by the incoming tide. In monoecious species the male areas are segregated from the female areas. The former are yellowish white or white and the latter purple. The carpospore has a stellate chromatophore.

Germination of Carpospores The liberated naked carpospore becomes amoeboid. It exhibits slow amoeboid movements of 2 or 3 days. It then comes to rest, becomes spherical, secretes a cell wall and germinates to produce a branched filamentous structure which is some species (*P. umbilicalis*) resembles an alga known as *Conchocelis rosea*. The *Conchocelis* stage produces monospores whose function is unknown. They may serve to multiply the conchocelis stage or germinate to form the plate-like *Porphyra* thallus.

Taxonomic Position:

Division	:	**Rhodophyta** (Rhodophycophyta)
Class	:	**Rhodophyceae**
Sub-class	:	**Bangiodeae** (Protofloridae)
Order	:	**Bangiales**
Family	:	**Bangiaceae**
Genus	:	*Porphyra*
Species	:	*umbilicus*

CERAMIUM RUBRUM

Class : Rhodophyceae

Order : Ceramialis
Family : Ceramiaceae

This plant is among the commonest of all the British marine Algae, occurring all round the coasts, chiefly in rock pools of the midshore zone, attached to rocks, stones or to other Algae. The plants can be recognized under a hand lens by the incurved (connivent) tips of the apices of the filaments (Fig. 13.12). The plants usually grow in tufts, varying from 2 to as much as 30 cm. in length, and rather irregularly dichotomously branched. The fronds are clear red in colour when living in deep sea, or reddish brown or yellowish green when living in shallow tidal water.

The plants are apparently dioecious, the antheridia being formed in sessile patches on the upper branches of the male plants, while the cystocarps are formed at the nodes of the females. The asexual plants resemble the sexual ones, the tetrasporangia-being developed in cells of the cortex.

Fig. 13.12 *Ceramium*–Habit

Structure of Thallus

The apex of each filament is formed of a single apical cell which cuts off segments basally. Branching begins of the formation of two obliquely intersecting walls in the apical cell, each segment thus cut off becoming the apex of a new branch. This oblique division of the apical cell usually occurs after about a dozen axis segments have been formed, and the further elongation of the axis is due to the expansion of these cells. Before elongation of the branches have a characteristic incurved form, and each elongating branch is crowned with a pair of thorns, which are the beginnings of the next youngest branches.

The segments cut off basally from the apical cell become the central cells of the axis. Each of these very soon cuts off a single layer of small pericentral cells, usually seven in number, beginning on the outer (convex) side of the curved filament and proceeding regularly around to the opposite side. These cells are cut off at the upper end of each central cell so that they form nodal rings. Each pericentral cell now produces four cortical cells which develop and divide, forming a close external layer of irregularly shaped cells around the node. As the cortex is forming, the central cells elongate and increase in thickness. In *C. rubrum* the cortical cells divide rapidly enough to keep pace with this growth, so that the filament retains a continuous cortex when mature. In some other species, however, the central cells elongate to such an extent that they separate the nodal hands of cortex from each other, and the equatorial belt of each central cell is left uncovered and exposed.

The outer surface of the cortex is covered by gelatinous material and at certain seasons there are produced, from the cortical cells large numbers of long unicellular hairs. In some species, though no in *C. rubrum*, spines may also be developed at the nodes in connection with the primary cortical cells. These spines consist of taperine filaments of about four or five cells. Secondary spines may be developed above and below the primary spines. These consist of only about two cells, and project upwards and downwards at an angle with the cortex.

The organ of attachment in *C. rubrum* develops from the base cell of the germinating tetraspore or zygote respectively, which elongates and then divides to form a number of rhizoids, each which becomes multicellular, and forms a flat plate of tissue by means of which the plant is firmly fixed to the substratum of rock. Frequently a number of separate shoots arise from a common basal attachment.

Sexual Reproduction

The sex organs are developed on different plants. The female sex organs are developed terminally on short lateral branches. The groups of carpospores are partly protected by a number of short branches which are formed laterally and grow up to surround, at least partially, the carpospores. No cystocarp envelope is formed. The antheridia are not produced on special organs but are scattered in patches over the surface of the thallus.

The Antheridium

The antheridia arise by the division of cortical cells. In the younger regions every cortical cell may function, but in the older parts of the thallus antheridia formation is limited to the cortical cells of the nodes. Each cortical cell divides first by the formation of periclinal walls and then by anticlinal ones, thus cutting up the cell into four or five cells, each of which functions as an antheridium mother cell. This cell next elongates and forces its terminal part through the gelatinous covering, forming a protruberance which is gradually abstricted as its base. This is the antheridium. It is enclosed in a rather thick gelatinous wall which increases greatly in thickness at the apex. The antheridium itself contains a single nucleus enclosed in rather dense cytoplasm. A split occurs in the apex of the antheridial wall and a single spermatium slips out.

Meanwhile a projection of the antheridial mother cell grows out and forms a second antheridium, and before the discharge of the second spermatium a third projection may arise. This process appears capable of successive repetitions, until after the production of about six spermatia the mother cell becomes exhausted.

The spermatium contains a single nucleus; it is oval, and the blunt apex is occupied by the nucleus while the rest is largely filled with vacuolate cystoplasm. It is enclosed in a delicate wall.

The Carpogonium

The procarp originates very close to the apex itself. One of the segments cut off by the apical cell divides to form a group of pericentral cells. The first of these to be formed is the procarp branch. It elongates laterally and becomes the basal cell. This cell cuts off a four-celled procarp branch, the terminal cell of which is the carpogonium. In *Ceramium rubrum*, unlike other species of *Ceramium*, only one procarp is cut off, either to right or left of the basal cell; in others two procarp branches are formed. The carpogonium elongates and forms a tubular trichogyne at its distal end. After fertilization the basal cell cuts off on the opposite side to the procarp branch a large auxiliary cell. The zygote nucleus from the carpogonium passes into this and the cell then begins to cut off a series of short gonimoblast filaments, from the tips of which carpospores are formed. These filaments are often simply short lobes of the auxiliary cell. The carpospores germinate to produce the *tetrasporic* plants.

Asexual Reproduction

The tetrasporic plants do not differ morphologically from the sexual ones. The tetrasporangia arise in the cortical cells in the region of the nodes. The details of the division have not been fully investigated, but it appears probable that they are formed by the division of certain of the pericentral cells in the nodal region, which each cut off externally a single cell which enlarges and pushes between the cortical cells. Not all the pericentral cells at the same level behave similaily, but it is quite common to find transverse bands of from four to about six tetrasporangia formed together at one node. Successive nodes generally behave similarly so that tetrasporangia occur in tiers on the branches. The nucleus of each tetraspore mother cells divides twice, during which meiosis occurs, resulting in the formation of four monoploid nuclei which become arranged around the periphery of the cell. Wall formation is effected by the ingrowth of the walls varies somewhat, both tetrahedral and cruciate arrangements occurring.

Paraspores are formed in some species, though not in *Ceramium rubrum*. They arise from undivided superficial cells of the cortex and are only formed on the tetrasporic plants. Their germination has no been followed.

The tetraspores are liberated through the gelatinous covering of the thallus as oval bodies which, on reaching a suitable rocky sub-stratum, divide into two cells. The lower or basal cell elongate and forms a rhizoid, whose terminal extremity becomes thickened and flattened and develops into a tiny adhesive disc. The upper cell functions as an apical cell, cutting off a row of cells which forms the central axis of the new sexual plant.

Ceramium rubrum, therefore shows the same *diplobiontic* alternation of generations, as in other members of the Rhodophyceae.

IMPORTANT QUESTIONS

Essay Type

1. Briefly describe the main characteristics of the class Rhodophyceae.
2. Describe the range of vegetative structure and reproduction in Rhodophyceae.
3. Give a brief account of the characteristics and classification of Rhodophyceae.
4. Describe the general characters of the class Rhodophyceae. Give an outline of the classification of the class.
5. Describe the thallus structure and reproduction in *Porphyria*.
6. With the help of labelled sketches, describe the thallus structure of *Porphyridium*.
7. Describe the different types of life cycles met within Rhodophyceae. Discuss their interrelationships and evolutionary tendencies.
8. Discuss in detail the reproduction in Rhodophyceae.

9. Discuss in brief the main characters of Myxophyceae, Pheophyceae and Rhodophyceae.
10. Describe the main distinguishing features of Chlorophyceae, Phaeophyceae, Rhodophyceae and Cyanophyceae.
11. Describe the cell structure in Rhodophyceae.
12. Describe the reproduction in Rhodophyceae.

Short Answer Type

13. Write short notes on the following:
    (a) Spores in Red Algae.
    (b) Carpospores
    (c) Pigments in Rhodophyceae.
    (d) Gonimoblast filaments.
    (e) Carpogonium.
    (f) Carposporophytes.
    (g) Cystocarp
    (h) Distinguishing features of Phaeophyceae and Rhodophyceae.
    (i) Carposporophyte of Red Alga.
15. Give the salient features of class Rhodophyceae. Why the members of the class are known as Red algae?
16. Differentiate between
    (a) Rhodophyceae and Cyanophyceae
    (b) Rhodophyceae and Chlorophyceae
    (c) Rhodophyceae and Phaeophyceae.
17. Write economic importance of Red algae.
18. Write explanatory notes on
    (a) *Porphyria*
    (b) *Porphyridium*
    (c) Rhodophyceae
    (d) Alternation of generations in Rhodophyceae.

Objective Type

19. Select the correct answer:
    (i) Reserve food material in Rhodophyceae is
        (a) Starch
        (b) Floridean starch
        (c) Cyanophycean
        (d) Laminarin.
    (ii) Members of class Rhodophyceae are commonly called
        (a) Green algae
        (b) Blue algae
        (c) Red algae
        (d) Blue-green algae.
    (iii) Red colour of the red algae is due to the pigment named
        (a) Chlorophyll
        (b) Pycoerythrin
        (c) Phycocyanin
        (d) All of the above.
    (iv) In Rhodophyceae non-motile male gametes are called
        (a) Spermatozoids
        (b) Anthrozoids
        (c) Zoospores
        (d) Spermatia.

(v) Algae belonging to Rhodophyceae are
    (a) Only marine
    (b) Only fresh water
    (c) Both marine as well as fresh water
    (d) Fresh water and terrestrial.

(vi) Cystocarp consists of
    (a) Oogonium
    (b) Carpogonium only
    (c) Carpogonium and gonimoblast filaments
    (d) Carpogonium, loosely arranged gonimoblasts and carposporpangia.

(vii) In higher red algae, Carpospores are usually
    (a) Haploid
    (b) diploid
    (c) neither haploid nor diploid
    (d) may be haploid or tetraploid.

(viii) In *Porphyridium* cell, the axial stellate chromatophore contains
    (a) no pyrenoid
    (b) a single pyrenoid
    (c) two pyrenoids
    (d) many pyrenoids.

(ix) Which of the following is a unicellular red alga?
    (a) *Chlamydomonas*
    (b) *Hydrodictyon*
    (c) *Porphyridium*
    (d) *Polysiphonia*.

QUESTIONS

Essay Type

1. Describe the thallus and reproduction in *Batrachospermum*. Mention its features of significance.
2. Describe the structure of thallus and development of cystocarp in *Batrachospermum*.
3. Give an account of the life history of *Batrachospermum*.
4. Give an illustrated account of post fertilisation changes in the life cycle of a fresh water red alga studied by you.
5. Describe the mode of sexual reproduction in *Batrachospermum*.
6. Give an account of the thallus structure and life cycle of *Batrachospermum*.
7. Compare life cycle patterns of *Batrachospermum* and *Polysiphonia*.
8. Compare the process of sexual reproduction in *Ectocarpus* with that of *Batrachospermum*.
9. Bring out clearly with a series of sketches ad brief explanation the differences between the modes of reproduction of *Dictyota* and *Batrachospermum*.
10. What is chantransia stage? How does it help n multiplication and production of plant body in *Batrachospermum*? Illustrate the development of cystocarp in *Batrachospermum*.
11. Describe the structure of sex organs and process of cystocarp formation in *Batrachospermum*.
12. How would you illustrate that *Batrachospermum* exhibits an example of advanced oogamy, while maintaining simple plant body?
13. Give detailed account of development of cystocarp in *Batrachospermum*.
14. Compare the life history of *Sargassum* with that of *Batrachospermum*.
15. Describe the structure and reproduction in *Batrachospermum*.

Short Answer Type

16. Write short notes on:
    (i) Chantransia stage

(ii) Carposporophyte or cystocarp in *Batrachospermum*.
(iii) Structure of Chromatophore in *Batrachospermum*.
(iv) Post fertilisation in *Batrachospermum*.
(v) Gonimoblast filaments.
(vi) Carpogonium.
(vii) Carpospores.
(viii) Sexual reproduction in *Batrachospermum*.
(ix) Cystocarp formation in *Batrachospermum*.
(x) Branching in *Batrachospermum*.
(xi) Carpospore in *Batrachospermum*.
(xii) Somatic phase of *Batrachospermum*.

17. (a) Where does meiosis occur in the life cycle of *Batrachospermum*?
    (b) Draw a labelled diagram showing structure of cystocarp.
18. Name the reserve food materials found in the cells of red algae.
19. Draw neat and labelled diagrams of
    (a) Cystocarp of *Batrachospermum*.
    (b) Thallus of *Batrachospermum*.
    (c) Sex organs and chantransia stage of *Batrachospermum*.
20. Give name and systematic position of a freshwater red alga.
21. Give the most important characteristic of *Batrachospermum*.
22. Name an alga which produces tetraspores.
23. Throw light on the significance and formation of chantransia stage.
24. Throw light on the development of cystocarp in *Batrachospermum*.
25. Describe chantransia stage in *Batrachospermum*.

Objective Type

26. Fill in the blanks:
    (i) Dwarf branches of *Batrachospermum* form a thick clusters of branches at node called the ....
    (ii) In *Batrachospermum*, the gonimoblast filaments, carpospores and carpogonium constitute the ....
    (iii) The terminal cell of gonimoblast filament in *Batrachospermum* functions as ....
    (iv) The red colour in red algae is due to the pigment named ....
    (v) *Batrachospermum* belongs to the order ....
    (vi) Chantransia stage is characteristic of the red alga ....
    (vii) The fresh water red alga you have studied is ....
    (viii) The life cycle of *Batrachospermum* is ....
    (ix) Adult plant of the red alga *Batrachospermum* is ....
    (x) Carpospores after germination produce Protonema like stage that resembles the alga ....

27. Select the correct answer:
    (i) *Batrachospermum* is
       (a) Red alga of sea   (b) Red alga of fresh water
       (c) Green alga of fresh water   (d) Brown alga of sea.

(ii) The adult plant of red alga *Batrachospermum* is
   (a) haploid
   (b) diploid
   (c) Batrachospermum
   (d) Porphyria.
(iii) Charansia stage is the characteristic stage of the alga
   (a) *Polysiphonia*
   (b) *Sargassum*
   (c) *Batrachospermum*
   (d) *Porphyra*.
(iv) The branching in *Batrachospermum* is
   (a) Monopozdial
   (b) Sympodial
   (c) Both of the above
   (d) None of the above.

QUESTIONS

Essay Type

1. Describe the life history of *Polysiphonia*.
2. Illustrate the phenomenon of alternation of generations in the life history of *Polysiphonia* and compare it with that of the Archegoniatae
   OR
   Explain th alternation of generations with reference to *Polysiphonia*.
3. Give an account of life history of *Ploysiphonia* with necessary diagrams and point out the special features in it.
4. Describe the reproduction of *Polysiphonia*.
5. Describe the process of cystocarp formation in *Polysiphonia*.
6. Where *Polysiphonia* is found? Describe morphology of the thallus and explain formation of tetrasporophyte.
7. Give a detailed account of post fertilisation changes in *Polysiphonia* till the formation of cystocarp.
8. Compare life cycle patterns of *Batrachosperum* and *Polysiphonia*.
9. Define alternation of generation. Explain it with the help of life cycle of *Polysiphonia*.
10. Explain the changes that take place after male and female gametes fuse in *Polysiphonia*. Illustrate your answer with labelled diagrams.
11. Give a diagrammatic presentation of *Polysiphonia* life cycle.
12. Describe briefly the illustrated life history of *Polysiphonia*.
13. Give an illustrated account of structure and reproduction in *Oedogonium* or *Polysiphonia*.
14. Describe the reproduction in red alga studied by you.
15. With labelled diagrams, describe the structure of thallus and sexual reproduction in *Polysiphonia*.
16. Compare life history of *Polysiphonia* with that of a fern.
17. Give a diagrammatic representation of the life cycle of *Polysiphonia* illustrating the relative lenghts of haploid and diploid phases.
18. Illustrate the phenomenon of homologous alternation of generation from the life history of *Polysiphonia*.
19. (a) Give a comparative account of the mode of reproduction of Chara and *Polysiphonia* and discuss the evolutionary tendencies in them.
    (b) Describe sexual reproduction in *Polysiphonia*.

Short Answer Type

20. Identify the following by giving one important character and give their systematic position
    (a) *Polysiphonia*
    (b) *Porphyridium*
21. (a) Compare ceramiales with Nemalionales

    OR

    (b) Compare *Polysiphonia* with *Batrachospermum*
22. Draw well labelled sketches of the following:
    (a) Cystocarp of *Polysiphonia*
    (b) Thallus structure of *Polysiphonia*
    (c) Filament of *Polysiphonia* with male sex organs.
23. Write short notes on:
    (a) Cystocarp of *Polysiphonia*
    (b) Sexual reproduction in *Polysiphonia*
    (c) Reproduction in *Polysiphonia*
    (d) Female reproductive organs in *Polysiphonia*
    (e) Post fertilisation changes in *Polysiphonia*
    (f) Sex organs in *Polysiphonia*
    (g) Structure of thallus in *Polysiphonia*
    (h) *Polysiphonia*
    (i) Tetra sporophyte in *Polysiphonia*
    (j) Alternation of generation in *Polysiphonia*
    (k) Development of cystocarp in *Polysiphonia*
    (l) Cystocarp
24. Name the type of braching found in trichoblasts of *Polysiphonia*
25. Name the semi parasitic species of *Polysiphonia*
26. What type of thallus is seen in *Polysiphonia*
27. Explain the tetrasporophyte in *Polysiphonia*?
28. Differentiate between carpospore and tetraspore.
29. Where does reduction division takes place in *Polysiphonia*?
30. Where would you find *Polysiphonia* in India?

Objective Type

31. Fill in the blanks:
    (i) Heterotrichy means having .... and .... branches.
    (ii) Tetraspores in *Polysiphonia* germinate to produce ....
    (iii) The alga which yields agar agar is ....
    (iv) Gonimoblast filaments in *Polysiphonia* is ....
    (v) Semiparasitic species of *Polysiphonia* is *Polysiphonia* ....
    (vi) The alga which produce tetraspores is ....
    (vii) The alternation of generation in *Polysiphonia* is ....
    (viii) Reduction division in *Polysiphonia* takes place during the formation of ....

(ix) The adult plant of *Polysiphonia* is ....
(x) In *Polysiphonia* short branches of limited grouwht are also called ....

32. Select the correct answer:
   (i) Which of the following alga is marine?
      (a) *Vaucheria*
      (b) *Polysiphonia*
      (c) *Chalamydomonas*
      (d) *Spirogyra.*
   (ii) Heterotrichy means having
      (a) Long and short branches
      (b) Prostrate and erect branches
      (c) Branches modified into leaves and air bladders
      (d) Rhizoidal and photosynthetic branches
   (iii) Agar Agar is obtained from
      (a) *Ectocarpus*
      (b) *Fucus*
      (c) *Polysiphonia*
      (d) *Gelidium.*
   (iv) The tetraspore germinates to produce
      (a) tetrasporophyte
      (b) sporophyte
      (c) Carposporophyte
      (d) gametophyte.
   (v) The Red Sea gets its name because of the presence in its water red coloured algae belonging to class
      (a) Rhodophyceae
      (b) Phaeophyceae
      (c) Chlorophyceae
      (d) Cyanophyceae.
   (vi) The male gamete in *Polysiphonia* is called
      (a) Spermatium
      (b) Spermatozoid
      (c) Androspore
      (d) Zoospore
   (vii) Cystocarp of *Polysiphonia* is
      (a) Gametophyte
      (b) Sporophyte
      (c) Saprophyte
      (d) Lithophyte.
   (viii) In *Polysiphonia* the life cycle is
      (a) haplonitc
      (b) haplo-diplonitc
      (c) haplo-haplobiontic
      (d) haplo diplodiplontic
   (ix) The sterile sheath of the carposphorophyte in *Polysiphonia* is called
      (a) Epicarp
      (b) Procarp
      (c) Pericarp
      (d) Cystocarp
   (x) Reduction division in *Polysiphonia* takes place during the
      (a) formation of carpospores
      (b) formation of tetraspores
      (c) germination of carpospores
      (d) germination of tetraspores.

14 Class: Myxophyceae (Blue-Green Algae): General Characters and Type Study

GENERAL CHARACTERS OF MYXOPHYCEAE

Occurrence

Members of Myxophyceae have a unique capacity to live in diverse conditions. They are very common as freshwater forms, occurring in a wide variety of habitats. Some of them form planktons e.g., *Microcystis* and others occur on wet soils forming felt-like mass. Some species are marine growing in the intertidal zones of temperate seas.

Freshwater Myxophyceae may grow so abundantly in permanent or temporary waters, as to cause *'water blooms'*. Some of the blue-green algae occur as terrestrial species. A number of them found on damp soils of rice-fields, have been found to be greatly useful in nitrogen-fixation. Sometimes, they occur even below the surface layers of the soils. Some species may form extensive coating on the soil.

Many of these members are the first plants to inhabit rocky areas and they have been instrumental in paving the way for other types of vegetation (xerosere succession).

Some species of Cyanophyceae are the useful algal partners in the *lichens*. They occur as space parasites in the tissues of coralloid roots of *Cycas*, and other plants like *Anthoceros* and *Azolla*. They occur in extraordinary conditions as in frigid lakes (*Phormidium*), hotsprings (*Mastigocladus*), rushing torrents and on barren rocks. A few of them occur on salt marshes.

Thallus

Most of the members of Myxophyceae are filamentous, a few are unicellular or colonial. Very often, daughter cells formed after cell division do not separate, but remain together in a copious mucilage forming irregular colonies, so that unicellular forms are formed (*Gloeocapsa*).

In the filamentous forms, cells divide only in transverse plane. In these species, a trichome must by clearly distinguished from filament. The filament is formed of the basic structural unit called *trichome* consisting of a row of cells surrounded by a gelatinous sheath. Thus, in *Oscillatoria*, the filament shows distinct mucilage sheath. In some, several trichomes may be in a common gelatinous sheath *(Micrococcus)*. Sometimes, trichomes may be irregularly present embedded in spheres of mucilagenous masses, as in *Nostoc*.

The trichomes may not show differentiation into base and show diffuse growth. But in some *(Rivularia)*, the trichomes are be broad at the base, tapering towards the tip and even terminate in a colourless hair; in such forms, *trichothallic* growth is due meristematic cells present at the base of the hair. Trichomes may unbranched *(Oscillatoria)*, truly branched (Stigonematales), or show false branching *(Scytonema)*. Some members of Stigonematales exhibit heterotrichy with disk-like prostrate system and a filamentous portion.

Cell Structure (Fig. 14.1)

Studies with electron microscope have revealed the details of cell structure. The Cyanophycean cell lacks a definite nucleus, Golgi bodies and mitochondria. The endoplasmic reticulum is lacking but ribosomes are present. But these are 70s ribosomes unlike the 80s ribosomes of other cells. There are no chloroplasts. Such cell is described as *prokaryotic* unlike the eukaryotic cells of algae. The cells of bacteria are also prokaryotic like those of green algae. The prokaryotic cell of Bacteria and Myxophyceae also known as Moneran cell.

A Cyanophycean cell has a mucilage sheath of variable thickness. It absorbs and retains water. It may be sometimes stratified *(Gloeocapsa)*. In some, the gelatinous sheath may be tinted brown, yellow or red. The cell wall is inner to the sheath and it is double layered. The inner layer contains

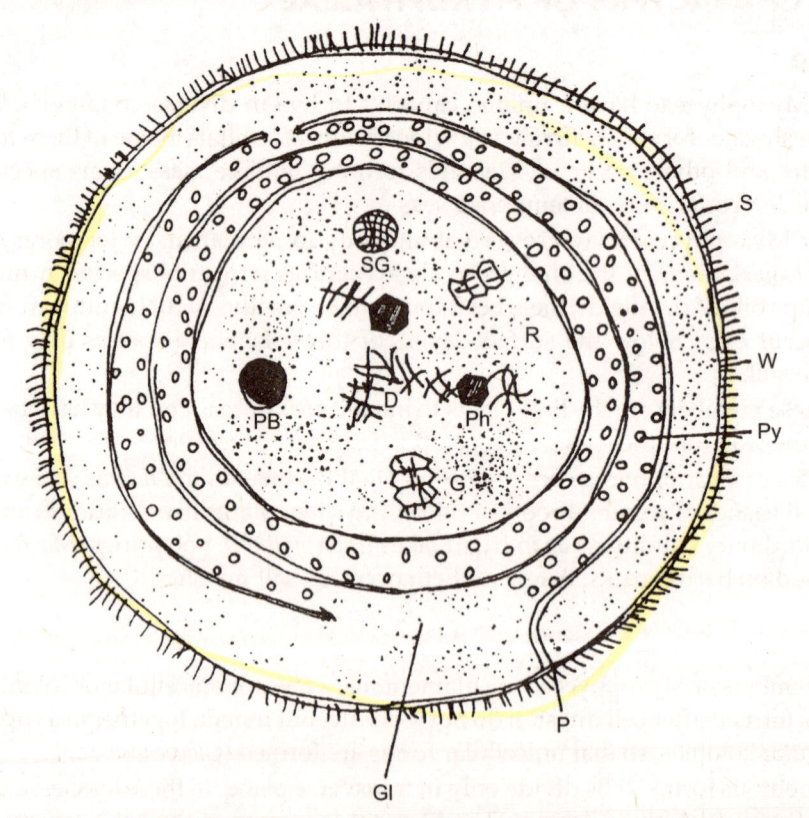

Fig. 14.1 Fine structure of blue-green algal cell (Myxophyceae). D.-DNA fibrils; G.-Gas vesicles; Gl.-Glycogen granules; P.-Plasmalemma; PB.-Polyphosphate body; Ph.-Polyhedral body; Py.-Phycobilisomes; R.-Ribosomes; S.-Sheath; SG.-Structured granules (Cyanophycin granule); W.-Wall.

mucopeptide which is found in bacterial cell walls. The cell membrane is inner to the cell and it is a lipoprotein membrane.

The cytoplasm is distinguished into a peripheral pigment *chromatoplasm* and an inner colourless *centroplasm*. The chromatoplasm has a complex lamellar system associated with the photosynthetic pigments. These lamellae or thylakoids are not delimited into chloroplasts. The centroplasm was, in the past, described incipient nucleus. But there is no nucleus. The central region contains DNA, but as it is not associated with histones or protamines (proteins), chromosomes are not formed as in most of the organisms.

The important cell inclusions are reserve foods and pseudovacuoles. The Cyanophycin granules are of the nature of proteins whereas Cyanophycean starch represents polyglucose molecules.

Pseudo-Vacuoles

These gas vacuoles are numerous in planktonic species making them buoyant. They appear as dark dots under low power. According to Fogg (1970) these vacuoles contain metabolic gases. They are produced in low light intensity so that osmotic pressure of the cells increases causing the collapse of the gas vacuoles.

Photosynthetic Pigments

The pigments are chlorophyll a, β-carotene, c-phycocyanin, c-phycoerythrin, myxoxanthophyll, and myoxanthin. Other pigments like oscilloxanthin, zeaxanthin and lutein may be found in smaller amounts. These pigments are not located in chloroplasts but they are found in the thylakoids of chromatoplasm. Generally, these algae are blue-green due to phycocyanin and chlorophyll pigments occurring in higher proportions. But in forms growing in low light intensity, more of phycoerythrin is present so that they are reddish. Some species of *Oscillatoria* are green in red light, bluish-green in yellow light and reddish in green light. Such a change in colour according to the light is known as chromatic adaptation or *Gaidukov phenomenon*. The change in colour is due to the variation in the proportion of the pigments. The chromatic adaptation enables the alga to utilise the available light to the maximum. Several heterocystous forms like *Nostoc*, *Anabaena* and *Aulosira* can fix atmospheric nitrogen due to the presence of enzyme-complex nitrogenase. These species greatly contribute to the fertility of rice fields.

Movements

Many of the Oscillatoriaceae show pendulum like movements of the trichomes. They glide forward or backward rotating around their axes. The actual mechanism of this movement is not definite. Several algologists consider that the movement is due to the secretion of mucilage through pores in the cell wall.

Cell Division

Cell division commences with the formation of a median constriction in the cytoplasm. This is followed by the centripetal growth of cell membrane and inner layer of cell wall, resulting in completing the cell division.

Reproduction

In *Myxophyceae*, reproduction takes place by vegetative means. A number of species show formation of spores. Sexual reproduction is completely absent.

Vegetative Reproduction—Cell Division

In *Chlorococcales,* the only method of reproduction is by cell division. After cell division, the daughter cells may remain together in a common gelatinous envelope. The colonies reproduce by accidental breaking of the colonial envelope.

Hormogonia. In filamentous forms, the trichomes are capable of indefinite growth, but usually it breaks up into two or more parts. The breaking of the filament may be accidental or by special means. A filament breaks into short pieces are separated by death of some cells. In certain, filamentous forms having heterocysts, separation may be affected due to weaker adhesion between a heterocyst and a vegetative cell abutting on it. In some, discs of gelatinous material which are biconcave are formed between the vegetative cells. These discs gradually become thin by absorbing water. This results in the separation of the filament into hormogonia. The hormogones move away from the parent plant by gliding movement. Some consider that the movement is related to the production of mucilage. The nature of this movement is not fully understood.

Hormospores. Hormogones formed at the tips of trichomes have thick-walled cells. These spore-like multicellular bodies are known as hormospores. They give rise to new filaments (*Lyngbya* spp.).

Spore formation. It takes place by production of non-flagellate spores.

1. *Akinetes* are common in Nostocaceae and Rivulariaceae. They may occur singly or in chains anywhere or next to the heterocyst. These are large, thick-walled cells rich in reserve foods. They grow into new filaments under favourable conditions.
2. *Endospores* are formed in species which do not produce hormogonia. The contents of an enlarged vegetative cell divided into a number of naked endospores. They germinate immediately after they are liberated.
3. *Exospores* are rare and produced only in two genera e.g., *Chamaesiphon*. The cell wall ruptures and the protoplast is extruded to be cut off as an exospore.
4. *Nannocytes* are formed in *Gloeocapsa*. They are just like endospores and are formed in vegetative cells of normal size.
5. *Heterocysts*. Most of the filamentous Cyanophyceae, except Oscillatoriaceae have heterocysts. They may be basal (*Rivularia*), intercalary or terminal (*Nostoc*). Geitler observed that in some species of *Anabaena* heterocysts may germinate forming new filaments.

Affinities of Blue-green Algae with Bacteria

Like most groups of algae, the Cyanophyta contain several colourless forms (e.g., *Beggiatoa*). These bear a strong superficial resemblance to some bacteria. Because of this Cohn (1872) suggested that both bacteria (Schizomycetes) and blue-green algae (Schizophyceae) be regarded as two classes of the phylum, Schizophyta. Stanier and van Niel (1941) concluded that the two groups are closely related by, their absence of nuclei, absence of sexual reproduction and absence of plastids. Pringsheim (1949) questioned these three relationships and are found to be negative. But recent electron microscopic studies reveal more common features between them like: procaryotic cell structure, biochemical features like ornithine biosynthesis and their sensitivity to certain antibiotics. Echlin and Morris (1965) warrant their separation from all other cellular organisms from the standpoint of several distinctive common features between bacteria and blue-green algae.

HETEROCYSTS IN CYANOPHYCEAE

Most of the filamentous members of Cyanoptlyceae, except Oscillatoriaceae form heterocysts. The heterocysts are generally intercalary in *Nostoc* and basal in *Rivularia*. They are intercalary as well as lateral in *Stigonema*. The heterocysts are larger than the vegetative cells and have a thickened cell wall. An intercalary heterocyst has a pore at both ends (bipolar). But when the heterocyst is basal or terminal, the pore is present only on the end which is in contact with the vegetative cell. Protoplasmic connections extend into the heterocyst through the pores from the vegetative cells. The cell wall is thickened around the pores forming polar nodules.

Recent studies with electron microscope reveal the sequence of changes during the differentiation of the heterocyst from a vegetative cell. It includes cell enlargement, gradual loss of photosynthetic pigment, reorientation of thylakoids, decrease in granular inclusions and thickening of cell wall. The photosynthetic lamellae lack the normal lipid components but contain *glycolipid* and *acylipid* not found in vegetative cells. Mature heterocysts lack phycocyanin. The Cyanophycin granules are almost absent and the ribosomes decrease in number.

Various views have been put forward to explain the significance of heterocyst. They are:

1. Heterocysts are related to the production of akinetes. This is based on the observation that in many species akinetes are formed only close to the heterocyst.
2. Heterocysts, according to some algologists, represent the points of breakage of a trichome into hormogones. This is because, in many forms, fragmentation occurs at the heterocyst.
3. Heterocysts are considered to be archaic reproductive structures which have lost their capacity to reproduce. This is based on observations that they can germinate into new filaments in some species of *Anabaena* (Geitler).
4. It is the site of nitrogen fixation. The nitrogen fixing ability is due to the presence of enzyme-complex nitrogenase. Most of the blue-green algae which fix atmospheric nitrogen are heterocystous forms. But the enzyme nitrogenase is also found in vegetative cells of *Anabaena cylindrica*. But *Gloeocapsa* sp., which is a non-heterocystous form, also fixes atmospheric nitrogen and therefore, some consider that nitrogen fixation is not the monopoly of heterocysts. But reports of nitrogen fixation by non-heterocystous forms are scanty whereas about 45 species of heterocystous forms possess this capacity. Evidences given by Fay (1968) suggest a dominant role for the heterocyst in nitrogen fixation.
5. Heterocysts may store food material.
6. They may contain enzymes.
7. It is the site for the production of a substance which at one stage controls the cell division and growth of the vegetative filament, and at another stage controls sporulation. In *Anabaena cylindrica*, the vegetative cells adjacent to a heterocyst sporulate earlier than other cells. But when the heterocyst is removed, sporulation is prevented. So, the heterocyst is supposed to have a role in sporulation.

Following type study in *Myxophyceae* are discussed here in detail.

TYPE STUDY

1. Gloeocapsa, 2. Rivularia, 3. Oscillatoria, 4. Nostoc, 5. Gloeotrichia, 6. Scytonema.

GLOEOCAPSA

Class : Myxophyceae

Order : Chroococcales

Gloeocapsa belongs to the order *Chroococcales*. It is subaerial growing on wet rocks, walls and moist soils. It may be associated with fungi forming lichens. *Gloeocapsa* occurs as a unicellular or palmellod colonial (Figs. 14.2 & 14.3). The irregular palmelloid colonies are formed as a result of the daughter cells formed from the parent cell remaining within a common mucilage sheath. Each colony consists of few more or less sperical cells which have colourless or blue or brown stratified mucilage sheaths. In some cases, the mucilage may be homogeneous.

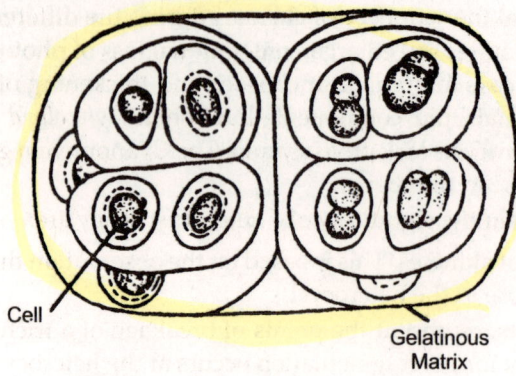

Fig. 14.2 *Gloeocapsa*—A colony.

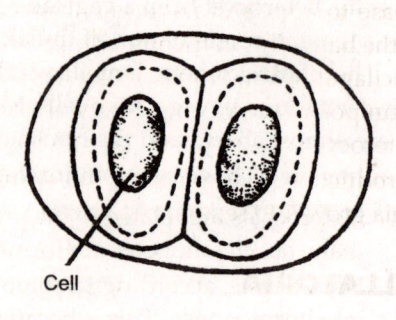

Fig. 14.3 *Gloeocapsa*—Cells of a colony.

RIVULARIA

Order : Nostocales
Family : Rivulariaceae

Rivularia belongs to the order Nostocales. Colonies of *Rivularia* occure as bluish-green gelatinous masses on aquatic plants growing in ponds and pools. *R. bullata* is a marine species. If a small portion of the gelatinous mass is examined, a number of trichomes of *Rivularia* which are radially arrainge can be seen (Fig. 14.4). These trichomes are unbranched and they have tubular mucilage

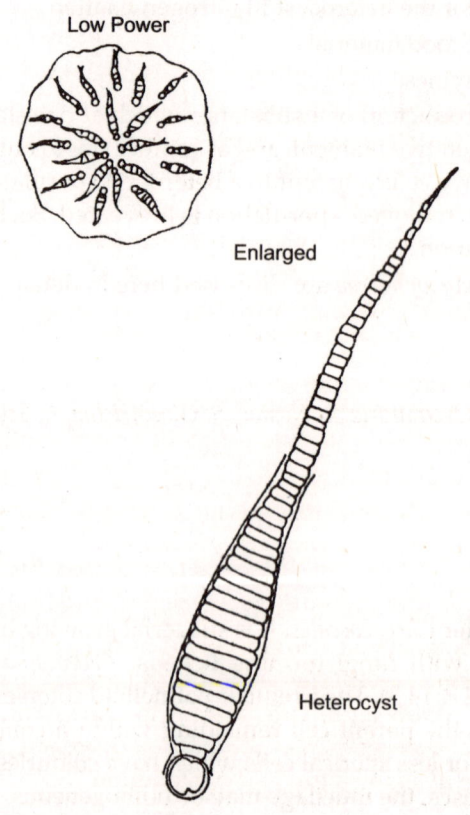

Fig. 14.4 *Rivularia*—Habit and enlarged.

sheaths which may or may not enclose the basal heterocysts. The trichome is gradually attenuated from base to apex and its terminal part is like a colourless hair. The meristem which is present below the hair helps in trichothallic growth. The terminal hair may be naked or may be enclosed by the mucilage sheath. The *heterocyst* has a pore where it is in contact with the vegetative cell. So, it is called unipolar heterocyst. Species of *Rivularia* may show false branching when there are intercalary heterocysts.

Reproduction takes place by the formation of hormogonia below the colourless hair. The hormogonia grow into separate filaments.

OSCILLATORIA

Order : Oscillatoriales
Family: Oscillatoriaceae

Occurrence

Oscillatoria spp., are mostly present in abundance in stagnant freshwaters. They also occur in subareal environments. *O. bonnemaisoniae* occurs in littoral zones. Yellow green *O. chlorina* is a saprophytic form (occurring in putrifying bottom deposits). Prat recorded a species of *Oscillatoria* from hot springs. The frequent occurrence of oscillatorias in polluted waters is a proof of its capacity to live in habitats lacking oxygen.

Structure

It exhibits a very simple type of construction having unbranched *trichomes* (Figs. 14.5 & 14.6). The trichome consists of a row of identical cells, and it is surrounded by a sheath of mucilage.

In some species of *Oscillatoria,* the mucilage envelope is not distinct and in some others, it is well-defined. All the cells are capable of division and growth. In some species, the terminal cells are conical or rounded, often showing attenuation into *capitate* structures. The outer surface of the terminal cell of the trichome may be covered by a thickened cap (calyptra) which is protective in function. The terminal part of the trichome is bent, in some species, towards one side.

Fig. 14.5 *Oscillatoria*—Habit in low power.

The cells show typical Cyanophycean structure. They are prokaryotic and they lack nucleus, mitochondria, Golgi bodies, chloroplasts and endoplasmic reticulum. But ribosomes are present. The central colourless region contains DNA. The peripheral pigmented chromatoplasm shows lamellae with which the photosynthetic pigments are associated. The reserve foods are Cyanophycin granules and Cyanophycean starch. They are found in the peripheral portion of cytoplasm.

The pigments are chlorophyll *a*, β-carotene, c-phycocyanin, c- phycoerythrin and some xanthophylls like oscilloxanthin, myxo-xanthophyll. Different species of *Oscillatoria* vary in colour due to the occurrence of these pigments in different proportions. The pigmentation seems to depend on the quality of light in which they grow. Some species of *Oscillatoria* are given in red light, reddish in green light and bluish-green in yellow light. This capacity to change colour according to the colour of the light is known as chromatic adaptation or *Gaidukov phenomenon*. This is advantageous since it enables the alga to absorb the available light to the maximum.

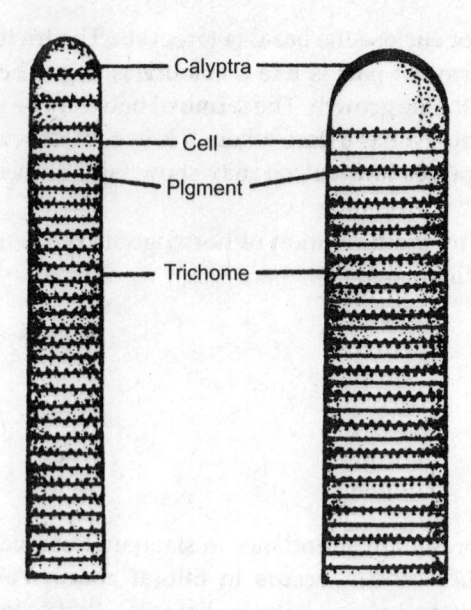

Fig. 14.6 *Oscillatoria*—Enlarged filaments.

Movements

The movement observed in *Oscillatoria* is probably helpful in dispersal of *hormogonia*. The terminal parts of *Oscillatoria* filaments exhibit somewhat slow and jerky oscillations. This is said to occur, only when the terminal end is not in contact with the substratum. Increase of temperature causes acceleration in the rate of movement. In some species of *Oscillatoria*, it is accompanied by rotation. The movement is usually along a curved path and the direction of curvature is dependent on the rotation.

The mucilage envelope of *Oscillatoria*, probably formed by modification of cell-sheath according to Fritsch, has no bearing on the mechanism of movement. But other investigators like Schmid, Ullrich have regarded secretion of mucilage as directly connected with movement. Krenner has shown that only trichomes of minimum length exhibit movement. The movement of the filaments, in diverse species of *Oscillatoria* has been shown to be influenced by light.

Reproduction

The only method of prolific reproduction in *Oscillatoria* is by vegetative means. The trichome breaks up into short pieces called hormogonia. *Oscillatoria*, formation of hormogonia is facilitated by the formation of biconcave separate discs.

A separation disc is formed when a cell becomes moribund and becomes biconcave due to the pressure exerted by adjacent cells. They degenerate, when the trichome breaks up into hormogonia. The hormogonia grow into new trichomes.

NOSTOC

Order : Nostocales
Family : Nostocaceae

Occurrence

Nostoc is a filamentous form and the filaments are closely crowded together forming jelly-like masses which are spherical or ellipsoidal. In some species, they reach the size of a hen's egg (*N. pruniforme*). *Nostoc* is found as a free-floating form in ponds and pools. However, the gelatinous lumps of *Nostoc* filament may also occur on tree trunks, on wet soils and on rocks. Some species are subterranean.

Some species of *Nostoc* are endophytes. *N. punctiforme* is found in *Cycas*, *Zamia* and *Gunnera*. *Nostoc* species occur in the mucilage cavities of *Anthoceros*. This alga is an important constituents of the Lichens. *Nostoc* species, which occur in paddy fields, are useful in nitrogen fixation.

Structure (Fig. 14.7)

The soft jelly-like gelatinous lumps contain numerous filament of *Nostoc* which lie irregularly. The gelatinous masses of *Nostoc* are like spherical or ellipsoidal structure or may be flat and lobed as *N. commune*.

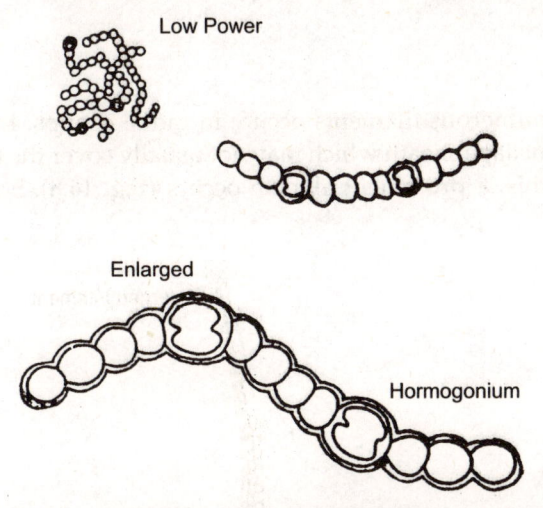

Fig. 14.7 *Nostoc*—Habit and enlarged filaments with hormogonia.

The unbranched filaments of *Nostoc* are much contorted. The moniliform cells give it a beaded appearance. The heterocysts are intercalary in position, but sometimes they are terminal. Each hetrocyst is slightly larger than a vegetative cell and it has a thick wall composed of two layers. The wall of the intercalary heterocyst shows two pores, one at each end. The nature of the heterocyst is much in doubt for a long time, but Geitler observed in *N. commune* heterocysts germinating into new filaments. The cells have typical Cyanophycean structure and they are prokaryotic.

Reproduction

It takes place purely by vegetative means. Asexual reproduction and sexual production are entirely absent. Filaments of *Nostoc* simply break up into small pieces the *hormogonia*, which grow individually. The heterocyst helps in fragmentation of the filament into harmogonia since the adhesion between the heterocyst and vegetative cell is weak.

Akinetes

In *Nostoc*, akinetes are formed in contact with the heterocysts. These may be isolated or may be in series. The ordinary vegetative cell enlarges and accumulates reserve food and subsequently, the wall gets thickened appreciably, accompanied by differentiation into an outer exospore and an inner endospore. Such a structure is known as *akinete*. These are useful in crossing over unfavourable periods *(N. muscorum)*. When conditions are favourable, the akinete germinates and gives rise to a new filament.

GLOEOTRICHIA

Class : Myxophyceae

Order: Nostocales
Family: Gloeotrichiaceae

Occurrence

It occurs as globose mucilaginous colonies of appreciable dimensions. These colonies occur in freshwater ponds and pools attached to various plants and other substrata. They may occur on moist soils.

Structure

In each globose colony, numerous filaments occure in radial groups. Each filament consists of a trichome enclosed in a mucilage sheath which may not usually cover the heterocyst. The heterocyst is basal and just above this, a prominent akinete occurs (Fig. 14.8). Sometimes, heterocysts are intercalary.

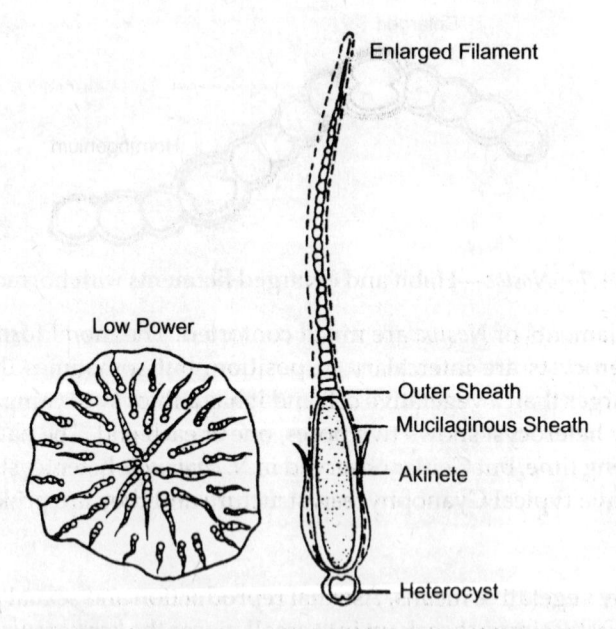

Fig. 14.8 *Gloeotrichia*—Filaments in low power and enlarged showing heterocyst and akinete.

Thus, *Gloeotrichia* chiefly differs from *Rivularia* in producing akinetes. The trichome is broad towards the heterocyst but is whiplike tapering into a colourless multicellular hair. The trichomes grow by means of a meristem of short cells situated at the base of the hair. Such an intercalary growth is described as *trichothallic*. The cells of the trichome show the typical structure characteristic of Myxophyceae.

Some filaments of *Gloeotrichia* with intercalary heterocysts exhibit *false branching*.

Reproduction

The sexual reproduction, so prevalent in other algae, is characteristically absent, as in other Myxophyceae.

Reproduction takes place by the trichomes accidentally breaking into few-celled bits called *hormogonia* which grow into new trichomes.

The *akinete* is present next to the heterocyst. They are enlarged, thick-walled, vegetative cells with accumulated reserve food. They are useful in tiding over the unfavourable period. They germinate into new filaments when conditions become favourable.

SCYTONEMA

Class : Myxophyceae

Order: Nostocales
Family: Scytonemataceae

Occurrence

Scytonema is a filamentous form showing a characteristic false branching (Fig. 14.9). It occurs in damp places.

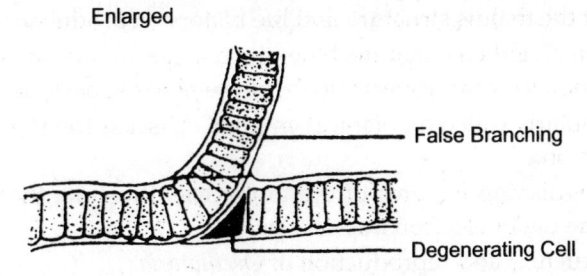

Fig. 14.9 *Scytonema*—Habit enlarged showing false branching.

In shallow waters, loose-lying aegagropilous forms occur. Some are *saxicolous* forms i.e., they grow on the exposed hard rocky surface (*S. myochrous*). Filaments of *Scytonema* may produce compact felt-like mucous growths, in regions of high humidity, on the rocks and trunks of trees.

Structure

Filaments of *Scytonema* have no distinction into base or apex distinctly and the trichomes are more or less of uniform thickness. The trichomes show characteristic false branching and the heterocysts are intercalary. The trichome is enclosed in a thick mucilaginous sheath which is usually stratified.

The cells of trichome show the typical structure of Myxophyceae.

False branching commences with the trichome breaking at a certain point. A trichome may break due to degeneration of some intercalary cells so that a triangular space will be left between the branches. In some, vegetative cells modify into biconcave, colourless *separating discs*. The separating discs degenerate so that the trichome will be broken. Thus, a trichome breaks by any of these methods and the broken ends of one or both the branches grow out piercing the parent sheath. These branches secrete their own mucilage sheaths. Branches will be paired (germinate) when both the broken ends grow into the laterals. This type of growth is known as *false branching*. Sometimes, false branches may appear adjacent to the heterocyst due to degeneration of vegetative cells near to it and the heterocyst itself has no role in the branching.

Reproduction

It takes place by hormogonia. Hormogonia are small fragments of trichome and they arise by the breaking of tips of trichomes. These hormogonia are small trichomes with few cells and are free at both ends. They grow into new trichomes by repeated cell division.

In *S. velutinum*, structures with single cells enclosed in stratified envelopes are formed from the trichomes. They are comparable with the hormocysts of other related algae.

IMPORTANT QUESTIONS

Essay Type

1. Describe the range of thallus structure in cyanophyceae.
2. Describe the different modes of reproduction met with in cyanophyceae.
3. Give an account of cell structure and reproduction in cyanophyceae. Illustrate your answer with neat labelled diagrams.
4. Give an account of the habit and habitat of *Nostoc*. Explain how its form and structure is adapted to the environmental conditions.
5. Give an account of the thallus structure and life history (reproduction) of *Nostoc*.
6. List the features which indicate that the blue-green algae are primitive and ancient members of the plant Kingdom. In what respects do they resemble the bacteria?
7. "Heterocyst is popularly called a botanical enigma". Discuss the statement in the light of its structure and function.
8. Describe the distinguishing features of cyanophyceae. Illustrate the structure of a cell of a cyanophycean algae under electron microscope.
9. Discuss habitat, structure and reproduction of *Oscillatoria*.
10. Give an account of the habitat, structure and reproduction of *Nostoc*.
11. Identify the following giving their important characters with the help of diagrams:
    (a) *Oscillatoria*   (b) *Nostoc*
    (c) *Rivalaria*   (d) *Gloeotrichia*
    (e) *Scytonema*
12. What are heterocysts? Describe their structure and function in detail.
13. Describe the structure of thallus and modes of reproduction in *Oslillatoria* or *Rivularia*.
14. What do you know about the distinguishing characters of cyanophyceae? By means of labelled diagrams only show the structure of cyanophycean cell under light and Electron microscope.
15. Give the distinguishing characters of Cyanophyceae and describe the structure of the cell and reproduction in any type which you have studied.
16. What the cell type of Bacteria and blue green algae called? Give an account of Morphological features, physiological functions and methods of reproduction in any of the type studied by you.
17. Describe the various methods of reproduction in *Nostoc* and *Oscillatoria*.
18. Give an illustrated account of the ultrastructure of any prokaryotic algal cell.
19. Write an essay on economic importance of Blue green algae.

20. Describe the different modes of reproduction in Cyanophyceae.
21. Give an account of the ultra structure of a cell of a blue green algae studied by you. Why are blue green algae considered to be most primitive of all algae?

Short Answer Type

22. Write short notes on:
    (i) Movements in *Oscillatoria*
    (ii) Akinete
    (iii) Branching in *Scytonema*
    (iv) Structure of a cell in *Oscillatoria/Nostoc*
    (v) Structure of mature heterocyst
    (vi) Hormogonium of *Oscillatoria*
    (vii) Reproduction in Cyanophyceae.
    (viii) Cell structure in Blue green algae.
    (ix) Economic importance of Blue green algae.
    (x) Heterocysts
    (xi) Nuclear structure of *Nostoc*
    (xii) Ultrastructure of Cyanophycean cell
    (xiii) *Gloeotrichia*
    (xiv) Endophytic Blue green algae
    (xv) Nitrogen fixation by Blue green algae
    (xvi) Use of Blue green algae as fertilisers
23. Draw labelled diagrams of the following:
    (i) False branching in *Scytonema*
    (ii) A fertile trichome of *Gloeotrichia*
    (iii) Ultra structure of Myxophycean cell
    (iv) Structure of *Oscillatoria*
    (v) Cell structure in Cyanophyceae
24. In what respects do cyanophyceae resemble bacteria?
    (i) *Oscillatoria* and *Gloeotrichia*
    (ii) *Nostoc* and *Oscillatoria*
    (iii) Chlorophyceae and Cyanophyceae
    (iv) Scytonema and *Rivularia*
26. Give reasons for the following:
    (i) It is not necessary to add nitrogen fertilizers to paddy fields.
    (ii) Although all the blue green algae have a blue pigment, they appear as red or yellow brown.
    (iii) Blue green algae are considered primitive of all the algae.
    (iv) Blue green algae can thrive in relatively dry places.
27. Explain hormogones and Heterocysts in *Nostoc*.
28. Blue green algae can thrive in relatively dry places compared to other algae. What structural peculiarities make this possible?
29. Describe the general characteristics of Cyanophyceae
30. Describe the plant body of *Oscillatoria*

286 A Textbook of Algae

31. Describe the reproduction in *Oscillatoria/Nostoc*.
32. Discuss briefly the thallus structure of *Scytonema*.

Objective Type

33. Fill in the Blanks:
    (i) The commonest mode of Reproduction in *Oscillatoria* is by ....
    (ii) In Cyanophyceae, the site of nitrogen fixation is ....
    (iii) False branching is characteristic feature of ....
    (iv) Water blooms are generally formed by ....
    (v) Myxophyceae are so named because of ....
    (vi) Plastids are absent in ....
    (vii) *Nostoc* does not reproduce ....
    (viii) In Algae, the prokaryote cell is present in the member of ....
    (ix) The name of an alga which helps in nitrogen fixation in ....
    (x) The procaryotic alga having heterocyst is ....
    (xi) The genetic apparatus is not bound by membranes in members of ....
    (xii) The two principal components of cell wall of blue green algae are .... and ....
    (xiii) The reserve food material in Cyanophyceae is ....
    (xiv) The alga in which basal heterocyst and akinete are present is ....

34. Select the correct answer:
    (i) Which of the following has a single polar nodule in the heterocysts?
        (a) *Volvox* (b) *Oscillatoria*
        (c) *Gloeotrichia* (d) *Scytonema.*
    (ii) The photosynthetic apparatus, the respiratory apparatus and the genetic apparatus is not bound by membranes in
        (a) Chlorophyceae (b) Myxophyceae
        (c) Rhodophyceae (d) Phaeophyceae.
    (iii) Which of the following has prokaryotic cells?
        (a) *Chlamydomonas* (b) *Volvox*
        (c) *Polysiphonia* (d) *Nostoc.*
    (iv) The alga in which basal heterocyst and akinete is present is ....
        (a) *Oscillatoria* (b) *Gloeotrichia*
        (c) *Nostoc* (d) *Chlamydomonas.*
    (v) False branching is characteristic of :
        (a) *Scytonema* (b) *Gloeotrichia*
        (c) *Nostoc* (d) *Anabaena.*
    (vi) Vegetative reproduction through hormogonia is present in
        (a) *Oscillatoria* (b) *Scytonema*
        (c) *Gloeotrichia* (d) *Volvox.*
    (vii) Plastids are absent in the members of class.
        (a) Chlorophyceae (b) Phaeophyceae
        (c) Rhodophyceae (d) Myxophyceae.

(viii) Phycocyanin-c is the dominant pigment in
- (a) *Volvox*
- (b) *Ectocarpus*
- (c) *Polysiphonia*
- (d) *Oscillatoria*

(ix) The member of the class Myxophyceae are characterised by
- (a) Presence of motile spores and gametes
- (b) Sexual reproduction
- (c) Fusion and meiosis
- (d) Absence of motile reproductive bodies and incipient nucleus.

(x) The algaologist who has done extensive research on cyanophycean algae is
- (a) F.E. Fritsch
- (b) R. N. Singh
- (c) M.O.P. Iyengar
- (d) R.S. Randhawa.

(xi) The dominant pigment of blue green algae is
- (a) Phycoerythrin
- (b) Xanthophyll
- (c) Phytocyanin
- (d) fucoxanthin.

(xii) *Scytonema* usually reproduces by
- (a) hormogonia and endospores
- (b) hormogonia and akinetes
- (c) heterocysts and akinetes
- (d) heterocysts and endospores.

(xiii) The outer portion of the protoplast in a cyanophycean cell is called chromoplast because of the presence of
- (a) Coloured plasma membrane
- (b) Chromosomes
- (c) Pigment
- (d) Chromoplast.

(xiv) Psuedovacuoles in Cyanophyceae help as organs to
- (a) Regulate buoyancy
- (b) Store food
- (c) Store oxygen
- (d) Store CO_2

(xv) In *Rivularia* growth of the filament is due to the activity of
- (a) an apical cell at the apex of the hair
- (b) Intercalary meristem at the base of the hair
- (c) all cells of the trichomes
- (d) basal heterocysts.

Appendix I— Life Cycles of Algae

LIFE CYCLE OF *PANDORINA*

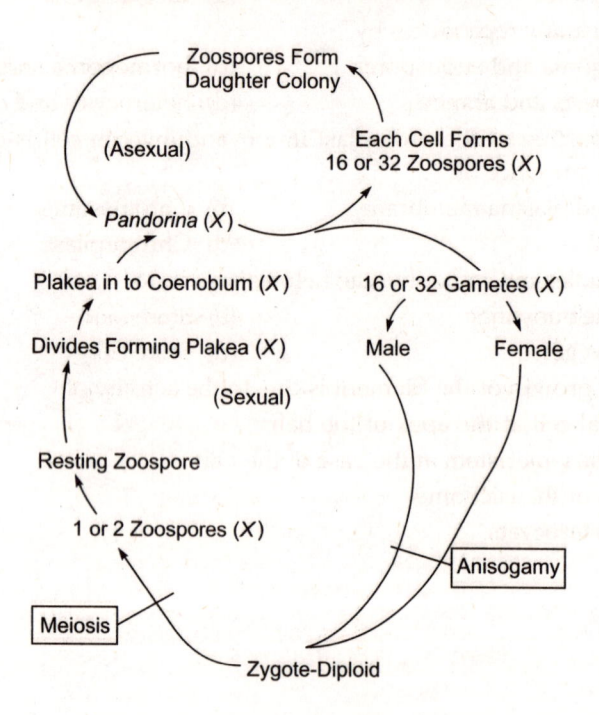

- Zoospores Form Daughter Colony
- Each Cell Forms 16 or 32 Zoospores (*X*)
- (Asexual)
- *Pandorina* (*X*)
- Plakea in to Coenobium (*X*)
- 16 or 32 Gametes (*X*)
- Divides Forming Plakea (*X*)
- Male
- Female
- (Sexual)
- Resting Zoospore
- 1 or 2 Zoospores (*X*)
- Meiosis
- Anisogamy
- Zygote-Diploid

Appendix I—Life Cycles of Algae 289

LIFE CYCLE OF *VOLVOX*

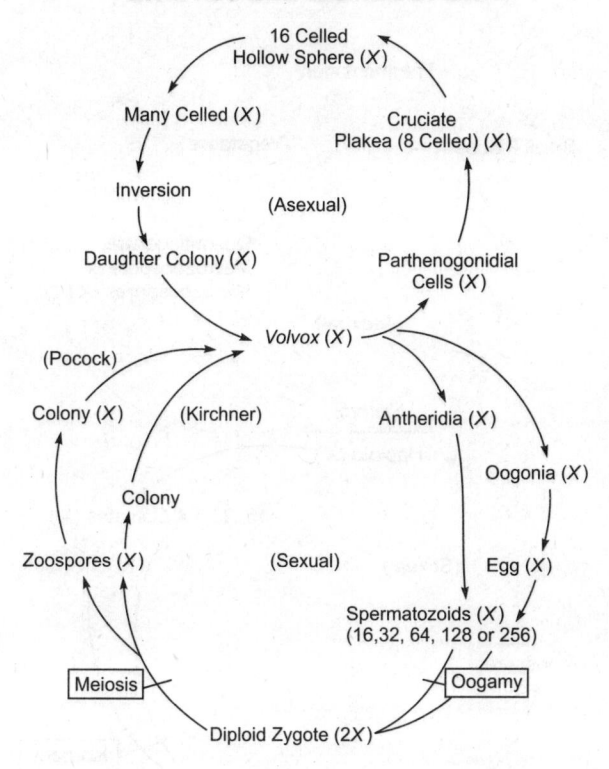

16 Celled Hollow Sphere (X)
Many Celled (X)
Cruciate Plakea (8 Celled) (X)
Inversion
(Asexual)
Daughter Colony (X)
Parthenogonidial Cells (X)
Volvox (X)
(Pocock)
Colony (X)
(Kirchner)
Antheridia (X)
Colony
Oogonia (X)
Zoospores (X)
(Sexual)
Egg (X)
Spermatozoids (X) (16, 32, 64, 128 or 256)
Meiosis
Oogamy
Diploid Zygote (2X)

LIFE CYCLE OF *HYDRODICTYON*

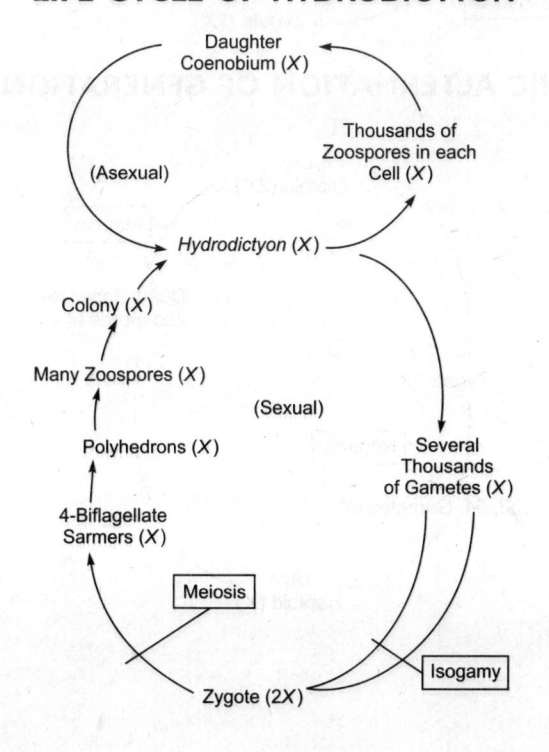

Daughter Coenobium (X)
Thousands of Zoospores in each Cell (X)
(Asexual)
Hydrodictyon (X)
Colony (X)
Many Zoospores (X)
(Sexual)
Polyhedrons (X)
Several Thousands of Gametes (X)
4-Biflagellate Sarmers (X)
Meiosis
Isogamy
Zygote (2X)

LIFE CYCLE OF *ULOTHRIX*

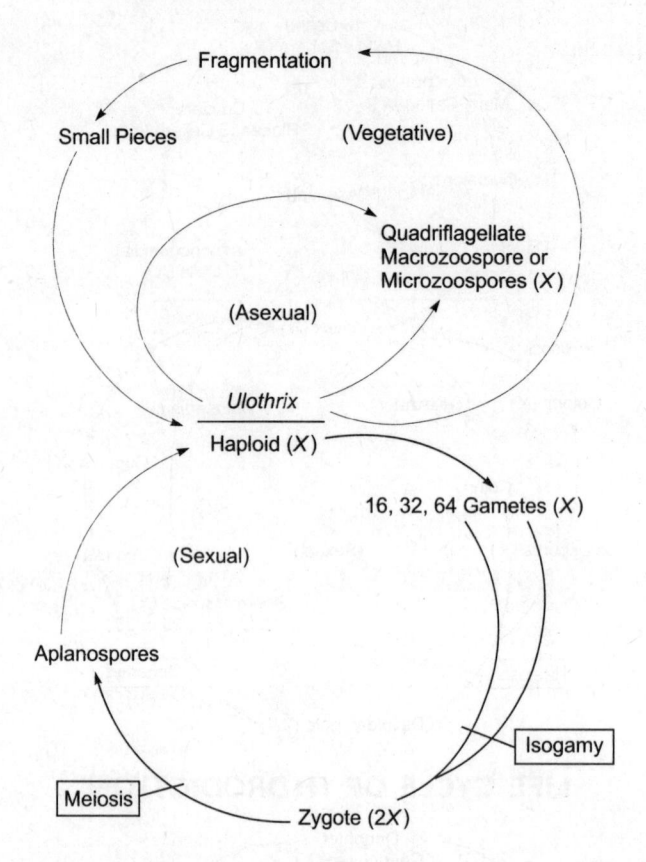

- Fragmentation
- Small Pieces
- (Vegetative)
- (Asexual)
- Quadriflagellate Macrozoospore or Microzoospores (X)
- *Ulothrix* Haploid (X)
- 16, 32, 64 Gametes (X)
- (Sexual)
- Aplanospores
- Meiosis
- Isogamy
- Zygote ($2X$)

ISOMORPHIC ALTERNATION OF GENERATIONS IN *ULVA*

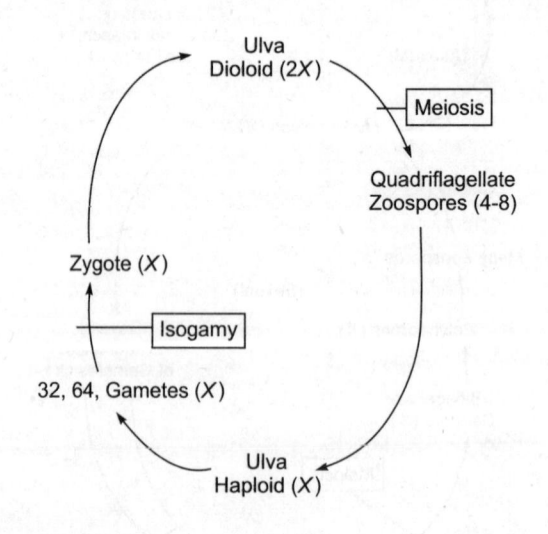

- Ulva Dioloid ($2X$)
- Meiosis
- Quadriflagellate Zoospores (4-8)
- Ulva Haploid (X)
- 32, 64, Gametes (X)
- Isogamy
- Zygote (X)

Appendix I—Life Cycles of Algae 291

LIFE CYCLE OF *CLADOPHORA GLOMERATA*

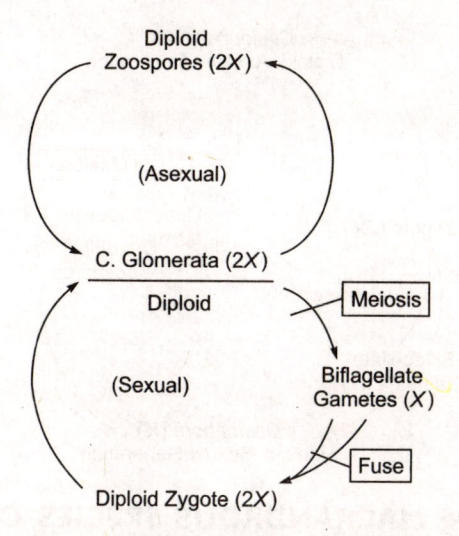

- Diploid Zoospores (2X)
- (Asexual)
- C. Glomerata (2X) — Diploid
- Meiosis
- Biflagellate Gametes (X)
- Fuse
- (Sexual)
- Diploid Zygote (2X)

OEDOGONIUM LIFE HISTORY OF NANNANDROUS SPECIES

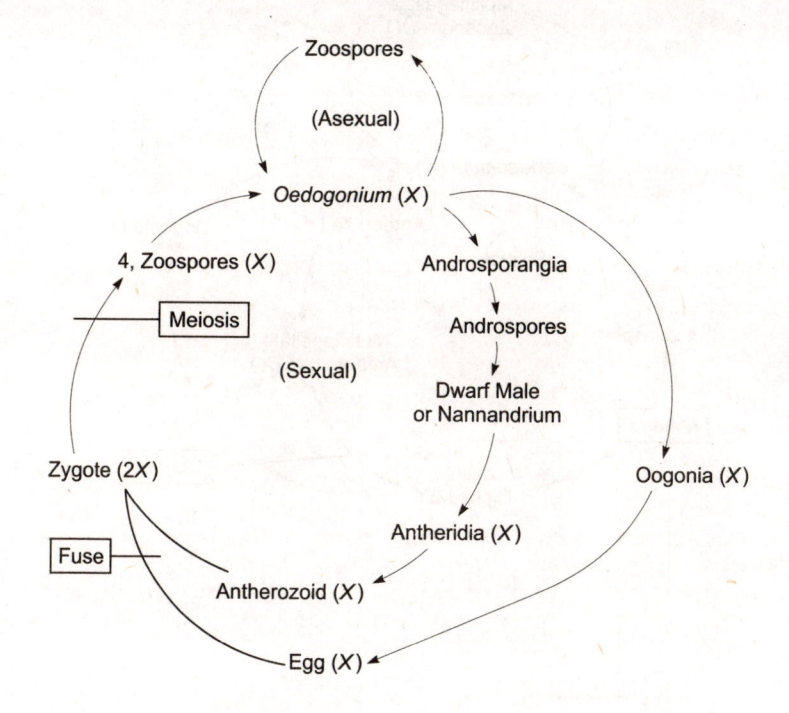

- Zoospores
- (Asexual)
- Oedogonium (X)
- 4, Zoospores (X)
- Meiosis
- (Sexual)
- Androsporangia
- Androspores
- Dwarf Male or Nannandrium
- Oogonia (X)
- Zygote (2X)
- Fuse
- Antheridia (X)
- Antherozoid (X)
- Egg (X)

ISOMORPHIC ALTERNATION OF GENERATIONS IN *CLADOPHORA*

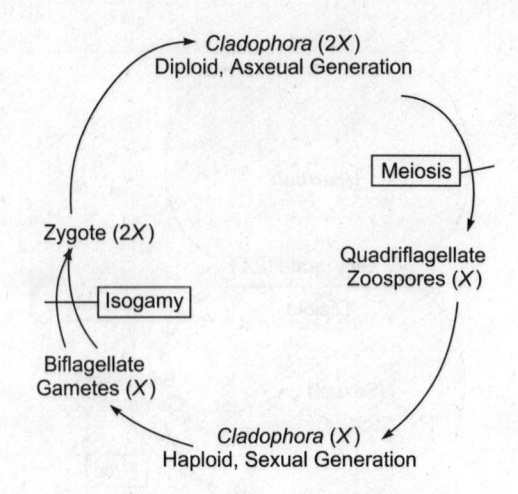

Cladophora (2X) Diploid, Asexual Generation → Meiosis → Quadriflagellate Zoospores (X) → *Cladophora* (X) Haploid, Sexual Generation → Biflagellate Gametes (X) → Isogamy → Zygote (2X) → *Cladophora* (2X)

LIFE CYCLE OF MACRANDROUS SPECIES *OEDOGONIUM*

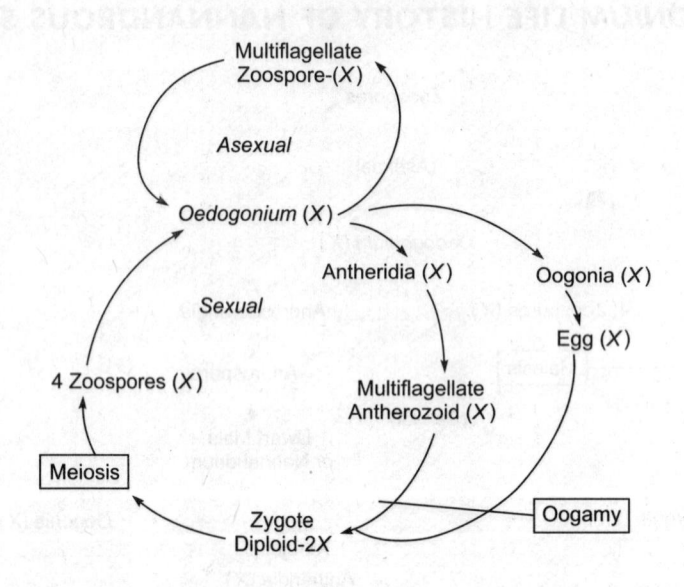

Asexual: *Oedogonium* (X) → Multiflagellate Zoospore-(X) → *Oedogonium* (X)

Sexual: *Oedogonium* (X) → Antheridia (X) → Multiflagellate Antherozoid (X); *Oedogonium* (X) → Oogonia (X) → Egg (X) → Oogamy → Zygote Diploid-2X → Meiosis → 4 Zoospores (X) → *Oedogonium* (X)

Appendix I—Life Cycles of Algae

LIFE CYCLE OF *COLEOCHAETE*

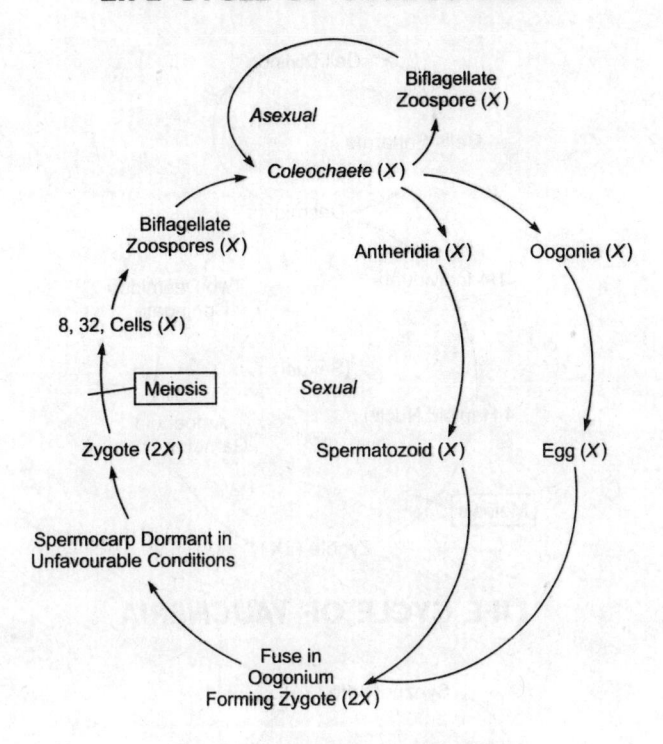

- *Coleochaete* (X)
- Asexual → Biflagellate Zoospore (X)
- Sexual: Antheridia (X) → Spermatozoid (X); Oogonia (X) → Egg (X)
- Fuse in Oogonium Forming Zygote (2X)
- Spermocarp Dormant in Unfavourable Conditions
- Zygote (2X) → Meiosis → 8, 32, Cells (X) → Biflagellate Zoospores (X)

LIFE CYCLE OF *ZYGNEMA*

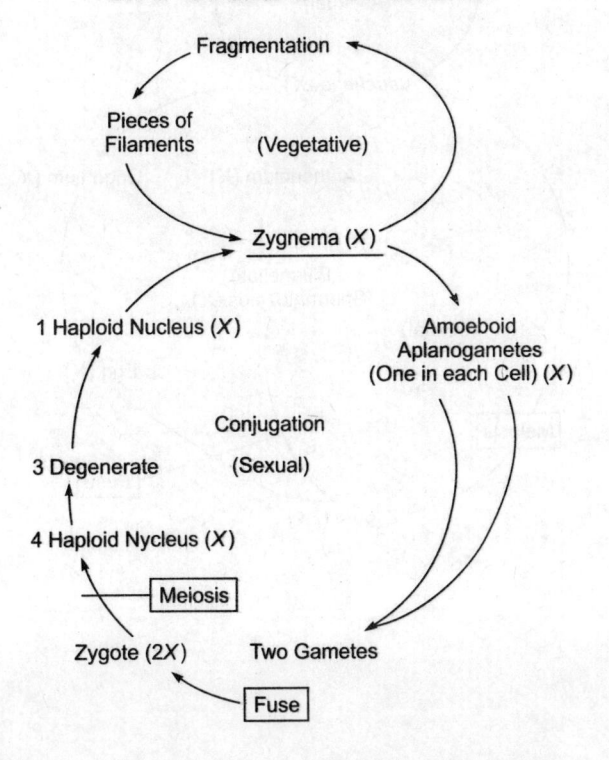

- *Zygnema* (X)
- Fragmentation (Vegetative) → Pieces of Filaments
- Conjugation (Sexual): Amoeboid Aplanogametes (One in each Cell) (X) → Two Gametes → Fuse → Zygote (2X)
- Meiosis → 4 Haploid Nucleus (X) → 3 Degenerate → 1 Haploid Nucleus (X)

LIFE CYCLE OF *DESMIDS*

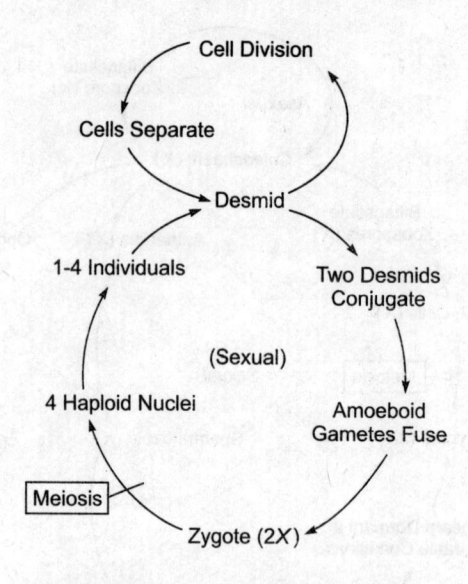

Cell Division → Cells Separate → Desmid → Two Desmids Conjugate → Amoeboid Gametes Fuse → Zygote (2X) → [Meiosis] → 4 Haploid Nuclei → 1-4 Individuals → Desmid

(Sexual)

LIFE CYCLE OF *VAUCHERIA*

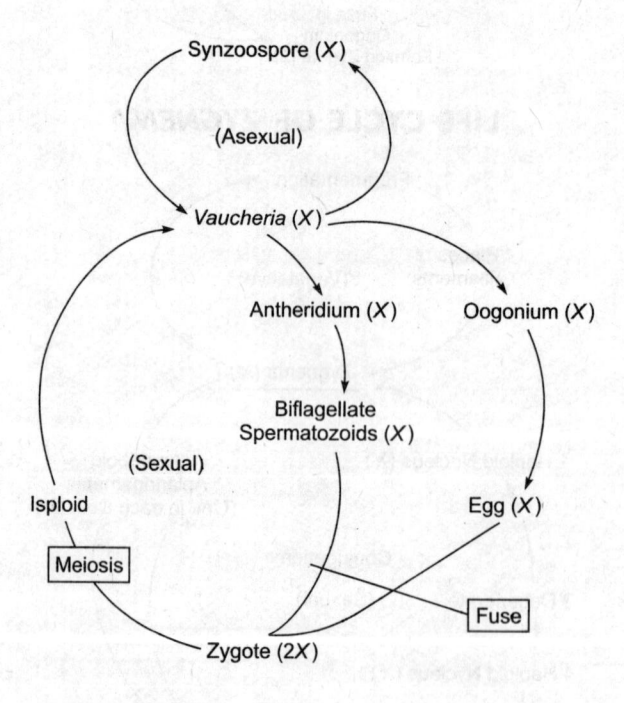

Synzoospore (X) → *Vaucheria* (X)

(Asexual)

Vaucheria (X) → Antheridium (X) → Biflagellate Spermatozoids (X)

Vaucheria (X) → Oogonium (X) → Egg (X)

[Fuse] → Zygote (2X) → [Meiosis] → Isploid → *Vaucheria* (X)

(Sexual)

Appendix I—Life Cycles of Algae

LIFE CYCLE OF *ECTOCARPUS*

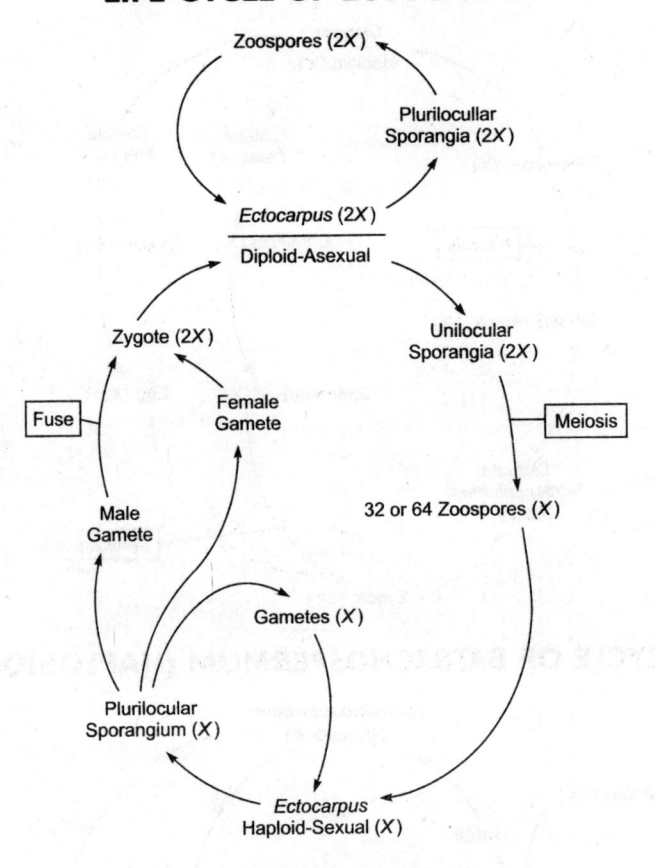

Zoospores (2X)

Plurilocullar Sporangia (2X)

Ectocarpus (2X)
Diploid-Asexual

Unilocular Sporangia (2X)

Zygote (2X)

Female Gamete

Fuse

Meiosis

Male Gamete

32 or 64 Zoospores (X)

Plurilocular Sporangium (X)

Gametes (X)

Ectocarpus
Haploid-Sexual (X)

LIFE CYCLE OF *SARGASSUM*

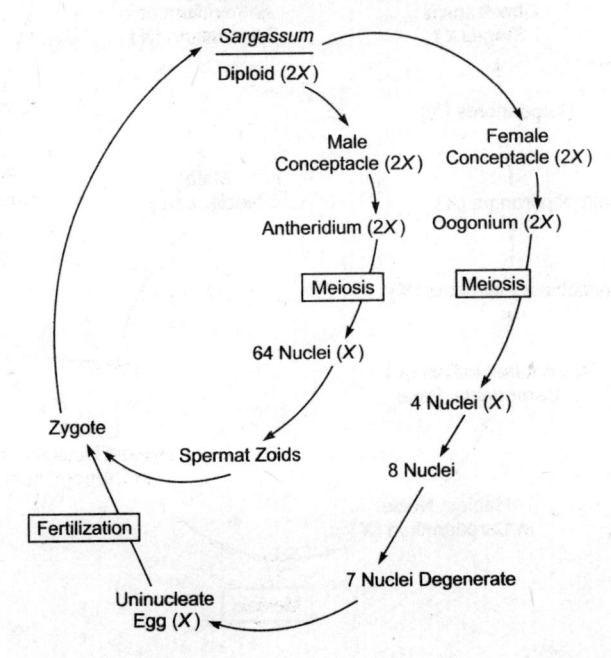

Sargassum
Diploid (2X)

Male Conceptacle (2X)

Female Conceptacle (2X)

Antheridium (2X)

Oogonium (2X)

Meiosis

Meiosis

64 Nuclei (X)

4 Nuclei (X)

Zygote

Spermat Zoids

8 Nuclei

Fertilization

7 Nuclei Degenerate

Uninucleate Egg (X)

296 *A Textbook of Algae*

LIFE CYCLE OF *DICTYOTA*

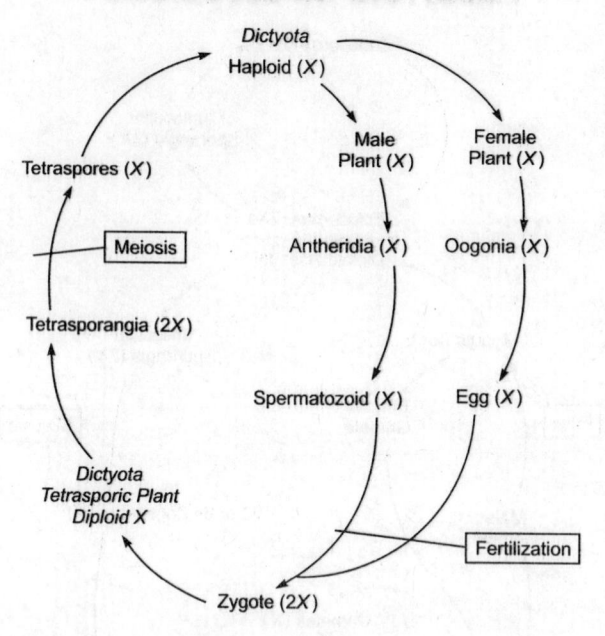

- *Dictyota* Haploid (X)
- Male Plant (X) → Antheridia (X) → Spermatozoid (X)
- Female Plant (X) → Oogonia (X) → Egg (X)
- Fertilization → Zygote ($2X$)
- *Dictyota* Tetrasporic Plant Diploid X
- Tetrasporangia ($2X$) → Meiosis → Tetraspores (X)

LIFE CYCLE OF *BATRACHOSPERMUM* (HAPLOBIONTIC)

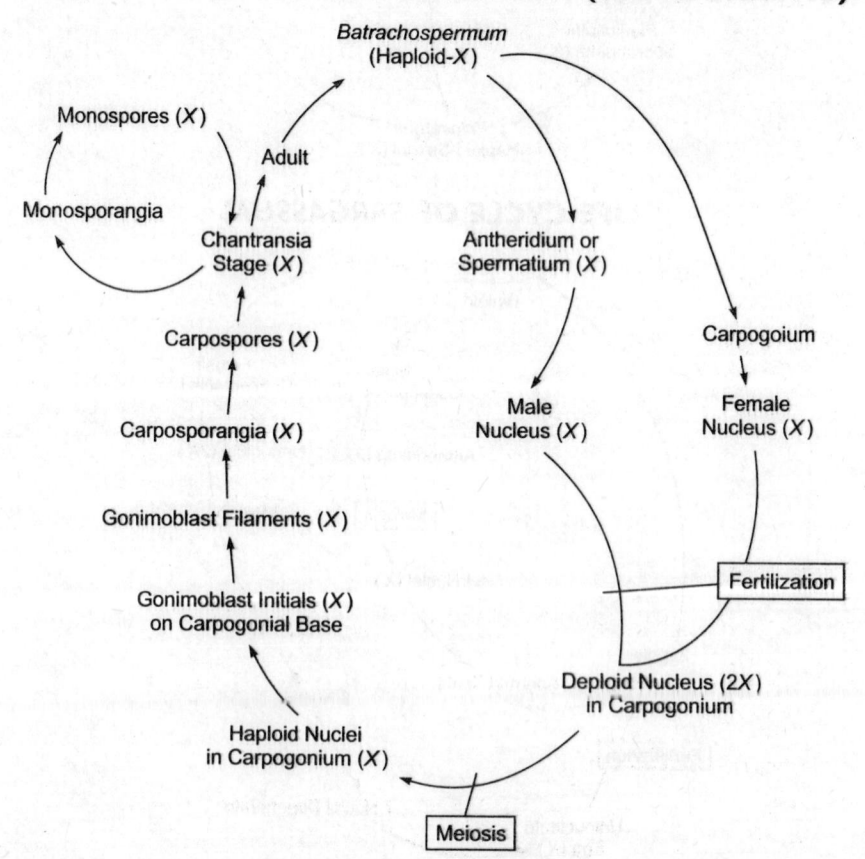

- *Batrachospermum* (Haploid-X)
- Adult → Chantransia Stage (X)
- Monosporangia → Monospores (X)
- Antheridium or Spermatium (X) → Male Nucleus (X)
- Carpogoium → Female Nucleus (X)
- Fertilization → Deploid Nucleus ($2X$) in Carpogonium
- Meiosis → Haploid Nuclei in Carpogonium (X)
- Gonimoblast Initials (X) on Carpogonial Base
- Gonimoblast Filaments (X)
- Carposporangia (X) → Carpospores (X)

Appendix I—Life Cycles of Algae **297**

LIFE CYCLE OF *POLYSIPHONIA* (DIPLOBIONTIC TYPE)

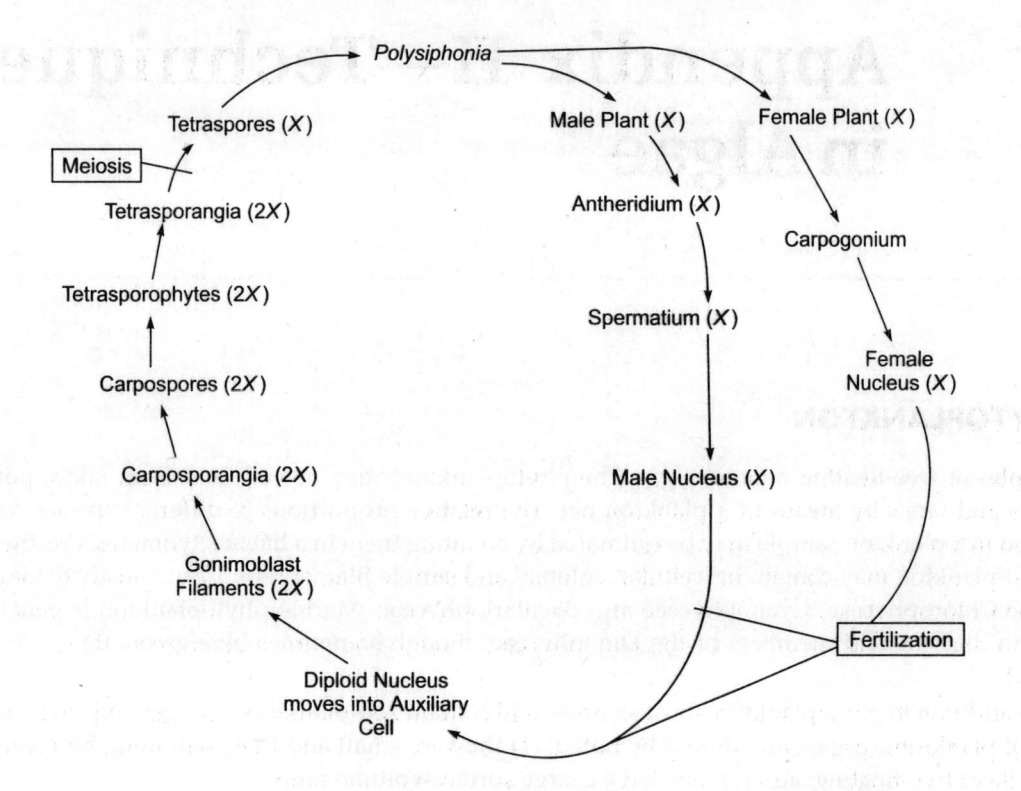

- Polysiphonia
- Tetraspores (X) — Meiosis
- Tetrasporangia ($2X$)
- Tetrasporophytes ($2X$)
- Carpospores ($2X$)
- Carposporangia ($2X$)
- Gonimoblast Filaments ($2X$)
- Diploid Nucleus moves into Auxiliary Cell
- Male Plant (X)
- Antheridium (X)
- Spermatium (X)
- Male Nucleus (X)
- Female Plant (X)
- Carpogonium
- Female Nucleus (X)
- Fertilization

Appendix II—Techniques in Algae

PHYTOPLANKTON

Samples of free-floating aquatic algae (the phytoplankton) may be collected from lakes, ponds, rivers and tanks by means of a plankton net. The relative proportions of different species represented in a plankton sample may be estimated by counting them in a haemactyometer. Freshwater phyto-plankton may contain unicellular, colonial and simple filamentous algae, mostly belonging to the Chlorophyceae, Cyanophyceae and Bacillariophyceae. Marine phytoplankton is generally rich in diatoms and members of the Dinophyceae, though sometimes blue-green algae are also found.

In addition to phytoplankton, most samples will contain zoo-plankton. Two general characteristics of planktonic organisms should be noted: (1) they are small and free-swimming by means of flagella or free-floating; and (2) they have a large surface-volume ratio.

Freshwater phytoplankton, collected from relatively 'clean', nutrient-deficient or oligotrophic waters, exhibit a great diversity of algal species, though the concentration of each species or of the algae as a whole is very low.

PRESERVATION

When it is intended to store algae in the laboratory for subsequent morphological studies, they may be killed and preserved in a 4% solution of formalin (prepared by adding 4 ml of 40% formalin to 96 ml of distilled water; 40% commercial formalin is regarded as 100% for purposes of calculation). For preserving aquatic algae, appropriate quantity of 40% formalin may be added directly to the sample so as to obtain a final concentration of 4%.

Many terrestrial or subaerial algae can be preserved dry and stored in paper envelopes for long periods without any apparent damage or loss of viability.

For maintaining the algae in their natural (green) colour, any one of the following solutions may be employed. The algae are immersed in the preservative for a few days and then transferred to formalin acetic alcohol (FAA) solution.

1. | | |
    |---|---|
    | 50% ethyl alcohol | 90 ml |
    | 40% formalin | 4 ml |
    | Glycerol | 3 ml |
    | Glacial acetic acid | 3 ml |
    | Cupric chloride ($CuCl_2$) | 9.5 gm |

	Uranium nitrate (UNO$_3$)	1.5 gm

Although suitable for most green algae, this preservative can also be used for blue-green algae, if 10 gm of copper acetate is substituted for the cupric chloride and uranium nitrate.

2.	Cupric sulphate (CuSO$_4$5H$_2$O)	0.25 gm
	Water	38 ml
	Glacial acetic acid	4 ml
	40% formalin	8 ml
	95% ethyl alcohol	50 ml
3.	Potassium chromealum	10 gm
	40% formalin	6 ml
	Distilled water	500 ml

PREPARATION OF HERBARIUM SHEETS

Herbarium sheets of most marine algae, especially those having mucilaginous or gelatinous thalli (e.g., *Delesseria, Plocamium, Gracilaria, Porphyra* and *Ulva*) can be prepared with great ease. The specimen is floated in an enamel tray containing freshwater or sea water, as the case may be, and a piece of moderately thick herbarium sheet inserted in the water below the specimen. The alga is then spread and the tray tilted gently while the specimen is held on the sheet. If the specimen is mucilaginous, it will stick to the sheet and any excess water on the latter may be drained off by keeping the sheet on an inclined plane for some time. The sheet is then placed between two dry newspapers or blotting papers and after a number of herbarium sheets of different algae have been prepared in this way, they are all pressed in a herbarium press. After 24 hours the wet newspapers are replaced by fresh, dry sheets and pressed again. Five or 6 such changes are usually sufficient to absorb all traces of water from the sheets and specimens may then be mounted and labelled.

This method may be modified for non-sticky algae by affixing the specimen on to herbarium sheet by means of a few small pieces of cellotape.

Permanent Preparations

Some of the more common methods of making permanent slides of algae are described here:

1. This method may be used for both freshwater and marine algae. It is suitable for morphological study and identification, but not for cytological study.
   (1) Place alga on a slide in a drop of freshwater or sea-water.
   (2) Add a small drop of 40% formalin to fix. Drain out excess water.
   (3) Place a small lump of glycerine jelly on the alga. (Glycerine jelly is prepared by dissolving 5 gm of gelatine in 30 ml of water by gentle heat and then adding 0.125 gm phenol and 35 ml of glycerol.)
   (4) Transfer the slide to an incubator or oven (at 60°C) for a few minutes for the jelly to melt. Spread the algal material appropriately.
   (5) Apply circular coverglass. Keep the slide again in the oven for a short while. Wipe of excess jelly from around the coverglass and seal it with Gold Size. Store flat.
2. Method employed for making permanent slides suitable for cytological studies.
   (1) Fix the material in any of the fixatives described in Schedule III, or in a mixture containing equal volumes of glacial acetic acid and absolute ethyl alcohol. Material should be immersed in the fixative for 10-15 min.

(2) Wash in 3 changes of water.
(3) Mordant in 1.5% aqueous iron alum solution for 10-60 sec.
(4) Wash thoroughly in running tap water for 2-3 min.
(5) Stain the material in a drop or two or acetocarmine on a slide, or, in the case of unicellular forms, in a centrifuge tube, warm the slide gently over a spirit flame and apply a coverglass.
(6) Dip the slide and coverglass horizontally in a petridish containing 95% ethyl alcohol and allow them to remain there till the coverglass separates from the slide. Carry the coverglass and slide through two other petri dishes containing absolute ethyl alcohol, mount in Euparal and apply the coverglass again.

STAINING TECHNIQUES IN ALGAE

(1) The fresh algal material may be fixed in any one of the following solutions:
   (i) *Chrome-acetic Fixative*

10% aqueous chromic acid	2.5 ml
10% aqueous acetic acid	5.0 ml
Distilled Water	92.5 ml

   (ii) *Dioxan Fixative*

Dioxan	50 ml
40% Formalin	5 ml
Glacial acetic acid	5 ml
Distilled water	50 ml

   (iii) *Chrome-osmo-acetic Fixative*

Chromic anhydride	1 gm
Glacial acetic acid	3 ml
1% aqueous osmic acid	1 ml
Distilled water	100 ml

2. Wash the fixed material several times in tap water. Dehydrate gradually by passing the material through a number of grades of ethyl alcohol (3%, 5%, 8%, 12% absolute alcohol, 2 changes in each grade). In the lower grades the material may be kept for 3 min., in the higher grades for about 5 min. After absolute alcohol the alga is mounted directly in a drop of Euparal on the slide and a coverglass applied.
3. If stained specimens are required, any one of the following stains may be used:
   (i) Aniline Blue (1% solution in 90% ethyl alcohol) used for filamentous green algae, stain for 5 min after 85% ethyl alcohol stage.
   (ii) Ertythrosine Bluish (1% solution in absolute ethyl alcohol) used for staining gelatinous envelopes or sheaths; stain for 30 sec after 95% ethyl alcohol.
   (iii) Light Green (0.2% solution in 90% ethyl alcohol) used for cellulose cell walls; stain for 30-60 sec after 85% ethyl alcohol.
   (iv) Congo Red (0.2% solution in absolute ethyl alcohol) used for staining mucilage sheaths of Cyanophyta; stain for 60 seconds after 95% ethyl alcohol.

General Staining

Different algae vary in their affinity for stains. More frequently employed stains are Methylene Blue, Gentian Violet or Acid Fuchsin (up to 1% aqueous solutions). The following simple procedure is recommended. Mount algae in a drop of water on a slide and apply a coverglass. Add a drop of the stain to one edge of the coverglass and let it diffuse to the opposite edge by removing water from the other side with a piece of dry blotting paper. In this way a spectrum of staining is obtained, the algae near one side of the coverglass are intensely stained and on the opposite side weakly stained.

Flagella

Most algal flagella cannot be seen under an ordinary light microscope without special staining. Following are the two commonly used methods to render flagella visible:
- (i) Add a few particles of lead of copying ink pencil to a drop of algal suspension, apply a coverglass and examine, and
- (ii) Fix algae in dilute IKI solution or by exposure to osmic acid vapour, apply coverglass and examine. If overstained, decolourize to desired degree by adding dilute sodium thiosulphate.

Gelatinous Envelopes and Sheaths

They are rendered conspicuous by staining with a very dilute aqueous solution of either Ruthenium Red or Methylene Blue. Alternatively, a trace of India ink or of Gurr's Negative Stain may be employed for the same purpose.

CULTURE

Certain morphological and reproductive stages can best be demonstrated by suitable manipulation of live cultures of algae. Algal cultures may either be obtained from a Culture Collection, or raised in the laboratory. The three major Culture Collection Centres are:
- (i) The Culture Collection of Algae. Department of Botany, Indian University, Bloomington, Indiana, U.S.A.
- (ii) The Culture Centre of Algae and Protozoa, Storey's Way, Cambridge, England.
- (iii) The Culture Collection of Algae and Microorganisms, Institute of Applied Microbiology, University of Tokyo, Bunkyoku, Tokyo, Japan.

Unialgal cultures may, however, be isolated without much difficulty from fresh material collected from nature. A number of culture media have been found suitable. The following three are likely to prove useful for the isolation and multiplication of most freshwater and soil algae.

1. Chu No. 10 (modified), for common freshwater and soil algae:

Calcium nitrate $Ca(NO_3)_2$	0.04 gm
Dipotassium hydrogen phosphate (K_2HPO_4)	0.01 gm
Magnesium sulphate $(MgSO_4.7H_2O)$	0.025 gm
Sodium carbonate (Na_2CO_3)	0.020 gm
Sodium silicate (Na_2SiO_3)	0.025 gm
Ferric citrate	0.003 gm
Citric acid	0.003 gm
*A_5 Trace elements stock solution (optional)	1.0 ml
Glass-distilled water	1000 ml

*The A_5 Trace elements stock solution has the following composition in grams per litre of glass-distilled water:

Boric acid (H_3BO_3)	2.86
Manganese chloride $(MnCl_2.4H_2O)$	1.81

Zinc sulphate (ZnSO$_4$.7H$_2$O)	0.222
Molybdenum trioxide (MoO$_3$(85%))	0.0177
Cupric sillphate (CuSO$_4$.5H$_2$O)	0.079

2. Allen and Arnon's Medium (modified) for nitrogen-fixing blue-green algae:

Magnesium sulphate (MgSO$_4$.7H$_2$O)	0.025 gm
Calcium chloride	0.05 gm
Sodium chloride	0.20 gm
Dipotassium hydrogen phosphate	0.35 gm
A$_5$ Trace elements stock solution	1.0 ml
Glass-distilled water	1000 ml

If 0.20 gm of potassium nitrate is added, this medium will also support the growth of many non-nitrogen-fixing blue-green algae.

3. ASM-1 Medium for fresh water planktonic algae:

Sodium nitrate	170 mg
Dipotassium hydrogen phosphate	17.4 mg
Disodium hydrogen phosphate	14.2 mg
Magnesium chloride (MgCl$_2$.6H$_2$O)	40.7 mg
Magnesium sulphate	49.3 mg
Calcium chloride	22.2 mg
Ferric chloride (FeCl$_3$.6H$_2$O)	1.1 mg
Sodium ethylene diamine tetraacetate (Na$_2$EDTA)	6.7 mg
Boric acid	2.5 mg
Manganese chloride (MnCl$_2$.4H$_2$O)	1.4 mg
Zinc chloride (ZnCl$_2$)	0.4 mg
Cobalt chloride (CoCl$_2$.6H$_2$O)	0.02 mg
Cupric chloride (CuCl$_2$.2H$_2$O)	0.00014 mg
Glass-distilled water	1000 ml

4. *Marine Algae Medium*

Smaller marine algae may be grown in either synthetic or natural sea-water culture media. The latter have, however, been found to be more suitable than the former. In inland places far away from sea, the synthetic media may be used if sea water cannot be procured. The composition of a suitable culture medium, both natural and synthetic, for growing marine algae is given here.

	Natural Seawater Medium	Synthetic Medium
Natural sea-water	1 litre	—
Glass-distilled water	—	1 litre
Potassium nitrate (KNO$_3$)	200 mg	200 mg
Potassium monohydrogen phosphate (K$_2$HPO$_4$)	35 mg	35 mg
Ferric chloride (FeCl$_3$.6H$_2$O)	3 mg	—
Manganese chloride (MgCl$_2$.4H$_2$O)	0.2 mg	—
Sodium chloride (NaCl)	—	30,000 mg
Magnesium chloride (MgCl$_2$.6H$_2$O)	—	5,000 mg
Calcium sulphate (CaSO$_4$.2H$_2$O)	—	1,000 mg
Sodium ethylene diamine tetra acetate (Na$_2$EDTA)	—	20 mg

Potassium chloride (KCl)	—	750 mg
Potassium bromide (KBr)	—	15 mg
Ferrous sulphate (FeSO$_4$.7H$_2$O)	—	0.7 mg
Aluminium sulphate Al$_2$(SO$_4$)$_3$	—	0.25 mg
Cobalt sulphate (CoSO$_4$.9H$_2$O)	—	0.03 mg
Cupric sulphate (CuSO$_4$.5H$_2$O)	—	0.005 mg
Lithium chloride (LiCl.H$_2$O)	—	0.05 mg
Sodium molybdate (Na$_2$MoO$_4$.2H$_2$O)	—	2.0 mg
Rubidium chloride (RbCl)	—	0.5 mg
Strontium chloride (SrCl$_2$.6H$_2$O)	—	5.0 mg
Zinc sulphate (ZnSO$_4$.7H$_2$O)	—	10.0 mg

(The pH of all the above media should be adjusted to 7.0- 7.5 for green algae and 8.5-9.0 for blue-green algae.)

In all these media, phosphate should be autoclaved separately from the other components and mixed aseptically upon cooling.

The media may be solidified with 1 or 1.5% agar. It is advisable to sterilize agar in a small volume of water separately from the mineral salts, to mix the two components after autoclaving, and then to pour in sterile petri dishes.

For liquid culture, a small volume of the mineral salts medium is taken in an Erlenmeyer flask or a test tube which is plugged with non-absorbent cotton wool and then sterilized at 15 lbs/in^2 for 15 min.

Appendix III—Glossary of Algae

Acropetal: toward the apex.
Aerobic: needing oxygen.
Agglution: chemical substance involved in the recognition of gamete of the opposite strain.
Agglutination: aherence of gametes of different mating types by their flagella tips.
Akinete: thick-walled resting spore.
Alkalinity: in water chemistry the total quantity of base in equilibrium with carbonate or bicarbonate that can be determined by titration with strong acid. Alkaline waters have a high pH.
Alpha Granule: protoplasmic structure containing myxophyceae starch in the Cyanophyceae.
Amphisema: outer covering of a dinophycean cell including the thecal plates, peripheral vesicles, and attendant microtubules.
Amylum Star: star-shaped aggregate of cells filled with starch that forms new plants in the Charophyte.
Anerobic: without oxygen.
Androsporangium: sporangium that forms androspores.
Androspore: spore that forms a dwarf male filament in the Oedogoniales (Chlorophyceae).
Antapical: opposite from the apex.
Antherozoid: male gamete.
Anticlinal: perpendicular to the circumference of the thallus.
Apical Growth: growth by means of an apical cell dividing to form the thallus beneath it.
Apochlorotic: colourless.
Aragonite: orthorhombic crystals of calcium carbonate.
Autospore: aplanospore with the same shape as the parent cell.
Autotroph: not needing an external source of organic compounds as an energy source. Energy is obtained from light or inorganic chemical reactions.
Auxiliary Cell: cell that receives the diploid nucleus from fertilization in the Rhodophyceae. The nutritive auxiliary cell provides nutrients for the developing carposporophyte. Whereas the generative auxiliary cell gives rise to gonimoblast filaments.
Axoneme: central two and nine peripheral doublet microtubules of a flagellum.
Basal Body: bottom part of a flagellum beneath the transition zone; basal bodies divide to perpetuate flagella.
Basipetal: toward the base.
Bathal Zone: ocean water over continental slope.

Benthos: organisms living on, and attached to, the bottom of aquatic habitats.
Bilaterally Symmetrical: symmetrical about a line.
Bioluminescence: emission of light by a living organism.
Biomass: the amount at any one time of living matter in a habitat.
Bisporangium: a sporangium forming two meiospores in the Rhodophyceae.
Bloom: heavy growth of planktonic algae in a body of water.
Brackish: saline water with a salinity less than that of seawater.
Bulbil: small plant formed on rhizoids in the Charophyta.
Calcification: deposition of calcium carbonate, usually in association with smaller amounts of other carbonates.
Clacite: rhombohedral crystals of clacium carbonate.
Callose: polysaccharide associated with pores in sieve cells.
Canal: rigid opening in some flagellates.
Capsule: oxygen-free, unsaturated, hydrocarbon carotenoid.
Carotenoid: yellow, orange, or red hydrocarbon fat-soluble pigment.
Carpogonium: female gametangium in the Rhodophyceae.
Carposporangium: carpospore-producing sporangium derived directly or indirectly from the zygote in the Rhodophyceae.
Carpospore: usually diploid spore produced by the carposporangium in the Rhodophyceae.
Carposporophyte: usually diploid generation in the Rhodophyceae derived from the zygote, and which forms the carpospores.
Carrageenin: red algal polysaccharide (phycocolloid) similar to agar but needing higher concentrations to form a gel.
Cellulose: polysaccharide composed of — 1, 4 linked glucose molecules that forms the main skeletal framework of most algal cell walls.
Central Nodule: a wall swelling that divides the raphe into two parts in the Bacillariophyceae.
Centric: type of ornamentation arranged around a central point in the Bacillariophyceae.
Centriole: equivalent to a flagella basal body.
Chemotaxis: the movement of a whole cell in response to a concentration gradient of a chemical substance. If toward higher concentrations it is positive; if away from higher concentrations, it is negative chemotaxis.
Chitin: polysaccharide made up of repeating units of N-acetylglucoseamine.
Chlorophyll: fat-soluble, green, porphyrin-type pigment.
Chloroplast: plastid with chlorophyll.
Chloroplast Endoplasmic Reticulum or Chloroplast E.R.: one or two membranes surrounding the chloroplast envelope; ribosomes are usually attached to the outside of the outer membrane.
Chromatic Adaption: change in the proportions of different photosynthetic pigments enabling optimum absorption of the available wavelengths of light.
Chromoplast or Chromatophore: a chloroplast with some other colour than green.
Chrysolaminarin or Leucosin: a liquid polysaccharide storage product composed principally of — 1, 3 linked residues of glucose.
Cilium: flagellum.
Cingulum or Girdle: transverse furrow in the Dinophyceae containing the transverse flagellum.
Circadian Rhythm: repeated sequence of events that occur at about 24-hour intervals.

Coccoid: spherical.
Coccolith: calcified scale in a coccolithophorid (Prymnesiophyceae).
Coenobium: colony of algal cells in a specific arrangement and number that does not increase once mature.
Coenocyte: large multinucleate cell without cross walls except where reproductive bodies are formed.
Conceptacle: cavity in a thallus where gametangia are produced.
Conjugation: fusion of two nonflagellated protoplasts.
Connecting Band: part of the girdle in the Bacillariophyceae.
Contractile Vacuole: vacuole fed by smaller vesicels that expels water and solutes rhythmically to the outside of the cell.
Coralline: calcified alga.
Cornutate Process: hornlike wall extension.
Corps De Maupas: a vesicular body in the Cryptophyceae used in the digestion of unwanted cell components.
Cortex: outer part of an algal thallus.
Cosmopolitan: occurring everywhere.
Costa: cell wall extension.
Cruciate: cross-shaped.
Cryptostomata or Cryptoblast: flasklike opening in the thallus with hairs.
Cuticle: coating on the outside of a cell wall.
Cyanelle: endosymbiotic blue-green alga.
Cyanome: host cell containing a cyanelle.
Cyanophage: virus in cells of the Cyanophyceae.
Cyanophycin Granule: polypetide storage granule in the Cyanophyceae.
Cyst: cell resistant to poor conditions with a thick or silicified wall, usually not including the original wall, if one was present.
Cystocarp: in the Rhodophyceae the carposporophyte and surrounding gametophytic tissue (pericarp).
Cytosome: cell opening used for ingestion of food particles.
Dendroid: a type of nonmotile colony that produces mucilage in one area, usually forming a stalk.
Diatomaceous Earth: a mineral consisting of the remains of the silicified frustules of the Bacillariopbyceae.
Dichotomy: division of a thallus into two equal branches.
Dictyosome: stack of vesicles in a Golgi apparatus.
Diffuse Growth: type of growth where most of the cells in a thallus are capable of division.
Dioecious: an organism that has male and female gamete borne on separate plants.
Disc or Thylakoid: membrane sac in the chloroplast.
Distichous: arranged in two equal rows.
Diurnal: daily.
Dorsiventral: flattened in one plane.
Dulse: a preparation of Rhodymenia (Rhodophyceae) used as food.
Ecad: a form of a plant species produced in response to a particular habitat, the modifications not being heritable.

Ecdysis: shedding of the theca in dinoflagellates.
Ecotype: a locally adapted variant.
Ectoderm: outer protoplasm next to the plasma membrane in the Charophyta.
Egg: large nonmotile female gamete.
Electron-Dense or Electron-Opaque: term used to describe a material that absorbs electrons and appears dark in electron micrographs.
Electron-Transparent: term used to describe a material that does not absorb electrons and appears light in electron micrographs.
Endemic: occurring in a specific area.
Endochite: inner wall of oogonium in the Fucales (Phaeophyceae).
Endoderm: inner protoplasm next to vacuole in the Charopyta, capable of cytoplasmic streaming.
Endolithic: living inside rock.
Endophyte: plant living inside another plant.
Endosome: nucleolus in the Euglenophyceae.
Endospore: asexual spore in the Cyanophyceae formed by the internal division of a cell.
Endosymbiotic: term that describes an organism living inside a host in a symbiosis.
Endozoic: living inside an animal.
Enucleate: without a nucleus.
Epicone: part of a cell above the girdle in the Dinophyceae.
Epidermis: outer layer of cells.
Epipelic: growing on mud.
Epiphyte: one plant living on another plant.
Epitheca: larger of the two halves of a diatom frustule.
Epizoic: living on an animal.
Estuary: the mouth of a river where tidal effects are evident, and where freshwater and seawater mix.
Eucaryotic: a cell having membrane-enclosed organelles such as the nucleus and mitochondria.
Eurythermic: term that describes a body of water that receives large amounts of nutrients, usually resulting in a large growth of algae.
Exochite: outer wall of an oogonium of the Fucales (Phaeophyceae).
Exotoxin: toxin secreted into the medium.
Extant: living today.
Extinct: not living today.
Eyespot: red to orange area in a cell, composed of lipid droplets.
Facultative Heterotroph or Autotroph: organism that is able to live as a heterotroph or an autotroph.
Facultative Parasite or Saprophyte: organism that is able to live as a parasite or a saprophyte.
False Branching: in the Cyanophyceae breakage of a trichome through a sheath, giving the appearance of a branch.
Filament: in the Cyanophyceae, one or more trichomes enclosed in a sheath.
Floridean Starch: red algal storage product composed of 1, 4 and 1, 6 linked glucose residues.
Florideoside: primary product of photosynthesis in the Rhodophyceae.
Foliose: leaflike.
Fragmentation: type of asexual reproduction where a thallus breaks into two or more parts, each of which forms a new thallus.

Frustule: silicified cell wall in the Bacillariophyceae.
Fucoidin: polysaccharide in the cell wall and mucilage of the Phaeophyceae composed of sulfated fucose units.
Fucosan or Phaeophycean Tannin: a colourless acidic fluid in brown algae giving a characteristic red colour with vanilin hydrochloride.
Fucoserraten: a sex attractant secreted by the eggs of *Fucus* (Phaeophyceae).
Fusiform: spindle-shaped.
Gametangium: structure forming gametes.
Gamete: cell capable of fusion with another to form a zygote.
Gametophyte: plant generation that forms the gametes, usually haploid.
Gas Vacuole: a collection of gas vesicles.
Gas Vesicle: hollow cylindrical gas-filled structures in the Cyanophyceae.
Generative Auxiliary Cell: cell in the Rhodophyceae forming the gonimoblast filaments.
Geotaxis: movement of a cell away (negative geotaxis) or toward (positive geotaxis) gravity.
Girdle: the transverse groove containing the transverse flagellum in the Dinophyceae.
Gliding: active movement of an organism in contact with a solid substrate where there is neither a visible organ responsible for the movement nor a distinct changes in the shape of the organism.
Globule: male reproductive structure in the Charophyta.
Glycocalyx: sticky polysaccharide secreted by the zygote in the Fucales (Phaeophyceae).
Glycopeptide or Glycoprotein: polysaccharide composed of sugars and amino acids or peptides.
Glycoside: a polysaccharide composed of glucose.
Gonidium: a cell that divides to form a daughter colony.
Gonimoblast: usually diploid cells that form the carposporangia in the Rhodophyceae.
Gonoid: type of ornamentation that is dominated by angles in the Bacillariophyceae.
Hair: appendage on a flagellum; colourless elongate cell.
Hapteron or Holdfast: bottom part of an alga that attaches the plant to the substrate.
Hemathochrome: red or orange lipid bodies occurring outside the chloroplast.
Hermaphroditic or Homothallic: producing both male and female gametangia on the same thallus.
Heterocyst: thick-walled, hollow-looking enlarged cell in the Cyanophyceae.
Heterokont: having fiagella of unequal length.
Heteromorphic Alternation of Generations: having haploid and diploid generations of different morphology.
Heterothallic: producing male and female gametangia on different plants.
Heterotrichous: term used to describe division of a plant into an erect and a prostrate part.
Heterotrophic: needing an external source of organic compounds as an energy source.
Histone: basic protein.
Holdfast: part of an alga that attaches a plast to a substrate.
Holophytic or Autotrophic: needing only light and inorganic substances for growth.
Holozoic or Phagocytic: absorbing food particles whole into food vesicles for digestion.
Homothallic: producing male and female plants on the same plant.
Hormogonium: short pieces of a trichome in the Cyanophyceae that become detached from the parent filament and move away by gliding, subsequently developing into new filament.
Hydrophilic: water-attracting.

Hydrophobic: water-repelling.
Hypha: long slender cell in the medulla of Laminariales (Phaeophyceae).
Hypnospore: aplanospore with a greatly thickened cell wall.
Hypogenous Cells: in the Rhodophyceae those cells under the carpogonium.
Hypothallus: lower part of the thallus composed of large cells in the coralline reds.
Hypotheca: smaller half of a diatom frustule.
Intercalary: in between two cells or tissues.
Intercalary Bands: bands between the valve and girdle band in the Bacillariophyceae.
Internode: part of axis between nodes.
Interstitial Water: that water trapped between particles of soil or mud.
Intertidal: occurring between the low- and high-tide marks.
Inversion: phenomenon in the green algae in which a colony turns itself inside out through a pore.
Iridescence: the play of colours caused by refraction and interference of light waves at the surface.
Isoenzyme: enzymes having the same function but of a somewhat different structure.
Isogamy: fusion of similar gametes.
Isokont: cell with flagella of the same length.
Isomorphic Alternation of Generations: generations that are morphologically alike.
Karyogamy: fusion of two gamete nuclei.
Kelp: a member of the Laminariales (Phaeophyceae); also used for the burnt ash of plants of the Laminariales.
Kerogen: yellow-brown amorphous organic matter in sedimentary deposits.
Kombu (Japanese): vegetable made from Laminariales (Phaeophyceae)
Laminarin: food storage polysaccharide in the Phaeophyceae composed principally of β-1, 3 linked glucose residues.
Laminate: flat.
Laver or Laver Bread: similar to Japanese nori, vegetable made from dried *Porphyra* (Rhodophyceae).
Lentic: related to a pond or lake.
Leucoplast: colourless plastid usually having a large number of starch grains and few thylakoids.
Leucosin or Chrysolaminarin: food storage polysaccharide of golden-brown algae composed mostly of-l, 3 linked glucose residues.
Lithophyte: plant growing on rock.
Lithotrophic or Autotrophic: needing only light and/or inorganic compounds for growth.
Litoral Zone: zone from the water's edge to a water depth of about 6 m or the maximum depth of rooted vegetation, if any exists.
Loculus: hexagonal chamber in the wall of the Bacillariophyceae.
Lotic: related to rivers or streams.
Lysosome: single-membrane-bounded cytoplasmic particle containing destructive enzymes.
Mannan: polysaccharide composed of mannose residues.
Mannitol: sugar alcohol, $C_6H_{14}O_6$; primary product of photosynthesis in the Phaeophyceae.
Mantle or Valve Jacket: part of a valve in the Bacillariophyceae that is bent inward.
Marl: deposits of calcium and magnesium carbonate.
Meiospore: spore formed by meiosis.

Meristem: dividing tissue that forms new cells.
Meristoderm: dividing layer of cells in the Phaeophyceae.
Mesochite: middle wall of an oogonium of the Fucales (Phaeophyceae).
Metachromatic Granule: protoplasmic body containing stored polyphosphate in the Cyanophyceae.
Microaerophilic: with small amounts of oxygen.
Micrometer (μm): 10^{-6} meter. 1 μm equals 1 micron.
Microtubule: submicroscopic tubule in the protoplasm.
Mixotroph or Facultative Heterotroph: photosynthetic organism capable of using organic compounds in the medium.
Monoecious or Homothallic: having male and female gametangia borne on the same plant.
Monopodial: having one main axis of growth.
Monosporangium: a sporangium that forms a monospore in the Rhodophyceae.
Monospore: asexual spore that germinates to reform the parent in the Rhodophyccae.
Mucopeptide: polysaccharide of the walls of the Cyanophyceae, composed of sugars and amino acids.
Multiaxial: having an axis with a number of apical cells that give rise to a number of nearly parallel filaments.
Multiseriate: with more than one row of cells.
Myxophycean Starch: storage polysaccharide of the Cyanophyceae, similar to glycogen.
Nannoplankton or Nanoplankton: plankton smaller than 75 μm.
Nanometer (nm): 10^{-9} meter, 1 nm equals to angstrom units.
Necridium or Separation Disc: a cell that dies in a trichome or the Cyanophyceae resulting in the formation of a hormogonium from part of the trichome.
Nemathecium: wartlike surface elevation containing the reproductive structures in the Rhodophyceae.
Nitrogen Fixation: the intracellular fixation of nitrogen gas from the atmsophere to ammonia in Cyanophyceae or bacteria.
Node: part of thallus that bears branches.
Nori or Laver: vegetable made from dried *Porphyra* (Rhodophyceae) in Japan, similar to laver bread.
Nucule: female reproductive structure in the Charophyta.
Oogamy: fusion of a large nonmotile egg with a small motile sperm.
Oogonium: single-celled female gametangium.
Oospore or Zygospore: thick-walled zygote with food-reserves.
Organelle: a membrane-bounded part of a cell.
Ostiole: an opening to the outside in a conceptacle.
Ovum or Egg: nonmotile large female gamete.
Palmelloid: term describing colony of an indefinite number of single, nonmotile cells in mucilaginous matrix.
Pantonematic or Tinsel Flagellum: flagellum with hairs attached to the surface.
Paraflagellar or Paracrystalline Body: photoreceptor in the Euglenophyceae consisting of a crystalline swelling in one of the flagella.
Paraphysis: sterile structure found with sporangia or gametangia.
Parasite: heterotrophic organism that derives nutrients from a living host.

Parenchyma: a tissue formed of thin-walled living cells produced by division in three planes.
Parietal: peripheral.
Parthenogenetic: germination of an egg without fertilization to form a new plant.
Pedicel: supporting structure of a reproductive tissue.
Peduncle: tail-like process with a central core of microtubules.
Pelagic: living in the open ocean or oceanic region; in some definitions, living at or near the surface of the open sea.
Pennate: term describing type of ornamentation arranged on either side of a central line in the Bacillariophyceae (bilateral symmetry).
Pericentral Cell: a small cell formed around a central axis.
Periclinal: parallel to the circumference of the surface.
Periphyton: organism attached to submerged vegetation.
Phaeophycean Tannin or Fucosan: colourless, acidic fluid in physodes in the Phaeophyceae.
Phagotrophic or Holozoic: ingesting solid food particles into a food vesicle for digestion.
Phialopore: hole in an inverted daughter colony in the Volvocales (Chlorophyceae).
Photoreceptor: the part of the cell that receives the stimulus in phototaxis, usually a dense area in a flagellar swelling.
Phototaxis: movement of a whole organism toward (positive) or away from (negative) light.
Phragmoplast: wall formation by the coalescence of Golgi vesicles between spindle microtubules.
Phycobiliprotein or Phycobilin: water-soluble blue-green or pink pigment in the Cyanophyceae, Rhodophyceae, and Cryptophyceae.
Phycobilisome: an aggregation of phycobiliproteins on the surface of a thylakoid.
Phycobiont: algal partner in a lichen.
Phycocolloid: polysaccharide colloid formed by an alga.
Phycocyanin: blue-green coloured phycobiliprotein.
Phycoerythrin: pink-coloured phycobiliprotein.
Phycophaein: black oxidized phaeophycean tannins.
Phycoplast: type of cell division in which the mitotic spindle disperses after nuclear division with the two daughter nuclei coming close together, another set of microtubules arising perpendicular to the former position of the microtubules of the mitotic spindle, and the new cell wall forming along these microtubules.
Physode: vesicle containing phaeophycean tannins in the Phaeophyceae.
Phytochrome: photoperiod regulating chemical.
Phytoplankton: plants that float aimlessly or swim too feebly to maintain a constant position against a water current.
Pit Connection: a continuous area between two red algal cells consisting of an aperture in a cross wall, a plug, and a plug cap.
Plakea: flat plate of cells in the Volvocales (Chlorophyceae).
Plankton: organisms that float aimlessly or swim too feebly to maintain a constant position against a water current.
Planogamete: motile gamete.
Planospore: motile spore.
Planozygote: motile zygote.

Plasmodesma (plural Plasmodesmata): the minute cytoplasmic threads that extend through openings in cell walls and connect the protoplasts of adjacent living cells.

Plasmogamy: fusion of protoplasm without fusion of nuclei (karyogamy).

Plastid: double-membrane-bounded organelle usually containing the photosynthetic apparatus or some part of it.

Plurilocular Sporangium: many-chambered sporangium in the Phaeophyceae, each chamber forming one swarmer.

Pneumatocyst or AIR Bladder: expanded part of thallus containing gases.

Polar Nodule: wall swelling near the end of a cell in the Bacillariophyceae.

Polyglucan Granule: protoplasmic structure containing the storage product in the Cyanophyceae.

Polyhedral Body: protoplasmic structure in the Cyanophyceae associated with DNA microfibrils; it may contain the carbon dioxide-fixing enzyme ribulose-1, 5-bisphosphate carboxylase.

Polymorphic: having more than one shape.

Polyol: sugar alcohol.

Polyphosphate Body: protoplasmic structure containing stored phosphate in the Cyanophyceae.

Polysiphonous: term describing thallus made up of vertical files of parallel cells.

Pore: a single hole.

Poroid: pore occluded by a plate in the Bacillariophyceae.

Procarp: association of carpogonium and auxiliary cells in the Rhodophyceae.

Procaryotic Cell: a type of cell lacking membrane-bounded organelles.

Propagulum: branchlets that fall off and form new plants in the Sphacelariales (Phaeophyceae).

Pseudoparenchyma: densely packed filaments resembling parenchyma tissue.

Pseudoraphe: unornamented area on a theca of the Bacillariophyceae where a raphe would occur.

Puncta: opening in the frustule in the Bacillariophyceae, either a pore or a loculus.

Pyrenoid: proteinaceous area of the chloroplast associated with the formation of storage product.

Radial: occurring around a central point.

Ramulus: reproductive branch.

Raphe: longitudinal slit in the valve of some Bacillariophyceae.

Receptacle: swollen tip of thallus containing conceptacles in the Fucales (Phaeophyceae).

Resting Cell: cell with the same morphology as vegetative cell in the Bacillariophyceae, but with a large amount of lipid and reduced size of organ cells.

Resting Spore: thick-walled cell resistant to unfavourable environmental conditions.

Reticulate: netlike.

Rhizoid: rootlike filament without vascular tissue.

Saprophyte: heterotrophic organism living off dead material.

Separation Disc or Necridium: a cell that dies in a trichome of the Cyanophyceae, resulting in the separation of a hormogonium from part of the thallus.

Septum: cross wall.

Sheath: extracellular mucilage.

Sieve Plate: end wall in a sieve cell with pores through which the cytoplasm is continuous (plasmodesmata).

Siphonaceous or Siphonous or Coenocytic Cells: large multinucleate cells without cross walls except when reproductive bodies are formed.

Sirenine: a sex attractant in *Ectocarpus* (Phaeophyceae).

Slime Layer: extracellular mucilage.
Somatic: vegetative.
Sorus: cluster of reproductive bodies.
Sperm: male gamete.
Spermatangium: male gametangium forming one spermatium in the Rhodophyceae.
Spermatogenesis: formation of sperm.
Spermocarp: zygote plus an enclosing layer of cells in *Coleochaete* (Chlorophyceae).
Sporangium: spore-producing structure.
Spore: cell that germinates without fusing to form a new individual.
Sporeling: young plant arising from a spore.
Sporocyte: reproductive cell.
Sporophyte: diploid plant that forms spores.
Statospore or Cyst: resting spore.
Stellate: star-shaped.
Stigma or Eyespot: group of pigmented lipid bodies that are associated with photoaxis.
Stipe: organ between a holdfast and blade.
Suffultory Cell: cell to which the dwarf male filament attaches in *Oedogonium* (Chlorophyceae).
Sulcus: longitudinal groove in the hypocone of Dinophyceae.
Supporting Cell: a cell that bears the carpogonial branch in some Rohodophyceae.
Suture: area of fusion of two adjacent structures.
Swarmer: motile cell.
Symbiosis or Reciprocal Parasitism: two organisms living together to the mutual benefit of each.
Sympodial Axis: axis formed from successive dichotomous branches in which one branch is shorter than the other, giving the appearance of a simple stem.
Syngamy: fusion of gametes.
Systole: contraction of a contractile vacuole.
Terrestrial: growing on soil.
Tetrasporangium: a sporangium producing four tetraspores, usually by meiosis.
Tetraspore: spore formed in a tetrasporangium, usually by meiosis.
Tetrasporophyte: usually diploid plant forming tetraspores in the Rhodophyceae.
Thallophte: plant lacking roots, stem, and leaves.
Theca: outer covering of the Dinophyceae and some Chlorophyceae.
Thecal Plate: a plate in a vesicle under the plasmalemma in the Dinophyceae.
Thermocline: layer in a thermally stratified lake where the temperature changes suddenly with depth.
Thermophiles: organisms that grow at high temperatures.
Thylakoid: membrane-bound sac in a plastid.
Tinsel Flagellum: flagellum with hairs.
Trabeculae: wall ingrowths in some coenocytic green algae.
Trichocyst: projectile in the Dinophyceae and Rhaphdiophyceae.
Trichogyne: long colourless part of a carpogonium that receives the spermatium in the Rhodophyceae.
Trichome: a row of cells without the sheath in the Cyanophyceae.

Trichothallic: term describing intercalary meristem producing a hair in one direction and the thallus in the other direction.
Trophocyte: vegetative cell.
Trumpet Hyphae: drawn-out sieve cells wider at the cross walls than in the middle of the cells (Laminariales, Phaeophyceae).
Uniaxial: having a main axis consisting of a single row of usually large cells.
Unilocular Sporangium: sporangium composed of a single cell producing zoospores usually by meiosis.
Uniseriate: having a single row of cells.
Unisexual: having only one type of gametangium formed on one plant.
Upwelling: an area of the ocean where nutrient-rich bottom water rises to the surface.
Uronic Acid: a type of monosaccharide.
Utricle: inflated branchlet.
Valve Jacket or Mantle: part of the valve in the Bacillariophyceae that is bent inward.
Velum: wall extension over a loculus in the Bacillariophyceae.
Volutin Granule: protoplasmic body containing stored polyphosphate.
Water Column: a vertical section of a body of water.
Whiplash Flagellum: flagellum without hairs on its surface.
Xanthophyll: a carotenoid composed of an oxygenated hydrocarbon.
Xylan: polysaccharide composed of xylose sugar residues.
Zoochlorellae: Chlorophyceae living inside invertebrate animals.
Zoosporangium: sporangium that forms zoospores.
Zoospore: flagellated planospore.
Zoosporogenesis: formation of zoospores.
Zooxanthellae: nongreen algal cells, usually Dinophyceae, living inside invertebrates.
Zygospore: thick-walled resting spore.
Zygote: product of the fusion of two gametes.

References

Bold, H.C. and Wynne M.J. 1978. Introduction to the Algae. Structure and Reproduction. Prentice Hall of India Private Ltd. New Delhi.

Chapman, V.J. 1962. The Algae. Macmillan, London.

Desikachary, T.V. 1969. Cyanophyta. I.C.A.R. New Delhi.

Fritsch, F.E. 1935. The structure and Reproduction of Algae. Vols, I and II. Cambridge University Press, Lond.

Godward, M.B.E. 1966. Chromosomes of Algae. Cambridge Univ. Press. Lond.

Govindji and Mananty, P.K. 1970. Photochemical aspects of photosynthesis in Blue-Green Algae. First inter. Symp. on taxonomy and biology of Blue-Green Algae. Univ. of Madras.

Kamiya, N. 1981. Physical and chemical aspects of cytoplasmic streaming. Rev. Plant Physiol. 32: 205-236

Kumar, H.D. and Singh, H.N. 1962. A text Book of Algae. East West Press, New Delhi.

Lobban, S.C. & Wynne, M.J. The Biology of Sea Weeds (ed.) Blackwell Scientific Publications. London. 1981. 783pp.

Misra, J.N. 1966. Phaeophyceae in India. I.C.A.R. New Delhi.

Morris, I. 1968. An Introduction To The Algae. Second ed. Hutchinson and Co. London.

Prasad, B.N. and Singh, Y. 1982. On Diatoms as indicators of water pollution. Ind. J. of Bot. Soc. 61: 326-336.

Singh, V.P. and Trehan, K. 1976. Recombination in Algae.

Recent Adv. in Bot. P.N. Mehra Comm. Vol. 1-9.

Smith, G.M. 1950. Freshwater Alge of United State.

Second. ed. Macmillan. New York.

Smith, G.M. Cryptogamic Botany. Vol. I Algae and Fungi. Mc Graw Hill Book Co. New York, 1955.

Index

Acanthophora 28
Acetabularia 74, 75, 78
Achrochaetium 27
Acronematic 98, 277
Aegagrophilous 154
Aerophytes 22
Agar agar 67
Agarth C.F. 2
Ahnfeldita 70
Akinetes 23
Algal cytology 78
Alginic Acid 20, 70
Allophycocyanin 72
Amansia 68
Amori 68
Amphiroa 28
Anabaena 23, 69, 71
Anabaenopsis 29
Ancyclonema 22, 25
Androsporangium 155
Anisogamous 277
Antheraxanthin 19
Anthoceros 23
Antithamnion 49, 72
Aphanizomenon 69
Aphanotheca 24
Aplanospores 277
Aquatic algae 22
Arabinose 21
Archothrix 31
Arthrothamnium 68
Ascophyllum 26, 63, 71, 92
Ascorbic acid 63
Asparagopsis 28, 60
Autospores 23, 24
Auxospore 23, 24, 280
Avarainvillea 72
Azolla 25, 27
Azotobactor 23

Bacillariophyceae 17, 277
Bangia 20
Batophora 74
Batrachospermum 2, 23, 91, 155, 298, 301
Beggiota 276
Benthic algae 24
Biliproteins 19
Biotin 64
Bodonella 298
Bold HC 277
Boodlea 26
Botrydium 279
Botryococcus 31
Bromophenols 62
Bryopsis 26, 72, 74, 75, 279
Bulbochaete 22, 25
Bullock 74

Calothrix 25, 29, 32
Campesterol 52
Carborundum 71
Caroigibua 299
Carotenoids 18, 48
Carpogonium 303
Carposporangia 91
Carposporophyte 299
Carrageen 67
Carragenann 21
Carteria 25
Caulerpa 4, 26, 32, 43, 55, 72, 73
Caulerpol 73
Centrales 279
Cephaleuros 22, 25
Ceramium 28, 65, 312
Chaetoceros 277
Chaetomorpha 68
Chantransia stage 91
Chapman 5
Chara 3, 31, 98, 278

Charales 30, 277
Chlamydomonas 2, 25, 30, 94, 98, 277
Chlorella 2, 23, 29, 71, 155
Chlorococcum 98
Chlorogloea 29
Chlorophyceae 15, 21
Chlorophylls A, B, C, D, E 18, 47
Cholesterol 52
Chondra 26, 63, 55
Chondrus 30, 34, 51, 67, 70, 77
Chorella 69
Chrococcum 98
Chromophyta 92
Chromulina 78
Chroococcus 69, 80
Chrysamoeba 79, 80
Chrysophyceae 15
Cladophora 2, 26, 50, 63, 71, 74, 75, 81, 90, 98, 154, 154
Closterium 154
Coccoid habit 98
Codisterol 53
Codium 27, 53, 65, 72, 74
Coenobia 112
Coilodesma 84
Coleochaete 2, 81, 93, 98, 155
Collinbsiella 94
Colpomenia 28
Commercial cultivation of algae 32
Conceptacles 308, 281
Conferva 24
Contophora 92
Corallina 298
Cortication 279
Cosmarium 277, 155
Costaria 87
Cryophytes 22, 25
Cryptoblasts 308
Cryptomonas 30
Cryptophyceae 20
Cryptostomata 308
Cutleria 76, 94, 298
Cyanophyceae 273
Cycas 25, 273
Cyclorinus 31
Cyclotella 4
Cylidrocystis 25
Cylindrocapsa 4
Cylindrocystis 277, 277
Cymbella 277
Cynidium 47
Cynophages 29

Cystocarp 91, 299
Cystophora 62
Cystophyllum 28

De Jussieu 2
Dendroid forms 80
Derbesia 26, 72
Desmidium 154
Desmids 155
Diatomaceous earth 70, 71, 281
Diatomite 70
Diatoms 30, 278
Diatoxanthin 19
Dictyopteris 26
Dictyota 28, 73, 298, 304
Diplobiontic 91, 266
Diplontic 90
Diterpenes 56, 73
Draparnaldia 23, 277
Draparnaldiopsis 4, 81, 154
Durvillea 67
Duvaliella 30

Ecballocystis 80
Ectocarpus 25, 68, 81, 91, 298, 299
Elatol 56
Endolithic 25
Endoplasmic reticulum 74
Endospore 24
Endospores 276
Endozoic algae 25
Engler & Prantl 4
Enteromorpha 51, 68, 75
Epilithic 25
Epiphytes 22, 25
Epiphyton 31
Epitheca 278
Equisetoid 278
Erythrocladia 27
Erythrotrichia 298
Estabularia 28
Euastrum 277
Eucaryotic 20
Euchetum 37
Eucheuma 34, 37, 42
Eudorina 154
Euglena 71
Exospores 276

Fatty acids 59
Flexilin 56

Floridae 27
Fossil algae 30
Fragilaria 30, 68
Fritsch FE 3, 5, 92
Fritschiella 4, 93, 154
Fucodin 20
Fucosterol 52
Fucoxanthin 19, 50, 72
Fucus 30, 63, 86, 279
Fumoria 70
Fumorin 70
Furcellaran 21
Furcellaria 70

Gaidukov phenomenon 275
Galactose 21
Gelidium 27, 67, 70
Giffordia 27
Gigartina 70
Gigartinine 64
Gilidiella 32
Globule 277
Gloeocapsa 22, 273
Gloeopeltis 40, 42, 70
Gloeotrichia 71
Glucronic acid 20
Godward MBE 4, 78
Golgibody 20
Golgicomplex 76
Gongrosira stage 279
Gonimoblast filaments 299
Gonium 154
Gracilaria 26, 28, 32, 40, 70, 73, 299
Gracilariopsis 32
Grammatophora 277
Grateloupia 28, 70
Griffithsia 299
Gymnodidum 24
Gynogonidia 154
Gyrosigma 30

Haemotococcus 22
Halicystis 72
Halidrys 63
Halimeda 30, 31, 72
Halophytes 25
Haplobiontic 91
Haplodiplontic 277
Haplontic 90
Harder 69
Hassall 2

Helioplankton 1
Hertwig 2
Heterocysts 276
Heteromorphic 91
Heterotrichy 277
Hildebrandia 27
Homoplasy 155
Homothallic 277
Hormogonia 276
Hormospores 276
Hyalotheca 277
Hydra 23
Hydrilla 25
Hydrocarbons 59
Hydroclathrus 83
Hydrocoleum 68
Hydrodictyon 2, 98, 154, 277
Hypnospores 23, 278
Hypoglossum 28
Hypotheca 278
Hypoxanthin 50

Isogamous 15
Isomorphic 91
Iyengar MOP 3

Jurassic 30

Keck W.M. 42
Khiesulghur 70
Kraft G.T. 11
Kuckuck 4

Lagerheim 2
Laminaria 2, 23, 30, 32, 34, 35, 38, 42, 45, 67, 76, 298, 2772
Lateral conjugation 155
Laurencia 55
Lemaea 298
Lens paper Porphyridium 72
Lichens 273
Limnoplankton 1
Link H.F. 2
Linnaeus 4
Lipids 59
Liptothrix 70
Lithodinia 31
Lithophytes 22, 25
Lobban 74, 94
Lomasomes 77
Loroxanthin 72

Lutein 19, 72
Luther 4
Lyngbya 29, 70, 71

Macrandrous 277
Macrocystis 2, 23, 42, 89, 298
Manubrium 278
Marginosporum 60
Mari culture 43
Marine algae 22
Marpolia 31
Mastigocladus 24
Melobasia 31
Melosira 24
Mesozoic 31
Microbodies 74
Microdictyon 4
Microspora 68
Microsterias 154
Misra J.N. 4
Mitochondria 20
Mixophyceae 19
Mixoxanthin 19
Mixoxanthophyll 19
Mizzea 30
Mobbins 2
Monosporangia 91
Monostroma 65
Monoterpenes 54
Morris 92
Mougeotia 4
Mougeotipsis 2
Myxophyceae 273

Nannandrous 277
Nannocytes 276
Narayan Rao, S.R. 4
Navicula 277
Nelumbo 25
Nemalion 49, 72, 81
Neoxanthin 72
Nereocystis 70, 88
Netrium 277, 155
Niacin 64
Nif genes 29
Nitophyllum 28
Nitrogen fixation 29, 69
Nitroglycerine 71
Nitschjia 277
Nodularia 69, 71
Nori 33

North W.J. 42
Nostoc 29, 68, 69, 81
Nostocales 29
Nucule 277

Obtusol 56
Ochtodes 54
Oedogonium 2, 22, 23, 25, 68, 90, 98, 154, 277
Oogamous 277
Origin of Soma 154
Oscillatoria 24, 30, 69, 273

Padina 26, 27, 85
Paleozoic 31
Palmella stage 23, 277
Pandorina 277
Pantonematic 20, 277
Papenfuss 5
Parenchymatous habit 81
Parisitic algae 25
Pediastrum 23
Pelagophycus 70
Pelvetia 65, 71
Penghu 33
Pennales 279
Petkoff 2
Phaeophyceae 17, 26, 298
Phenols 61
Phormidium 22, 29
Photolaxis 75
Photoplankton 1
Phragmoplast 74
Phycobiliproteins 48, 72
Phycobilisomes 1
Phycoerythrin 72, 275
Phycoerythrocyanin 72
Phycophages 29
Phycoplast 74
Phyllophora 67, 70
Phytoplankton 1
Pinnularia 279
Pithophora 24, 68
Placoderm desmids 277, 155
Plankton algae 24
Plectonema 29
Pleistocene 30
Pleodorina 277
Pleurocladia 298
Pleurococcus 23
Pleurotaenium 154
Plocarmium 54, 60

Index

Pluriliocular sporangia 299
Polyphloroglucinols 62
Polysiphonia 4, 23, 25, 28, 298, 303
Polysiphonous 303
Polysulphides 65
Porphyra 20, 33, 67, 71, 77, 81, 299
Porphyran 21
Porphyria 32
Porphyridium 47, 298
Postelsia 23, 298
Prasiola 22
Precambrian 30
Prescott G.W. 5
Pringscheim 2
Procaryotic 20
Protoderma 25
Pseudovacuoles 275
Ptilonia 60
Pyrenoid 279
Pyriform 277
Pyrrophyceae 31

Ralfsia 27
Raphe 278
Rapidonema 22
Receptacle 309
Rhibosomes 20
Rhododendron 25
Rhodomela 63, 71
Rhodophyceae 18, 26, 27, 77, 298
Rhodymenia 20, 28, 67
Rivularia 2, 22, 25, 276
Roth 2

Saccoderm desmids 277, 155
Sarcophycus 67
Sargassum 27, 28, 58, 70, 82, 90, 298, 299
Sarma YSRK 4
Scalariform conjugation 154
Scenedesmus 30, 71
Schmitz 2
Scotiella 25
Scott 74
Scytonema 22, 29, 69
Scytosiphon 93
Sesquiterpenes 55, 73
Sinus 154
Siphonaceous habit 81
Siphonaxanthin 72
Siphonous habit 277
Smith, G.M. 1, 5

Smithora 77
Solenospora 31
Spermatangia 299
Spermocarp 154, 155
Sphacelaria 75
Sphaerella 23, 277
Sphaerococcus 51, 56
Spihaeroplea 2
Spirogyra 68, 71, 81, 277
Spirotaenia 277
Spirulina 69
Spongomorpha 26, 65
Staphylococcus 71
Staurastrum 277
Steroids 51
Sterols 73
Stigeoclonium 277
Stigmasterol 52
Stigonematales 29
Struvea 28
Stypopodium 58
Subramaniam 3
Symbella 30
Symbiosis 23

Tamia 69
Tansley 4
Taonia 58
Taraxanthin 19
Terpenoids 53, 73
Tetraspora 80
Tetrasporangia 306
Tetraspores 91
Tetrasporophyte 304, 308
Thermal algae 24
Tiffany, L.H. 1, 3
Tilden J.E. 5
Tinsel flagellum 277
Tolypothrix 69
Transeau E.N. 3
Trentepohlia 155
Trichoblasts 299
Trichodesmum 24
Trichothallic growth 301
Triterpenoids 51
Turner 2

Udotea 72
Ulothrix 23, 25, 68, 75, 81, 93, 98, 98, 154
Ulva 26, 28, 50, 67, 75, 154, 154
Undaria 39, 40, 42

Unilocular sporangia 299
Urospora 75

Vallisnaria 25
Valonia 75
Valve 278
Vaucheria 22, 81, 278
Violacene 54
Violaxanthin 49, 72
Vitamin B$_{12}$ 63
Vitamins 63
Volvox 30, 93, 155, 277

West G.S. 3
Whiplash flagellum 277
Wildenow 3

Wittrock 2
Wrangelia 28
Wynne 74, 94
Wynne M.J. 11

Xanthidin 154
Xanthophyceae 15, 278
Xerosere 273
Xylan 20, 75

Zeaxanthin 275
Zobrasoma 68
Zonaroic acid 57
Zoochlorella 23
Zooxanthellae 25
Zygnema 154, 277, 277, 277
Zygogonum 3